KILLING ESCOBAR
AND SOLDIER STORIES
PETER MCALEESE

First published in Great Britain by Peter McAleese in 1993

Republished by Gadfly Press in 2021

Copyright © Peter McAleese 2021

The right of Peter McAleese to be identified as the author of this work has been asserted by him in accordance with the Copyright, Designs and Patents Act 1988. All rights reserved

No part of this book may be reproduced, stored in a retrieval system or transmitted in any form or by any means (electronic, mechanical, photocopying, recording or otherwise) without the prior written permission of the author, except in cases of brief quotations embodied in reviews or articles. It may not be edited, amended, lent, resold, hired out, distributed or otherwise circulated without the publisher's written permission

Permission can be obtained from gadflypress@outlook.com

This book is a work of non-fiction based on research by the author

A catalogue record of this book is available from the British Library

Typeset and cover design by Jane Dixon-Smith

SPELLING DIFFERENCES: UK V US

This book was written in British English, hence US readers may notice some spelling differences with American English: e.g. color = colour, meter = metre and = jewelry = jewellery

CONTENTS

Preface	1
Chapter 1 Hired to Kill Escobar	3
Chapter 2 First Principles	13
Chapter 3 Planning and Preparation	51
Chapter 4 Retirement	68
Chapter 5 Catharsis	77
Chapter 6 Interesting Work Abroad	83
Chapter 7 Trigger Time	120
Chapter 8 The Death of a Country	157
Chapter 9 High Standards Are the Only Ones Acceptable	200
Chapter 10 A State of Emergency	229
Chapter 11 Guardians of the Myth	254
Chapter 12 Hired by Medellín Cartel's Gacha	266
Chapter 13 Cali Cartel's Mission to Kill Escobar	294
Chapter 14 Escobar Surrenders	330
Chapter 15 Escobar's Prison Life	346
Chapter 16 Bombing Escobar's Prison	360
Chapter 17 Escobar's Death	368
Epilogue	391
Afterword: Meeting Pablo Escobar	395
Other Escobar Books by Gadfly Press	401

PREFACE

Before you start, I want to clear up a couple of points.

I've always been a professional operational soldier and not, repeat not, a 'career' soldier of the sort who wants only to keep his nose clean and worries about his promotion and pension prospects.

In fact, I don't think I made a bad depot soldier when necessary, and maybe my turnout has been smarter than most. I wonder how many soldiers nowadays bother to iron creases into their combat uniforms.

I've never soldiered just for profit. During all my service in three regular armies, my pay was unimpressive by modern standards. I went from one fighting zone to another but I receive no pension. Nor do I have funds from any other source. I've done it for the adventure, because I have always been a professional soldier, and because I love a fight. I've never been happier than when in action.

A British infantry officer said to me, "Staff McAleese, you are an extreme man. Your obsessions are your strongpoints, your qualities and your character. You have experience, you have drive, use it."

I think I have.

The press love extremes. You might think there is enough exciting material in my life to satisfy the most news-hungry of journalists, but there have been some terrible lies written about me and this book puts the record straight.

It's true that I've been a mercenary for a few months of my life, in northern Angola and in Colombia. But Gurkhas are mercenaries too, and so are all the British officers and men who served in the Sultan of Oman's Armed Forces.

I am proud to have served for years as a regular soldier in the world's finest airborne regiments in three armies: in the Parachute Regiment and 22 Special Air Service in Britain, in I Special Air Service Regiment in what was then Rhodesia, and in the Pathfinders of 44 Parachute Brigade in the South African Defence Force.

You, as the reader, must make up your own mind. Do not be manipulated by the press or second-hand stories you have heard. See what I've done and decide for yourself.

This, then, is my story.

Peter McAleese
Telford, 1993.

CHAPTER 1
HIRED TO KILL ESCOBAR

With Pablo Escobar's enemies finding him difficult to capture or kill, the godfathers of the Cali Cartel turned to their head of security for ideas. Over 6 feet tall, Jorge Salcedo towered over his peers. The softly spoken ex-army officer had short dark hair, thick eyebrows, a well-trimmed moustache and a firm gaze. He was a family man with degrees in mechanical engineering and industrial economics. Early in his career, he had designed forklifts and other machinery. His father was a retired Colombian Army general and a diplomatic figure. Regarding himself as more of an engineer than a soldier, Jorge had become an expert in electronic surveillance, which had drawn him into counterterrorism assignments.

In a previous job for the military, financed by the Medellín Cartel's Gacha, Jorge had worked with us mercenaries based in the UK. Enticed to Colombia on the pretext of fighting communism, we had found ourselves participating in a turf war with guerrillas over cocaine labs. Having been paid so well, we were enthused about doing more work in Colombia.

As detailed in his book, *Dirty Combat*, Dave Tomkins received a call from Jorge in February 1989. "Are you prepared to come back for another mission?"

"Yes, subject to terms and conditions."

With coiffured grey hair, dark brows, narrow features and a tattooed arm, Tomkins was a demolitions expert and soldier of fortune who gravitated towards the hotspots of the world. His career had started in Africa. In Afghanistan, he had fought with the Mujahideen, in Croatia with warlords, and in Uganda he was involved in a plot to assassinate President Idi Amin. In Angola, a mine had exploded on him. After a medical operation, his wounds had deteriorated into near gangrene. Back in the UK, he had been hospitalised. Two days later, when the nurses came to dress his wounds, they found that he had returned to the battlefield.

On February 13, Jorge greeted Tomkins at the Bogotá airport. "Your prospective clients," Jorge said, "are a group of businessmen whom Pablo Escobar has sworn to kill. The Medellín Cartel has engaged in a bombing campaign against their business interests. They live in Cali, which is Colombia's third-largest city. More than thirty bombs have exploded in their Drogas La Rebaja chain of 350 stores."

"Why has Pablo Escobar targeted them?" He pressed Jorge into revealing that the clients were the Cali Cartel. The express purpose of the mission was to kill Escobar. Although the mission had no official government backing, the authorities had blessed it.

In England, I received a visit from Tomkins. Although he was an excellent businessman, he wasn't ex-military, so I was needed to organise the tactical stuff. As a former paratrooper, SAS regiment soldier, South African sergeant major and Rhodesian SAS soldier, I was highly respected among the mercenaries. Back then, I was built like a tank, with hardly any neck or hair.

On February 24, we flew to Colombia. In Cali, we were situated in a heavily guarded five-star apartment. We had our own bar, en suite bathrooms and money-counting machines. After we showered, Land Cruisers took us to a little town outside Cali called Jamundí, which Cali godfather Pacho controlled. The four godfathers arranged to meet us in a private sports facility with a swimming pool, gym, sauna and running track.

After getting into a complex surrounded by sheet-steel fencing and through security checkpoints, we were escorted to the godfathers, who were sat around a table with drinks, wearing tailored shirts, designer trousers, Gucci shoes and modest watches made by Cartier or Rolex. In a Sergio Tacchini sports suit and with designer stubble, Pacho looked the youngest. Chepe wore overalls and no socks; a long-sleeved shirt hid his skin condition. As the godfathers spoke in Spanish and us in English, Jorge translated. Discussing the operation, the godfathers were relaxed and congenial, as if it were just another day at the office.

Tomkins extracted a cigarette and scanned the room for ashtrays, but there were none. "Can I?" Miguel shook his head and his face puckered. The godfathers had zero tolerance for substances being used during work. Jorge explained that Miguel was allergic to smoke.

At lunchtime, the front gates let in a 3-ton, open-back truck with over a dozen policemen. Their presence soothed our concerns about the authorities interrupting the meeting. They had come for lunch courtesy of the gentlemen of Cali.

Jorge got on a radio and equipment was brought for us to examine: night-vision goggles, low-cost bugging equipment, Desert Eagle guns, crossbows and a sniper rifle. Tomkins rolled his eyes at the inadequacies. Addressing the godfathers, he started to ask questions from a list he had formulated. As the cartel didn't know Pablo's whereabouts, Tomkins said that he couldn't act without real-time info on Pablo's location.

"How long will it take for you to find Pablo?" Gilberto asked.

"With what you know: how long do you think it will take?"

"A couple of weeks."

Disturbed by their ignorance of the mission's complexities, Tomkins restrained himself from shaking his head. He needed much longer than two weeks. "I'll bring a team with all the military skills required to complete this mission. These people are presently highly paid in various parts of the world. I can't calculate how long it will take to complete this mission, but we need to be paid in accordance with the risk. To finance the initial twelve-man team, I need a three-month advance payment. No matter how long the mission takes, all payments are to be for a minimum of three months, with an operational bonus payment to be discussed when the mission plan is formulated." After holding the gaze of each godfather, Tomkins asked for $1 million.

Without hesitating, Gilberto said that amount was no problem. If they killed Pablo, they would receive up to $3 million plus a bonus. Tomkins could barely contain his excitement. More weapons arrived in the night, including sub-machine guns and shotguns.

The next day, Tomkins and I were flown over Pablo's Hacienda Nápoles ranch. We took pictures of the huge building with terracotta roof tiles, an airstrip, two helipads, a lake, swimming pools, various other buildings for guests and bodyguards, a soccer field, a tennis court, aircraft hangars, a bullring and a massive satellite dish. The perimeter was protected by steel fencing and thatched-roof guard towers. Barriers to entry included trees, shrubs and lakes. Bodyguards lived in an L-shaped building, which contained a museum with the early 1930s Cadillac reminiscent of Al Capone's. Surveying the location made us realise the magnitude of killing Pablo.

We left Colombia to recruit a team, but kept the mission a secret from them. We chose some ex-members of the Special Air Service (SAS) – a special forces unit of the British Army – and a couple of ex-South African Reconnaissance Commandos, some of whom he had worked with. The younger ones were lean, fit and muscular. They weight-trained constantly and ran long distances. The older ones were stocky, worldly wise, scarred from bullets and shrapnel and hardened by experience, their bodies ravaged by combat and booze. One had a scar from knee to ankle from an ill-fated parachute drop.

In the UK, Tomkins went on a shopping spree. Cartel cash paid for bugging kits, radio scanners, frequency counters, direction-finding equipment, portable searchlights with infrared lenses, infrared markers with strobe lights visible to night-vision equipment, and medical equipment, including tracheotomy sets and inflatable splints. A security guard at Heathrow airport allowed his bulky luggage onto the plane without any questions asked. Jorge did the same for his luggage in Colombia.

Settling into apartments in Cali, we were warned to be vigilant as kidnapping was common. Soon given arms, we were advised not to take them outside. "We'll be moving to another place soon," I said, "where we can start our training. While we're here, don't go around in more than twos or draw any attention to yourselves. You must behave in a touristy manner." At the morning meeting, I chastised two men for getting drunk the previous night. "Our mission is to kill Pablo Escobar," I said. "It's called Operation Phoenix."

We analysed pictures of Pablo and aerial photos of Hacienda Nápoles. Using maps, we planned various reconnaissance routes. Jorge said that any reconnaissance in the vicinity of Hacienda Nápoles would be reported to Pablo, whose guards would swoop down on the pale-faced strangers. Using radios, the group tuned into conversations between traffickers and producers. The Spanish was translated by a skinny Colombian from New York who worked for the godfather Chepe.

Exotic women and disco bars in busy Cali distracted some of our men, so we relocated to a heavily guarded hilltop retreat frequented by one of Miguel's wives. It included a leisure complex set on 30 acres of gardens with ornamental bridges and streams. Maintenance men tended the Japanese garden. We swam in the pool, sweated in the sauna and had access to horse riding, tennis, indoor bowling and quad bikes. Our food was prepared by a female chef. Refraining from alcohol, we drank crates of Coca-Cola.

Worried about the luxuries softening our men, Tomkins and I enforced a strict discipline. At 6 AM, a mandatory run around the track commenced, followed by breakfast. Then came weight-training, volleyball and other activities. To divide the men into sections, a secret ballot was held to nominate leaders. The men were instructed to write down the name of the man they would most want to be in a foxhole with during a crisis.

After a month, Tomkins and I were transported to the cartel's offices in a suburb of Cali called 'Garden City'. In his typical low-key style, Gilberto ran the business from a complex of single-storey terracotta buildings. His weapons of choice were phones, fax machines, telex machines, typists and secretaries. We were escorted to a workshop which received incoming goods. Refrigerators were unpacked and their back panels removed to reveal weapons manufactured in America that had travelled via Mexico. They were hidden in the false floor of a van, which was activated by tilting the vehicle to one side.

We were delighted with the weapons' shipment: an assortment of guns, anti-tank rocket launchers, C-4 demolition charges, ammunition and night-vision equipment. With the new arms, we performed a week of drills until everything was second nature. We practised repelling attacks by helicopters and night-time raids by Pablo's men. At a remote hilltop farm, we did drills with live ammunition, machine-gunning down imaginary enemies.

On our way back with our weapons, we went to a village bar. A policeman on a motorbike noticed us and stopped. Approaching, he observed our black combat boots. After he spoke in Spanish, one of our escorts from the cartel intervened. Toting a mini-machine gun in a shoulder bag, Mario was ex-army, so he displayed his military ID. The unimpressed policeman yelled in Spanish, and both men argued. Some of us started to move towards the truck to get weapons. Jorge told us to stay put and to not do anything drastic. Mario produced a radio and made a call. A man showed up on a motorbike with a bag of cartel cash. Mario gave the bag to the policeman, who departed.

In the second month, one of the South Africans quit as he believed an attempt to kill Pablo was a suicide mission. Tomkins flew to England for a replacement.

The Cali godfathers hosted lunches for Tomkins, me and our two chaperones: Jorge and Mario. Gilberto assured us that Pablo had been seen in many places, but we viewed the information as stale and of no tactical value.

We stayed focused on Hacienda Nápoles. Jorge and Mario provided detailed information about the property, and military and police reports on attempts to capture Pablo. After processing the info, we realised that some of our earlier assumptions were wrong. Originally, we had intended to land on the soccer field, but as we examined the photos under a microscope, we spotted anti-helicopter wires on the field. The helipads were wired. The only option was to land on a tennis court.

A jungle cabin 50 miles west of Cali was transformed into a base camp. Built on wooden piles, it had a big room with tables and benches, a kitchen and bathroom. Animal skulls adorned the back wall. An electric generator pumped water from the Manguido River. The cabin was equipped with a cook, a barbecue, mosquito nets, mattresses, cooking appliances, and food and drinks by the time we arrived by helicopter. To avoid getting kidnapped by guerrillas, we took turns on lookout duty.

A Huey helicopter arrived for the assault, which we painted emerald and white to appear to belong to the Colombian police, complete with a Colombian flag under its tail rotor and Policía Nacional on both sides of the fuselage. The jungle drills included firing rockets and detonating grenades. On the beach, we shot at the river and killed imaginary enemies. We rehearsed landing at Hacienda Nápoles and bringing in additional helicopters by using coloured smoke. Yellow meant safe to land, whereas red indicated landing in hostile fire. Using maps, we chose emergency rendezvous points in case we had to evacuate the mission on foot because the helicopters had been damaged. If that happened, we would broadcast emergency radio transmissions to aircraft circling above.

We sprayed a Hughes 500 helicopter olive to match military colours and altered the doorway to accommodate a mounted machine gun. Previous raids from the air on Hacienda Nápoles had met with no resistance. Pablo's staff had even offered refreshments to the authorities. The plan was to deceive the staff with the helicopters, and to mow them down with machine guns before the mercenaries were spotted. Explosives would be dropped along the front of the main house. Upon entering the property, any locked doors would be blown open with C-4 explosives. We received black ski masks and combat vests with special pockets to hold equipment such as grenades and radios. Backpacks were stocked with emergency supplies such as civilian clothes and medical emergency shock packs with syringes and morphine.

By the fourth month, Pablo's location was still unknown. Torrential rain trapped us in the cabin, where we sweated blanketed by humidity. Tomkins and I were summoned to give a progress report to the godfathers at a ranch belonging to Chepe surrounded by cattle fields. Chepe's recently constructed mansion still had protective sheeting over its windows. We arrived at a massive front door with Doric columns. About sixty armed bodyguards were on duty. Others were playing soccer.

Gilberto revealed that although there had been no sighting of Pablo, he knew that he was going to hold a family party at Hacienda Nápoles. It would be large and he would definitely be there. Although pleased that the cartel finally had info about Pablo being at a specific location at a certain time, we declined the opportunity because there would be women and children attending and they would get killed by machine-gun fire.

In ski masks and camouflage gear, we resumed drills by the cabin. We practised launching assaults with rockets and grenades. From helicopters, we fired machine guns at the jungle. Practising so much gave us a permanent stink of guns and gunpowder. Equipment was adjusted after weaknesses were exposed and flimsy combat vests were replaced.

Months of rehearsing as if our lives depended upon it fine-tuned the operation but some of the men were exhausted. Jorge flew all of them except for me and Tomkins to Panama City for a break. They partied in clubs and bars and took a scuba-diving course.

With the team absent, we requested a meeting with the godfathers. We were transported to a luxury home, where we constructed a model of Hacienda Nápoles. We dangled aircraft from threads attached to the ceiling. We positioned helicopters on poles, so that we could replicate flights by hand. After we rehearsed the mission, we requested the godfathers.

The godfathers and four senior cartel members entered and sat down. Explaining what was about to happen, we spoke in English, which a translator converted. We demonstrated the assault and ended with Pablo getting killed and us leaving safely.

Gilberto stood and clapped, immediately joined by the rest, which delighted us. Speaking in limited English, Chepe said that he would pay us an additional million dollars if we fetched Pablo's head. For that price, we agreed to make room on the helicopter for a piece of Pablo.

On June 3, 1989, we awoke as usual to the cawing of jungle birds. We swore at each other in a comradely fashion. Armed with guns, we went to the river, which was our toilet.

Hours later, our radio came alive in coded Spanish: Pablo Escobar was at his swimming pool at Hacienda Nápoles. Gilberto gave the go-ahead for the operation, which required two Huey helicopters. As one of the Hueys was getting repaired in Brazil, Gilberto proposed that a smaller Hughes helicopter be used. As the lone Huey would have to be overloaded to compensate for the missing helicopter, Tomkins and the pilots objected on safety grounds. Having waited so long to get Pablo, Gilberto overrode the objections. He told the pilots to figure out how to handle the extra load. A pilot agreed that it could be done by redistributing the passengers and cargo. Gilberto ordered them to go as it might be their only chance and he announced another change. As soon as the mercenaries had landed and launched the attack, a plane was going to land on the opposite side of Pablo's ranch with over a dozen hit men sent by Pacho, who had a score to settle with Pablo.

The excitement was palpable as I told the team to get ready. The men cheered and grabbed battle gear. Steel clunked and rattled as each of them checked their equipment, which was double-checked by another man. They tested their radios. An hour later, I addressed the men with a final briefing. "I have colour-coded the areas of Pablo's mansion, which I've divided into three sections. The back of the house is black. The right of the house is red. The front is white. Once inside, the house team will advise me in the gunship which section they're in and they'll call out their position as they move. The support group will target the area one section ahead of the position of the house team. The gunship will act as an aerial stop group that can eliminate runners into or out of the building. I have 300,000 pesos emergency money for every man in case you have to make your way to an emergency rendezvous point."

Helicopter blades swooshed. The Hughes 500 buzzed like an angry bee. "Let's do it!" I said. Packed into the helicopters, we sat cramped amid weapons, explosives, ammunition and fuel tanks. Before noon, the helicopters set off on a two-hour journey to a refuel site, ten minutes from Hacienda Nápoles. We were quiet and focused as we fine-tuned our minds to the task of killing Escobar.

CHAPTER 2
FIRST PRINCIPLES

I could not describe the first person I killed, nor the last. There is nothing personal in that. I have never felt any personal animosity towards the enemies I have fought. It was just the way it happened, in the darkness and the chaos of fighting, but the first time was important to me, as I suppose it must be for anyone.

We were in the Aden Protectorate, about four hours' uncomfortable driving in a truck from Aden port over a rough graded road into the interior, which was an arid desert land of harsh rocky mountains. We lived in Habilayn, a dusty camp of tents, barbed wire and sand bags which the British Army had pegged out in neat military lines in the middle of a flat plane circled by hills, where our enemy lived. Every evening as the sun went down, the gunners fired the 105s into the mouths of wadis from which they thought the enemy would emerge to attack us. A group of us waited to go out on night patrol and I felt like a soldier on the Northwest Frontier, listening to the roar of the guns and watching the orange explosions far out on the darkening hillsides while the sky faded to deep blue. Stars appeared in the east and away across the guy ropes and poles of Habilayn Camp, I could faintly hear the garrison bugler sounding the Retreat. When the guns fell silent after nightfall, we walked out, in two half-squadron groups, in single file, through the barbed wire and pickets towards the dark jagged ridges to the west, under a black sky hanging with stars. The ground was brutal and rocky, like a moonscape, and we humped heavy rucksacks, old Para-type, with canvas sacks and A-frames. We carried Self-Loading Rifles (SLRs) and wore 'trousers and shirts OG' (olive green) and Clarks desert shoes because they were light and cooler than boots.

Our enemies were Arab tribesmen, the National Liberation Front (NLF), and their rivals the Front for the Liberation of Occupied South Yemen (FLOSY). They were hard men with dark faces, fine-looking, with long hair and dirty robes. They lived like the Saracens of centuries ago but they had twentieth-century weapons, from the huge Second World War arms dumps which the British Army had left behind in Egypt when we were booted out after Suez in 1956, only nine years before.

The thing I remember most about the march that night was being literally soaked with sweat. I followed Sergeant Dave Haley, our patrol commander – or rather the huge rucksack on legs in front of me which is all he seemed to be. He stopped us every hour for a break. When I sat down, I shivered as the sweat cooled on me in the night air and I worried about the amount of water I would need. I was carrying two 1-gallon heavy-duty plastic water containers in my rucksack and had two more water bottles on my belt kit. With that and ammunition, seven magazines and extra link for the gunner carrying the General-Purpose Machine Gun (GPMG), and food – which was last in order of priority – our rucksacks weighed nearly 70 pounds, excluding what we carried in pouches on our belts. It was tiring work and we needed to replace all the fluid lost.

In darkness, we crossed a small range of hills about 4 miles from Habilayn and then the moon came up. Six miles later, somewhere after midnight, we started to climb up the side of the Jabal Barash, which was very distinct, a sharp ridge with two lumps on it like a camel's hump. We slogged upwards, each one of us exhausted, drenched with sweat and almost mindlessly following the dark shape of the man ahead, until we finally breasted the top and moved into a saddle, on a reverse slope. We collected loose rocks lying about, of which there was no shortage, to make sangars with low protective walls, and then set up our shade for the day ahead. I stretched four netted face veils sewed together across the sangar, from wall to wall, held with rocks and supported in the middle with bits of stick that I always brought with me. There weren't any trees in those mountains, not that we would have been allowed to go out and cut them. We settled down to rest, sleep and lie up for the day.

The heat was really intense. And the Jabal was bare, like the moon. You could see for miles. Any movement in the open would have been seen by the enemy tribesmen, who kept lookouts sitting about on the tops of the mountains. So, before first light we always ran a piece of string out between the sangars tied to a rock which could be moved to attract the attention of people in other sangars. This meant sentry duty was done sangar by sangar, in pairs so that one person could not fall asleep on his own, and the sangars were mutually supporting. We used binoculars, binos, with the objective lenses covered with a piece of face veil, in case the enemy saw a glint of reflected sunlight.

The year before, a nine-man patrol from 3 Troop of 'A' Squadron had marched into the centre of the Radfan mountains and a goatherd had stumbled across them by chance. The enemy had gathered fast, pinned them down all day and killed two and wounded two others

before the patrol could make a fighting withdrawal to safety. We were sixteen and we had plenty of support, A41 radios to call up the 105mm artillery and Sarbe radios to bring in RAF ground-attack Hunter jets, and I could even see Habilayn Camp in heat haze far below us on the valley floor, but all that seemed a long way off and we kept very still in our sangars.

In the heat of the afternoon, we ate our meal – curry and rice – had a brew of tea, and then packed our rucksacks ready to go. As the sun faded, we had stand-down, a fine British military tradition. We crouched in our sangars, weapons ready, peering over the rocky hillside for about three-quarters of an hour while the sun set. This marked the transition from day routine to night routine. After dark, we took down the face veils, stowed them away and set about taking down the sangars, scattering the rocks about.

Then we started walking down the western side of the Jabal Barash into a small wadi which connected with the Wadi Mishwarrah. Halfway through the night, after about four hours' walking, Dave Haley stopped us at some muddy water and we filled our water bottles. It was filthy and brackish, but we were gasping. I filled my bottles without a second's thought and added the blue and white sterilisation tablets.

Suddenly, I smelled a scented sweetness on the still night air and the next moment a goatherd appeared. 'Shep' Shepherd was leading scout and Dave called me forward to speak to the goatherd. I had done an Arabic course in Aden three years before and was on the patrol to act as his interpreter.

"*Salaam alikum,*" I said softly; greetings to you.

"*Wa'alikum salaam!*" he replied in a shout; And to you!

When I asked him if he had seen any other British soldiers, he shouted, "*Aywa!*" Yes! About half an hour before.

"*Uskat, uskat!*" Be calm, I said, warning him to make less noise, and translated for Dave Haley. The goatherd talked as if he had seen the other half-squadron, which we knew was ahead of us and which sounded as if it had turned back up the Wadi Mishwarrah, while the goatherd had come into our tributary wadi behind them.

We assumed everything was going according to plan, and carefully moved on. The moon was not up yet and it was so dark in the narrow wadi that I could only just see the rock walls rearing up beside us. Five minutes later, Dave Haley whispered, "Pete, come up. There's some more chappies down here." I padded up to Dave to speak to them and made out the dark shapes of a group of shepherds, lying on the ground ahead. Just as I started to speak, I saw one of them shift on his side. It was probably the way he moved which warned me and I thought, *This is never right!*

The Arab shot Dave Haley, full in the chest.

He should have shot me, as I was closest, but I was standing to one side, ready to interpret, and the Arab was lying in such a way that, as he sat up, he was restricted.

Dave collapsed, badly wounded, and I shot the Arab. Six times. Then all hell let loose. The wadi walls came in close just there and tracer streamed past, pinning me to the warm rock at the side. The enemy had Degtyaryov Machine Gun (DPM) automatics with big pan-shaped magazines on the top; the darkness echoed with shouting and they started throwing old British No. 36 grenades.

The others behind me ran to the flanks at once, Mick Seale shouting he was going to one side and Shep to the other, climbing the steep rocks to find a higher position, where they poured fire down onto the enemy. Our medic, Jock Phillips, came up to help Dave Haley, who had crawled to one side.

I heard a vicious explosive crack at my feet and doubled back from my exposed position in the middle of the wadi to scramble up the rocks and join the flanking movement. As I climbed, I looked up and saw Shep hit in the shoulder. He was standing on a rock firing down into the wadi beyond and cartwheeled off the rock like a movie star in a Wild West film.

Jock Phillips and another of the guys, Dave Abbott, now had two wounded men to deal with. Jock was well trained, with weeks of experience in a hospital emergency ward back home, but he had none of the kit now issued to SAS patrol medics. What he really needed was plasma drips and giving sets (the needle and tube connecting an emergency plasma drip to the arm), but all he could do was pack on shell dressings to stop the bleeding (the gaping hole in the side of Dave's chest took six), jab in 100 ml units of penicillin and make them comfortable on the sand. Each man carried two syrettes of Omnopon, taped on string round their necks, but morphine can't be used on wounds to the head, chest or stomach, so Dave Haley did without.

I was extremely busy. Taking a position high up at the front once Shep was lifted down, I kept firing at the muzzle flashes of the Arabs' weapons and throwing grenades which the others passed up to me from below. I used white phosphorus grenades to light the wadi ahead and flush the Arabs out of cover – as the phosphorus blown into the air fell back on them behind the rocks where they were hiding – and high explosive American M26s, which were devastating on the hard open ground.

The explosions lit the night and boomed deafeningly in the

narrow valley, and the exchange of fire went on for a couple of hours. We were never too concerned about counter attack because we held the steep wadi flanks. After the firing died down, we lay there alert all night, until dawn at about five o'clock.

Come first light, several of us advanced down the wadi ahead. The Arabs had gone, leaving the fine silver sand and yellow rocks covered with blood. Another group of the guys picked up a trail of bloodstains which they followed up the hill above the wadi till they found an Arab who had somehow crawled up the steep slope and hidden under a rock. He was very badly wounded but refused to give up. He stupidly opened fire and they shot him dead where he lay.

In the wadi bottom, I looked round the scene of the battle. In their haste, the Arabs had left behind shoes, clothes and little bags of their belongings, including money and watches which they used as currency. Needless to say, these were pounced on by the chaps. We worked out there had been about twelve of them.

Behind us, in cover back up the wadi, our patrol signaller had been busy. As soon as early morning atmospheric conditions allowed, just before dawn, he had tapped out a signal to our base calling for helicopter casevac and just as it was light, we heard the beating din of a navy Wessex from the Royal Navy carrier standing off Aden port. Someone talked the pilot down on a Sarbe ground-to-air radio but the wadi was too narrow for the helicopter's big rotors. The pilot pulled off to circle round while we carried Dave and Shep as carefully as we could up the steep sides of the wadi. The sweat poured off us. The Wessex tried again, but was still unable to land. Finally, the pilot hovered about 50 feet up – his rotor blades terribly close to the rocks – the loadmaster winched them in and the Wessex pulled away en route for the hospital.

While all this was going on, the other half of our patrol picketed the heights above the wadi, just like the British Army had done high above the Khyber Pass. When we prepared to move on, I found the place where I had shot the first Arab. Typically, the others had dragged him off. The Arabs hated leaving their dead and I was certain I had killed him. The rock slab which had been behind him was covered in blood and there were six holes, in a good group, chipped deep into the stone. I also discovered that the sharp explosion at my feet the night before had been a British No. 36 grenade which had failed to explode. The detonator had fired, blowing the casing apart, but had failed to set off the main charge. I had been lucky. I had survived.

I felt good, I felt fit, I felt hard. This was the first time I had been in a contact and killed anyone. This euphoria was nothing to do with ending another person's life. I felt good because I had not panicked, I had not let down my friends, I had reacted as a professional soldier trained by professional soldiers, and the excitement of the firefight had been nothing short of fantastic. I've never taken drugs but I can't believe there's anything which can equal the thrill of a battle. I loved it.

This was no surprise. I was a very aggressive young man.

I've been told that there is a balance in everyone between what we learn from our families and what we are born with. If that's so, I think I was given more than my fair share of extremes in both cases. I'm making no excuses for myself, but you need to understand my background.

I was born on September 7, 1942, at Number 15 Kenmore Street, Shettleston, in Glasgow. My father and grandfather were miners and life in Glasgow was rough. We had moved twice by the time I was 5. First to Number 290 Carntyne Road in Carntyne, where three families shared our 'house', and then to Number 15 Lethamhill Road, Riddrie. This was a block of brick-built Victorian tenements which belonged to Barlinnie Prison and had been built as living quarters for prison warders. In 1947, they were empty and derelict, and because life was so cramped in Carntyne Road, a number of families moved in to Lethamhill Road as squatters. There was no caring Department of Health and Social Security (DHSS) then, and the council had nowhere else for us to go. Lethamhill Road was in sight of the prison, so it was not long before the prison authorities called the police to evict us. My father was away, either in the army or in prison (or both), but my mother gathered up her family – me; my elder brother Billy; Molly, who was only three; and her youngest, Rose, who had only just been born – and the police took us all off to the prison. They gave up with us soon enough, and let us stay in Lethamhill Road, but there aren't many – even in Scotland – who have the dubious claim to have been locked up in Barlinnie Gaol at 5 years of age.

I idolised my father, but it was not till later that I understood his life. He was a typical Glaswegian mixture of Catholic Scottish and Irish, and a miner. In the war, he was called up for the army, which promptly sent him to a Welsh regiment. As a result, he hated the Welsh and was constantly in trouble. He was in the army until 1953, so I saw little of him till I was ten. He was either away from home in the barracks in Brecon, or in prison, sometimes in Barlinnie, which had a military wing in those days. He had a cell overlooking Lethamhill Road which I could see from the street. He hung things out of his window as signals, like socks or a shirt of different colours, to show my mother what sort of mood he was in. They had known each other since he was six and she was four, and they adored each other. She must have loved him to have stayed with him and, in spite of all he did, there was never any doubt about the strength of feeling between them.

I grew up thinking that having a good time was getting blind drunk on Friday and Saturday nights and fighting. I remember when I was very young spying through the keyhole at my father and my uncle Billy shouting and battering each other with their fists in the front room. My uncle Billy was a vicious fighter. He would grab anything to hand to win, but my father got the better of him. He was a clean fighter, well respected as such in the neighbourhood, and Shettleston was as rough a part of Glasgow as any. The men in our area said the boys from the Gorbals came to Shettleston to learn how to scrap.

My father was leader of an illegal gambling school which he

ran on rough ground under a railway bridge near Carntyne dog track. It was well sited. They could carry on even if it rained, and if the police came, they had several routes to run away. He liked to wear smart suits, wide silk ties and had a fine pair of yellow shoes. He was the referee, or toller, nicknamed 'Kiter' as he flicked the pennies and took a portion of the money bet on the pitch and toss. Other men tried from time to time to beat him up and take his place. I remember the terrible shock when he came home one day and I could see in his eyes that he had been beaten. John Doran, a big man from another strong mining family in the area, had beaten him up. I hated to see how disgusted he was with himself and to see his sense of failure. He knew he had been knocked down because he had been in bad shape and he went to my grandfather for advice.

'Old Miles', as we all called my grandfather, was 60 years old then. He had been welter-weight champion of Scotland and was feared in his day. By then he was a steady man of experience, a fund of good sense, the sort of person who walked round all day in the house with his braces round his legs, and hoisted them over his shoulders to go out. He put my father through a punishing six-week training session, just as he had done himself at the boxing gym and in the army. The lesson was not lost on me for the future. I watched my father running and working out with weights, and the two sparring together. Then the challenge was issued to John Doran and all the men met on top of a slag heap, which we called a 'bing', to witness the fight.

I went and hung around on the edge of the crowd. I had to see what happened. There must have been 200 miners there, all our family and the Dorans in large numbers. My father, Kiter, and John squared up stripped to the waist and their seconds finalised the details. I watched them put up their bare fists and the fight started. Old Miles's training paid off. In moments, Kiter battered John Doran to the ground. He stayed down. In the pause, one of the Doran family remarked that the fight had been rigged. This was fuel to the fire. Without a second's delay, my uncle Billy drew a long bayonet from under his coat and stabbed the Doran man in the chest. At once, everyone started fighting. Maybe at first some were merely trying to separate the two, but in minutes the bing was covered with miners all wrestling and beating each other. But Kiter won.

He liked to keep ahead of the game. Someone came to tell him that a local villain called 'Crosby' was coming for him, to move him off his tossing school pitch. And, they said, Crosby had a gun.

'Fine,' said my father casually, and did nothing more about it.

In the early 1950s, few people – even where I came from – had guns and everyone waited, hanging out of the windows, to see what would happen. Hanging out of the windows was everyone's pastime. They could watch the neighbourhood and hear the gossip. Some claimed it was the only way to get fresh air.

Just as he said he would, Crosby walked over the derelict land to Kiter's pitch and toss group under the railway bridge. I was amazed as Crosby produced his gun, a small revolver, grinning fiercely with triumph. The other men backed off, waiting to see what Kiter would do. My father just opened his suit jacket both sides with both hands. Tucked in his belt were two guns, one on each side. Impressed and defeated, Crosby turned and walked off.

In addition to running the local gambling school, my father had other responsibilities. He was in charge of the tenants' association for the building we lived in. The families all put in half a crown a week to pay for repairs, but most people fell behind with their payments, and when a plumber was needed to unblock the one toilet serving sixty people, my father decided to do the job himself. The cause of the blockage was a miscarriage. Such was life in the slums. My father was a reluctant soldier. While in the Welsh Fusiliers he went absent, got caught and was remanded in custody in the guardroom, pending court martial. But in order to be court martialled, he had to be wearing uniform, and this he refused to do. On his second day of confinement, his civilian clothes were stolen while he was having a shower, and he was offered a uniform instead. But he refused to wear it, and for over a week, in mid-winter, this Gandhi-lookalike with shaven head and bare feet walked about the guardroom dressed only in a blanket.

During the Second World War, the army could not afford to have people like my father – people who continually questioned the system – but he certainly never lacked moral courage. And in more recent times he might even have been given credit for being an original thinker in his protests.

Unlike my father, my grandfather was a man with genuine military service in action. His stories had a profound effect on me. He too was a miner, a tough man who had joined the Argylls at sixteen and served his time in the trenches in the First World War. He told me that fighting in France was much the same as fighting in Glasgow, only he was allowed to shoot the opposition. He was wounded on three separate occasions, once badly in the thigh during the Battle of Loos, in Belgium. He lay down on a bank behind the trenches to wait for a horse-drawn wagon to take him to a field hospital and quietly lit his pipe. An officer passed by and when Old Miles did not move the officer began shouting at him to stand up and salute, and accused him of malingering. Old Miles merely took out his pipe, lifted his kilt and pointed with his pipe at the bloody hole in the top of his leg. Attitudes to suffering and survival were different then. Old Miles and the others in the First World War had no helicopters to swoop down and casevac them to hospital, no penicillin, no handy syrettes of morphine nor any of the modern drugs and well-stocked field surgical units for casualties on the battlefield.

Old Miles told me he respected the Germans. They fought hard and fair, and only once did he let himself go, when he heard that his brother, Alexander 'Sanny' McAleese, had been killed in the fighting around Neuve Chapelle. He got very emotional as he talked, sitting in our kitchen at home, his braces round his legs, and admitted to me that he had gone mad at that and killed some German prisoners. He said, "It was the only time. When I heard about Sanny, I let the war become personal."

Looking back, remembering the look in his eyes as he told me the story, I think he knew he had done wrong, but there was a time later, in Rhodesia, when I came to understand the same madness.

To his family and friends, Old Miles was a people's person with a strong sense of justice, and one story of his stuck firmly in my mind. He said there were trench police, like military police, whose job was to prevent desertion, and they were hated by the troops. One day, the Argylls had come through a town on their way up to the front and my grandfather and his mates had found a wine cellar in an empty house. There was no time to take any wine then, but one of them nipped back from the billet they took over that night nearer the front. He was caught going back by the trench police, who accused him of deserting. He admitted he was trying to steal wine, but no one believed him. He was caught in the system and my grandfather's face hardened as he said, "Desertion, they said it was. Before we knew it, he was ragged, tagged, bagged and shot by the firing squad." What a price for a drink. Come to think of it, the British trench police don't sound a lot different to the Iraqi special squads which Saddam Hussein used to stop deserters during the Gulf War.

My formal education, as they call it, was short, sharp and very strict. Being of a Catholic family, I was sent to St Thomas's Primary in Riddrie, which was run by nuns with a truly Catholic vision of what was right and wrong. We were all terrified of Sister Loyola, a tall, stern woman. All us kids reckoned she had been rejected by the Gestapo for being too harsh. When I was only six, Jimmy Dewar peed on me in the toilets so I peed on him. Most young kids have pee fights but we were caught and reported to Sister Loyola, who clearly had no idea what we had been doing. She was appalled. For her, this was proof of the degeneration of our minds and souls, and by the time I got home after school, she had told my mother that Jimmy Dewar and I had been "interfering" with each other in the toilets and recommended: "He should be stripped naked and beaten till he cries." My mother was a fine woman, a Catholic but a practical woman too. This was too much for her and she preferred to believe my less apocalyptic version of the story.

Most of the people in Riddrie were better off than us. The area was quite posh except for the tenement block slums in Lethamhill Road where we lived, and I was very self-consciously aware at school that all my clothes were handed down to me after they had been used by my older brother Billy. To be fair to my father, he did give us a new rigout once a year, on Easter Sunday for Mass: new trousers, new shirt, with a sleeveless grey pullover and 'sanies', sandshoes. However, most of the year I went to school in Billy's old clothes, and as I wasn't very big, I was always flopping about in baggy shorts, a worn jersey hanging off my shoulders and oversized shoes. Maybe this wasn't unique, but most of the other kids were better off, and their fathers had jobs. What really hurt was the meal-ticket system. All us children were given a ticket at the start of the week and a nun would punch a hole in your ticket each lunchtime, five times a week. That was all right, but an awful distinction was made. Children whose parents could pay for their lunches had buff-coloured tickets, but if you were 'on the parish', or your father was in jail, you were given red tickets. I always seemed to have a red ticket. It was like holding up a flashing red light, a beacon of shame, and I would hear people saying in low voices as I queued for meals, "They say his father's away working in the tunnels at Pitlochry, at the new hydro-electric plant," which meant they knew perfectly well he was in jail. To make matters worse, with a red card, I was given an old paper bag of rations every Friday to take home for the weekend!

We had hours of religious instruction from the nuns and I used to 'plonk' school to work on a milk float instead. One day, Father Brett, whose reputation was worse than Sister Loyola's, caught me on the milk round and made me come up on stage in front of the whole school.

"This child here chose to deliver milk," he shouted out to the entire hall, his Southern Irish accent making two syllables of the word 'milk'. "He chose to work rather than attend church!"

I cannot remember what else he said, but the mortal danger to my soul by missing religious instruction was nothing compared to the terrible humiliation of standing there feeling very small and exposed to ridicule in my worn clothes, with my hair too long and those great big shoes of Billy's banging about on the hollow wooden stage.

We had to go to church every Sunday and of course some of us plonked that too. One Monday morning, Thomas O'Donnell and I were told that Sister Loyola had noticed we'd been missing the day before and was gunning for us. She wanted to see us. That meant we would be subject to one of her famous interrogations and a certain hiding.

Terrified, I was called to her office. "Good day, Sister Loyola. God bless you, Sister Loyola," I said, holding my hands balled together on my chest as I walked in, my head bowed. God help us if we didn't always address the nuns and teachers this way.

"Peter McAleese!" she boomed at me sternly in her strong Irish accent. "You were not in the children's section at church yesterday morning!"

"Yes, Sister Loyola, I was," I lied, damned if I was going to take a beating without a fight.

The interrogation began. Sister Loyola wanted all the details. "Who was taking Mass?" She glared at me from behind her plain wooden desk.

"Father Courtney, Sister Loyola."

"What colour were his vestments?"

"Green, Sister Loyola."

"Who were the altar boys?"

"Tom McGrady and Pat McSweeney, Sister Loyola."

"Why weren't you on the register of those attending Mass?"

"I got there late, Sister Loyola, just before the consecration." Missing consecration was a mortal sin but, at that moment, Sister Loyola and her beatings were more terrifying.

She did not let up. She pressed me on details of the epistle, the gospel and Father Courtney's sermon. But I held my own. I had carefully learned all the answers from the other kids at playtime.

"Well done, Peter, you are a good boy," said Sister Loyola finally. Her bleak face broke into a smile. "You can have a coconut ice."

Thomas O'Donnell did not do so well. He told the truth, pleaded guilty and got a sound thrashing.

I don't know what happened exactly at St Thomas's but something went badly wrong there. I went back not long ago and felt the atmosphere of the place creep over me again. Sister Loyola and Father Brett were just names to the nuns there when I returned but they kindly showed me the big ledger with all the McAleese family names written in it. They looked very beautiful there, in that neat calligraphic handwriting, but it was hardly a true record of the reality of Lethamhill Road. Then I walked round the schoolyard and I was overcome with emotion. I hesitate to say this after all the things I have seen and done since, and some people who know me might find it hard to believe, but I could not stop myself crying. It wasn't just nostalgia or thinking of the one teacher, Miss Crawshaw, who I liked and who treated me as a person, and who was the only one to encourage me. It was the power of the place and the rigid Catholic discipline exerting its influence again. Perhaps Sister Loyola, Father Brett and the others were hard because they faced educating us in a hard, unforgiving society, but the iron rules of their education showed us no spark of freedom or opportunity to find a way out of the likes of Lethamhill Road. I spent seven years there, between the ages of 5 and 11, terribly formative years, and I think this place was as much a part of me as my family background and my own character.

In 1954, I left St Thomas's and went to St Roche's Junior Secondary School. In between the classroom, my education in violence continued and I became used to running round the streets in gangs. Like my father and uncle Billy before me, fighting rivals with sticks, iron bars and knives became second nature. I went round 'tooled up' with an axe or knife hidden under my jacket without thinking. I preferred a long bayonet, not for stabbing but whipping the 'enemy' about the head with the thick steel. Not surprisingly, Mr Kelly expelled me from St Roche's for fighting and I went instead to St Mark's for the last two years of schooling, which I finished aged fifteen.

To understand my behaviour as a boy, you have to understand the subculture of the slums. Success in that environment was not measured by the usual indicators – a good job, a nice home or a car. It was measured by how tough you were. Could you handle yourself, could you 'go ahead' (fight); were you capable of a 'comeback' (a return fight after you'd been beaten)? One local called Tommy Shaint was obsessed with the Westerns featuring a character called 'Edge', and used to use Edge's favourite weapon, a razor. He had a comeback in Kenmore Street, and I remember Tommy's adversary leaving the scene minus an ear. But after a comeback malice was uncommon. The pecking order was maintained or reshuffled and everyone then carried on as normal.

People hated the system that supported them, and felt it was there to be 'knocked' (abused). After joining the army, I came back on leave once and found one of my mates had strapped a vacuum cleaner hose to his gas meter in order to blow the numbers backwards. When I was growing up, a more common method was the potato trick. A small hole was drilled in the electricity meter and a needle inserted with a potato attached to act as a weight and stop the wheel from spinning. A character known as 'Johnny Boo' used to take the bulb from a streetlamp and attach a cable from his house, thereby enjoying the privilege of free electricity during the hours of darkness. Nowadays, things are more sophisticated, and professionals install switches into the system that allow electricity meters to be bypassed at will.

An outsider might be horrified by all this – the violence and the fiddling – but in the slums this was the norm, where the cramped conditions determined the values people lived by. However, my childhood memories are far from being uniformly grim. Easter Sunday was great fun. It meant new clothes and Easter eggs, and seemed to be the beginning of the Catholic social calendar; the weather picked up and school holidays were discussed. For myself, I used to work for a farmer called Sanny Mungall. The fresh air and milk which he used to let us have did us the world of good, and while he used to work us hard, deep down we knew he had a soft spot for us. Sadly, his farm was eventually eaten up by the encroaching housing estates.

Courting, or 'winching' as it was called, took the following form. The prospective candidate was 'eyed up' and normally approached in the local café owned by an Italian called Johnny Matteo. A date was made and the couple used to go to the pictures. With each visit they moved further back in the rows, and when you reached the back row you were winching. Soon after that, the couple would marry and move into a 'single end', a one-bedroom flat with an inset bed. As a rule, the marriages used to fare well in the beginning; there was employment for young people in the whisky bonds or the carpet factory. The problems began when the children came along: the income was halved, the single end became cramped and the men would start spending their time in the pub. Once the men reached the age of 21 they would lose their jobs, that being the age when employers would have to pay adult wages. The men were then on the 'buroo' (the dole) and the pressure was on. I don't claim that the whole of Glasgow was like this, but it is definitely the way it was in the East End suburbs of Shettleston, Parkhead and Carntyne. There were only two options if you wanted to escape this existence: move to England or join the army. I chose the latter.

I had wanted to be a soldier from an early age. Old Miles was a great influence with his stories of the war and I picked up everything I could about the army, especially about the Parachute Regiment. You might say I was obsessed. I saw the film *The Red Beret* seventeen times! I learned later that this was about the Bruneval Raid, but I can still remember Alan Ladd as Private MacKendrick, nicknamed 'Canada', Stanley Baker as the parachute jump despatcher Sergeant Breton, and Leo Genn as

Lieutenant Colonel John Frost, who was called Major Snow in the film.

And I laugh when I think of the day when I was 12 and I saw a Parachute Regiment soldier walking along Lethamhill Road going home on leave, kitbag over his shoulder, and I followed him through the streets. I was fascinated by his smart uniform, his brown and green camouflaged smock, his neatly pressed khaki trousers, his polished boots and immaculately rolled puttees, and most of all the shiny silver parachute badge on his red beret. I followed him for quite a long way and of course finally he lost his temper. He swung round on me and asked in utter amazement, "What the fuck are you doing?" It's a pity now that the Northern Ireland Troubles mean that soldiers proud of their uniforms daren't wear them in public.

I had a few years to kill before I could join up, so in 1957 I went to work in a Teachers' whisky bond (a bonded warehouse) for a while and then started as an apprentice plasterer. However, the following year my elder brother Billy and I left home. We went to work in Aberdeen and I found a job as a fish porter in the docks which I kept for two years, drifting back and forth to Glasgow, before I got sacked. Then Billy quit, went on the dole and was immediately picked up for his National Service. He went into the Ordnance Corps.

I joined the Parachute Regiment. On March 1, 1960, at seventeen, I signed on for nine years and began a crucial grounding in my soldiering career.

I was sent south, to Aldershot, where I was put in intake Number 197. I became very fit, and I must say I loved it. This was the first time I had been out of Scotland and the things that stick in mind are absurd, looking back. The accents from all over the country were utterly confusing to start with, the water was hard compared to the soft waters off the Scottish hills and English bitter tasted foul. However, the training was excellent and I absorbed myself in it totally. In the recruit company were my intake sergeant, George Brown, CSM Tovell, Company

Commander Major Maurice Tugwell and my officer was Lieutenant Duke Pirie, the first direct entry officer to the Parachute Regiment who was later 'B' Squadron Commander in the SAS, and quite coincidentally lived next door to Dave Tomkins, of whom more later. I took pride in my uniform and appearance. I was slow, but once I grasped the system, I think I did quite well. Lord Mountbatten took my passing out parade in June that year, and the following month I passed my parachute course at Abingdon, near Oxford.

My first posting was to I Para in Albuera Barracks in the old Wellington Lines at Aldershot and I was put straight into the Mortar Platoon in Support Company. This was unusual, in that most new

soldiers went to a rifle company first, and to begin with I was given a hard time by the older hands. Support Company men were normally the most experienced in any battalion, but I was beginning to learn the army system and settled in.

At the depot, I found out that while the Parachute Regiment is an elite among infantry troops, there was another regiment which demands higher standards of all its soldiers. I applied to join 22 Special Air Service and formed up to speak to WO 2 Johnson, the Support Company CSM.

"Permission to speak, Sergeant Major, sir!" I said, standing rigidly to attention in front of him. You could not speak to the CSM any other way.

"Yes?"

"I want to join the Special Air Service, sir."

He just looked at me, a faint mischievous smile on his face, and replied, "The SAS? We send all our shit there. All they want to do is go into the jungle and build tents." Then he just walked away and left me!

However, I applied and took selection in January 1962. There is a lot of rubbish written in the press about the SAS but it is true that selection over the Brecon hills is hard work. There are 400 miles of tabbing as fast as you can over the mountains, carrying a rifle and increasing weights up to 65 pounds. In a nutshell, people of all shapes and sizes pass simply because they want to, but it's sweaty work. I was in good shape, but I was only nineteen and my map reading was still shaky, so I scraped through on my fitness and determination.

My SAS continuation training was excellent, run by professionals such as Sergeant Tanky Smith, Sergeant Alec Spence and Captain George Morgan. It took seven months. Most of it was based in Hereford, though we spent an invaluable six weeks jungle training in Grik in Malaya.

I was posted to 'D' Squadron, where I was one of the youngest, and we went to train with the United States special forces in Fort Bragg, North Carolina. Here, I got into trouble. A group of us were off duty in a bar, when an argument started with an American who insisted on pressing me for a fight, so I obliged him. A squadron corporal took the American's side and I took him on too, twice, because later he tried to creep up on me in the middle of the night when he thought I was asleep. The squadron officers took a dim view of all this, perhaps because they were embarrassed for our American hosts. I certainly broke the rules, fighting with senior ranks, and I believe I would have done the same as they did faced by a person like me, but it would have been nice if the other side of the story had been aired too. Anyway, I found myself sent back to the Paras. This was bad news, but I make no excuses. We were out on the town, and, as I have said before, I was aggressive. Besides, the American and the corporal both deserved it!

My father had a more severe experience during the war. As I've said, he was posted to a Welsh battalion, and there he hit an officer. I can't remember the reason, but hitting officers is not encouraged, especially not in a national army in wartime. The discipline was not as severe as that practiced by Old Miles's trench police, but you can't have soldiers smacking officers and getting away with it. Kiter was sent away for two years. Barlinnie again. Even volunteering to fight in North Africa was no good. They made him serve his two-stretch. So, for me, going back to the Paras was not so bad.

At any rate, that's what I thought till I found myself in ' A' Company of I Para. The company sergeant major was a famous Parachute Regiment name, Nobby Arnold (later the regimental sergeant major), and the company commander was Major Jack Thorpe. Discipline was manic and our love-hate relationship with these two bound us all together. 'A' Company's reputation was such that, if the men in other companies stepped out of line, they were threatened with being posted to join us. Michael 'Beau' Geste himself would have cracked in 'A' Company!

At least we travelled and I saw something of the world when few people from my background could, and it was good military experience. In December 1962, we went to Bahrain in the Gulf, on standby to reinforce Kuwait in case Iraq attacked. I don't suppose Iraqis have changed much but they were not led by a megalomaniac then. They made a lot of noise but stayed their side of the border, so we spent several hot months over Christmas constructing Hamala Camp, breaking stones for the buildings with pick and sledge under the glaring sun, like prisoners, and we trained in Shajah in the desert. All this was a tough apprenticeship, not much like the plastering I had started in Glasgow, but these years gave me a solid base in soldiering.

I did one trip in a blue beret for the United Nations, right at the start of 1964. All I Para was on Christmas leave and I was in Glasgow when a telegram arrived on New Year's Day telling me to report back to Aldershot at once for an emergency deployment. Drinking was forgotten and I took the train south in great excitement. Hasty preparations were made and we were flown out to Cyprus with the UN. In my ignorance, I hoped for action. Instead, we spent three weeks in Dhekelia sitting about on our backsides in a tented camp wondering why we had been rushed back off Christmas leave. We concluded that the Paras didn't like us having Christmas at home.

Colonel John Woodhouse allowed me back to the SAS in

October that year and I suppose I was lucky he never made me do selection over again. However, I was very fit and had learned a lot in the intervening two years. I picked up all my old kit from Drag Rowbottom in the stores, where it had been kept in a container, and joined 'D' Squadron again on a training trip to South Arabia.

Being in Aden again brings me back to where I started, so I will explain something of the situation. The background is simply that Britain wanted out. This was officially called decolonisation, but on British terms. For this, in February 1959, the Conservative government invented the Federation of South Arabia, roping together a motley crew of sultanates, emirates and sheikhdoms, some of which had been enemies for generations. Aden port was supposed to be an economic jewel which would generate income and tie in the tribesmen of the interior. The British supported the new Federation's Arab leaders – the traditional rulers of these different groups – formed the British-officered Federal Regular Army and the Federal National Guard, and promised to keep troops there in support. This was the crucial part of the arrangement. HQ Middle Eastern Command was established in Aden in 1960, having been chased out of Egypt, Palestine and Cyprus, and Britain was keen to keep naval and air base facilities in Aden, which was still seen then as an important staging post to the Far East.

However, the tribes in the interior wanted total independence. I suppose they did not trust either the British or the town Arabs ruling Aden, and they were stirred up with revolutionary Marxism by Egypt, where Nasser was top dog and hated the British after the Suez debacle. Russia was behind him. They kept up a stream of anti-British radio broadcasts from Cairo and Sana'a in Yemen, and supplied plenty of arms carried over the border on camels.

Anyway, that was what we called 'the big picture'. On the ground, at troopie level, I went every year to Aden from 1964 to 1967, and each year things got worse. In 1964, there were only thirty-six terrorist incidents and thirty-six casualties, including two British servicemen killed. Then, while I was busy learning

Arabic on a six-week Arabic course in the Command Arabic Language School, Harold Wilson defeated the Conservatives. To start with, the new Labour government followed the existing policy, but it made no difference. Figures rose through 1965 to 286 incidents, 239 casualties with six servicemen killed and eighty-three wounded.

It's clear now that people in all the countries which had fallen under control of the British Empire wanted their freedom, but even when Labour announced independence for South Arabia and the withdrawal of all British forces, matters did not improve. In fact, the British announcement had the opposite effect. It completely undermined the existing Arab Federal leaders who depended on the British for their power. The strength of the tribes in the interior grew, and the terrorism increased in the streets of Aden itself, making some areas completely 'no-go' to British troops. The result? Soldiers and civilians died in greater numbers. At minimum, it became common for troops being trucked about the town on mere admin trips to expect to be fired on. So, by 1966 there were 480 incidents with 573 casualties, five servicemen killed and 218 wounded, and in 1967, the last year I went there and the year the British pulled out, there were nearly 3,000 incidents, with 1,248 casualties, forty-four servicemen killed and 325 wounded. It was an ignominious retreat, inevitable for more than two years before it actually happened, and a lot of soldiers died for nothing.

Compared with the line infantry, who had the thankless job of policing Aden, the SAS was fortunate in being given jobs with more purpose, based on what intelligence there was, with a reasonable chance of finding the enemy, like our Wadi Mishwarrah operation, and at least we had the opportunity to create a greater impression on the local situation through 'hearts and minds' work. So, I was sent to the fort at Ataq – about 200 miles from Aden near the Hadramaut border in the east, where a Federal National Guard battalion was based – to improve my Arabic, report what went on and show British support for the FNG.

We lived in a stone fort like something out of a Foreign Legion film – *Beau Geste* again – with sand, rocks, steep hills, goats, and thorny acacia trees around mud huts. The place was hot and dry as dust. However, my Arabic improved and I began to create a good rapport with the Arab FNG soldiers. They had FN rifles which they loved as they were heavy, looked the part and fired the powerful 7.62mm NATO round. One day on the range, which was just a piece of dry empty hillside nearby, they were firing a close quarter combat (CQB) practice and kept pushing me to join in. I had nothing more than a 9mm pistol but I could never resist a challenge. I suddenly drew my pistol, twisted round and fired at the target, a tin on a rock. To their delight and amazement, my first shot sent the tin flying and my second hit it again in mid-air. The Arabs put a great deal of store by good shooting and this really impressed them. They tried to persuade me to fire again but I coolly refused. I was just as amazed as them and I knew perfectly well it had been sheer luck, probably the product of good instinctive shooting, but I had no chance of repeating it and no intention of losing my advantage by trying.

Things did not always go so well. A local Arab came to me one morning with a sick goat. Its eyes were badly infected and seething with maggots. I had done a short medical course and decided to impress the Arabs with my concern for their goat. I explained this and they sat about watching in awe as I pulled out a syringe, drew off some penicillin and jabbed the goat. To these people, the height of good doctoring was to slap camel dung over a burn or a cut. Injections were 'white man's magic'. They watched fascinated as I cleaned up the wretched goat's eyes with swabs and antiseptic, until suddenly the goat leapt out of my hands and started jumping about like a dervish. The circle of Arabs backed off in alarm as the goat flung itself on the dusty ground, kicked its legs in the air and died. I had given it too much penicillin and the poor thing was allergic. It had gone into anaphylactic shock followed quickly by a total collapse of blood pressure and then death. The Arabs moved off in stunned silence, their confidence in white man's magic severely dented.

However, I got on well with them. They loved having their photograph taken and I often sent films of various characters in Fort Ataq down to base in Habilayn Camp for developing. Perhaps because I could use a camera, which they had never seen, they called me 'jaasoos', and for a long time I translated this as 'the elite' and felt rather pleased with myself. Then I discovered they actually meant I was a 'spy', mainly because these people tended to call any outsider a spy, but that was not how I saw myself at all.

However, even spies had their uses and when a dissident blew up the Khalifah's house in Ataq village, the FNG asked me to go on a patrol with them to level the score. The man had run away to a village called Bihan, a cluster of mud-brick flat-roofed houses some miles away in the hills. We drove there in Land Rovers over the bumpy desert and while the FNG picketed the high ground with watercooled Vickers machine guns, I was taken by a squad of Arab soldiers into Bihan and shown the culprit's house. At that time, I had never blown up a real target before, and I was conscious this was not the time to foul up. Recalling my demolitions instruction, my plan was to set off a fuel-air petrol bomb, vaporise the petrol all around the room and then ignite it. The ensuing conflagration uses up all the oxygen at once, creates a vacuum and implosion brings down the house. The FNG soldiers watched closely while I set up a charge attached to a petrol bomb dead centre in the main room of the mudbrick house. I ignited the safety fuse and shoved the curious Arabs out of the house, or I am sure they would have stayed to see what happened next. In seconds the charge blew, and the walls and roof lifted slightly and then totally collapsed inwards. This gave me a lot of satisfaction, seeing the theory work in practice. Frankly, I doubt it did much good for the cause of the Federation, but blowing up suspects' houses was the way it went then. Indeed, the Israelis still do it to Palestinians in Gaza.

Weakness was almost a sin for those Arabs. Perhaps it was the harsh, arid environment, but those who didn't measure up were rejected by their society. Outside the fort in Ataq, on the edge of the few drab village houses, the Arabs kept a man chained to a rock. There were no padlocks; the chains were just tied and knotted round him in thick coils. He was plainly a real villain, but when I asked what he had done, the FNG officers wagged their fingers and said warningly, "*Majnoon! Majnoon!*" He was mad.

In amazement, they agreed to let me take him for walks but on no account to untie his chains. So, every day, I got them to loosen him from his rock, and I gave him water and took him for walks in the desert on a length of chain, like a dog. Nick Angus was with me at the time and I suppose we made a bizarre sight, brown soldiers in shorts and Clarks desert shoes either side of this lunatic festooned in chains, but we practised our Arabic and he enjoyed the break from the awful monotony of his life.

I came back to Aden in February 1966. We did our build-up training at Sennybridge after Christmas and our troop commander was Captain Robin Letts. He had difficulty understanding that rifles zeroed in freezing damp conditions in mid-winter in Wales are not likely to keep on target in the blazing dry heat of Aden, but he was a good troop officer and took us on an operation in the Wadi Bila that year. The problem was always that the tribesmen seemed to know exactly what was going on in their areas, and we were always breaking new ground to keep ahead of them. Just creeping into the mountains by night was no longer any good in 1966. The Arabs had their own pickets on the tops of hills and every goatherd and small boy tending the animals was part of their lookout system. So, our squadron commander, Major Glyn Williams, known to us all as 'Punchy', devised a new scheme for the Wadi Bila operation which included a cunning deception plan.

The Wadi Bila was a deep and wide valley surrounded by steep, arid mountains. Two troops, 16 and 19 Troops, walked in at night, as we had on the Wadi Mishwarrah Op, and we hid on the tops of the mountains in rocky sangars on the reverse side of the slope, out of sight of the wadi. In 16 Troop, my patrol was led by Corporal Mick Welsh. Sergeant Don 'Lofty' Large, who was a veteran of Korea, hid 18 Troop out of sight to the north.

We went into a typical day and night routine. Every night, just after evening stand-down, we left our hilltop position and crept down into the wadi to ambush tracks used by dissidents moving under cover of darkness. In the small hours before dawn, we sweated back uphill to our sangars, where we lay up and rested, keeping a watch through the heat haze on the surrounding hills. Inevitably, very early one morning, an ambush position was compromised by a goatherd, who scuttled off to tell the dissidents.

At once, our deception plan was put into effect. Captain Letts called in the Scout helicopters to lift out all 16 Troop and, using one of our Tokai hand-held radios (which the SAS bought on the civilian market because the army had nothing suitable), he deliberately gathered us on open ground in full view of the whole of Wadi Bila. When four Scout helicopters came in to pick us up, all Arabia must have heard and seen what was going on. We were flown back to Almeila Camp, where we dumped our rucksacks in our billets, washed up, cleaned our weapons and sat about ready for the expected call out.

In the wadi, the dissident Arabs thought they had flushed out all the British soldiers and happily emerged from their hiding places, but 18 Troop were waiting for them and Don Large radioed back to our operations officer that he could see twelve dissidents approaching in line, Mark IV .303 rifles casually slung over their shoulders. The reply was, "Can you be sure they are the enemy?"

Don Large was not certain. Most Arabs in the region carried rifles, especially old British Army .303s, and there was no way of saying whether these men had been responsible for recent attacks on Habilayn Camp.

"Then report only and do not open fire," ordered the ops officer, conscious that shooting innocent locals could do nothing but harm to the British position. However, just in case, he ordered us to deploy at once.

Crossly, Don Large watched the twelve Arabs walk freely through the killing ground right in the centre of his ambush, while in Almeila, we all sprinted back to the helicopters.

Most SAS operations involve carrying ludicrously heavy rucksacks and it was a real treat to be going out with just belt kit and weapon. It was exciting too: the beat of the chopper blades, seeing the other three choppers flying with us, hugging the contours, cutting over the mountain tops, and the certainty of action. As we approached the Wadi Bila, Punchy's Scout heli climbed high into orbit so he could see the whole area and place the rest of us in the wadi as cut-off groups, so 18 Troop could sweep through. These were the same tactics used by the Rhodesians ten years later, and should have been used more often in Southern Arabia and Dhofar. The British Army doesn't seem to have picked up the real usefulness of helicopters in combat as fast as other armies. There is nothing better than good air support, especially helis, against dissidents hiding in difficult country. The fly-in was exciting. The enemy in the Wadi Bila were well stirred up by 18 Troop and I could hear bullets cracking round us as our Scout came in. I noticed our pilot was blissfully unaware of the danger. He was blocked off from reality by headphones and Punchy's voice ordering him where to land. He dropped us on top of a hill overlooking the wadi, then he pulled away in a steep curling climb to avoid enemy fire, and by the time he had turned through 360 degrees above us, he had lost his sense of direction. He glanced down, mistook us for a group of dissidents, and plunged back onto us, firing the general-purpose machine guns (GPMGs) mounted underneath the belly of his helicopter. It was a good thing these Scouts, early gunships by anyone's definition, were not so well set up as the US Huey Cobras later on in Vietnam, with mini-guns firing thousands of rounds a minute. We scattered for cover behind the tiniest rocks. I dived into a pouch on my belt kit for my red picket panel which I waved frantically. Thankfully, he recognised this piece of red cloth, an identifying air-marker panel, and he immediately pulled away with no damage done.

'Blue on blue' incidents, as they are called now, are more common than people think. An SAS vehicle patrol in Western Iraq during the Gulf War was bombed by US A-10 pilots, fortunately without casualties, and of course fourteen British infantrymen were killed by US 'friendly fire' some weeks later in the land battle for Kuwait. In Rhodesia, we got round this problem when sweeping through the bush by putting coloured patches on top of our hats. There can be terrible mistakes but no one could accuse the pilots of doing it deliberately. I see it as part of the risk of war which can be minimised but not eliminated.

The last three of 16 Troop, with Sergeant 'Smudge' Smith and Dusty Gray, flew suddenly over a ridge to find the twelve Arabs walking coolly in single file across a wide flat area. As the Scout turned over their heads, the Arabs opened fire on the chopper, which pulled round in a tight circle and dropped them in cover under the ridge. The three ran back to see where the Arabs had got to.

Far above them, I lay in cover behind rocks and watched the tiny figures moving in the wadi below. We fired on the Arabs as they ran away, but it was frustrating being well out of the actual fighting. I could see Smudge Smith's group and Captain Letts's group moving carefully along the floor of the valley, flushing out the enemy, and the sound of firing carried distantly. I saw one of the three fall, and minutes later I heard on the Tokai hand-held radio that Smudge had been badly wounded. At considerable risk of being shot down, one of the Scout pilots landed to take Smudge to hospital. On the way, he noticed blood spattering all over the instruments, blown about by the slipstream through the open doors, and realised it was his. He was lucky to survive.

I could also hear Punchy Williams high above in the Scout talking on the Tokai net, saying he had seen a group of enemies down one side of the steep spur I was on. I could see them with my binos and confirmed the sighting on my Tokai radio. Directed by Punchy above, Robin Letts's group gradually moved up the wadi and cornered several enemies at an impassable wall, where they killed two and wounded another four in close fighting with grenades. Trooper Dodds was awarded the Military Medal for his part in this engagement.

The fighting went on all day, with outbreaks of shooting and skirmishes, and we returned to Almeila Camp by helicopter in the afternoon, in time for a shower and the evening meal. For me, it was a good day, and unusual: in and out by chopper, showers and eating in the cookhouse. What a great way to soldier! The squadron had done well. By British standards of the time, and compared with most British counterinsurgency operations, two dead and four wounded was a great success. Don Large was cross he had not been allowed to shoot all twelve when he had them dead to rights in his ambush, which would have meant Smudge would not have been wounded too, but the ops officer had probably been right. I saw plenty of 'shoot first, ask questions later' in Africa and I do not think it helps any campaign.

The abiding recollection of Aden was the terrain, which was appal ling, in its way quite as bad as the jungle, and the whole episode illustrates the amount of sheer hard work, commitment, cost, resources and the professional techniques necessary to flush out and kill a very small number of the enemy.

Don't go away thinking that soldiers in the SAS are nothing but formidable fighting machines. I should mention an incident that occurred during my Arabic course, which illustrates a sense of the ludicrous to balance the fierce image of terrible firefights in the mountains. Much of our language training was done in laboratories wearing headsets with tape machines and repeating lessons into black, mesh-covered microphones. This was new, at the leading edge of technology in language training in those days, and the instructor – hard-bitten though he was – would have been appalled to see us adapting his shiny electronics to settle the winner of a farting competition. Ken Connor and Jake Allsop spent days eating nothing but peas and cabbage, and then one night, quite sober, we gathered round a microphone in the language lab to hear, and witness, the winning effort. The decibel level was registered on the tape-machine dial as, bare-arsed, Ken made a splendid contribution to science. These were real professionals. Jake was not to be outdone. He dropped his trousers, strained to swing the needle further, and shat all over the microphone.

Disgusting, I can hear you saying, but it amused us a lot, probably because we were living at the extremes. Hot, dusty days on hard, sweaty, dangerous patrols in unfriendly mountains, followed by more hot, dusty days in tented camps, where there was little to do after all the necessary weapon cleaning and admin had been done except rest, play sport and drink beer. That's army life. Certainly, SAS ops at best are a mixture of extreme excitement and action, followed by long periods of boredom. At the worst, it's all boredom and no action. I enjoy operations. I love the challenge and the commitment. I actually feel happy in action. But I am not so good with the boredom! I recall one afternoon in Habilayn Camp when Ken Connor and I had nothing to do and went to the Naafi Club, a large tent with folding tables and chairs on dusty coir matting. Much later, after dark, I said – as one does on these occasions – "Do you want another beer?"

Ken checked his watch and shook his head. "I've got to go to the Ops Room." He marched unsteadily out of the Naafi into the hot, dark night.

Two Guardsmen watched him leave. They were buying their first beers. One said to the other, "I've been looking forward to this all day."

The other wiped his sweating red face and nodded, gazing fondly at the beer being pushed across the counter towards him.

Their dream was interrupted by a distant whistling noise and a loud explosion on the far side of the camp, shaking the ground.

"Incomers!" I shouted, leaping to my feet and knocking over my chair behind me. "Mortars!"

"Fer fuck's sake!" muttered the Guardsmen as I ran outside.

I burst out of the lighted Naafi tent into the black night and saw the flash of explosions silhouetting the tents around. Enemy mortar rounds were landing inside the camp and the din increased as our 105s opened up on the hillsides where they thought the enemy were positioned. I blinked furiously to get my night vision and ran on. Only, instead of running straight down the path to the main avenue between the tents, I veered blindly sideways, tripped over a taut guy rope, and plunged head first into a deep slit trench. Most tents had trenches dug beside them, to shelter the occupants in such times of enemy attack, so maybe this was the safest place to be, but I landed on my head, split it open and passed out.

I came to with bright torchlight shining in my eyes. "Ere, Sarge! Wounded man!"

To my horror, I was lifted gently out of the trench by an immaculately dressed Irish Guards stretcher party, wearing highly polished steel helmets and shirts 'Khaki Drill' with shiny brass buttons, and laid on the stretcher on the ground.

"Let me up," I shouted, struggling.

"Hold him down, lads," the sergeant boomed, shining his torch over my bloody head. To me he said in a kindly voice, "The camp was mortared, lad. D'you remember? You've been hit. Nasty wound, by the look of it."

"No, I fucking well haven't!"

There was no arguing. The Guards had made up their minds. "You're concussed," said the sergeant knowledgeably. "That's common with head wounds."

And before I knew it, I was lying on an operating table in the full glare of the theatre lights and being inspected by a thoroughly outraged army surgeon in bloodied green overalls and rubber gloves, scalpel poised.

"This man is drunk!" he shouted. "Get out!"

I scrambled off the table and scurried off. I'm glad his training hadn't been entirely wasted, but what humiliation!

In 1967, during my last trip to Aden, 16 Troop did the British Army's first operational free-fall jump. I think it was also the first operational parachute jump of any kind since Suez in 1956, so you can imagine our excitement when the 'Sky God' free-fallers' talk actually turned into training for a real job. Robin Letts pushed for this op as a means of covert infiltration as, once again, we were trying to find new ways of getting patrols into the difficult terrain of the high mountains occupied by our enemy. Dropping silently out of the sky from 10,000 feet seemed worth a try. I presume staff officers with no experience of the desert picked the name: Operation Snowgoose.

I had been promoted by this time to corporal and the troop gathered for parachute training at Falaise Camp in Little Aden. For a change we lived in wriggly tin huts rather than tents. Para training is totally dependent on aircraft and the weather, and we were lucky. Headquarters authorised us full clearance to use Scout helicopters and a Beaver light aircraft flown by the Army Air Corps (AAC), and in nearly three weeks, we managed thirty-three training jumps. Modern free-fallers will not be impressed by this low number, but things were not very advanced then. First, we could only jump in the early morning and evening, as the winds picked up in the middle of the day. So, we rolled out of our pits at 5 AM for a couple of jumps before breakfast, finishing usually about 8 AM. Then we had to pack our 'chutes. Not for us the quick ten-minute field shake out and stuff back into the pack, which is all that is required for a modern square parachute. We jumped American T-10s, a parabolic-shaped 'chute which took over an hour to pack, with masses of complicated folds and rigging-line checks to make, all vital or you will pack in an automatic malfunction. So, after packing the morning 'chutes, we had lunch and a sleep before driving over to the aircraft for another couple of jumps between 4 and 6 PM. Later, we jumped with our rucksacks made up to a realistic weight and there is really nothing like a heavy bergen to spoil the pleasure of free-fall. Always at the end of the day there was more 'chute packing to be ready for the next morning. These were long, tiring days, but frankly I didn't care. I was too wound up with the chance of a real operational jump.

Captain Robin Letts gave the detailed operation orders in a stuffy tent in Habilayn Camp to a much smaller group than had started the build-up training. From a full troop, we had lost men through injuries, mostly twisted ankles and knees, so that by the time we were ready, there were only six of us left and I was acting troop sergeant. The team were Captain Robin Letts, Corporal

Peter McAleese acting Troop Sergeant, Tony Ball, Dave Gregory, Woody Woodford and Tony Dicker. A Royal Marine sergeant gave us the intelligence back ground. Our aim was to ambush a track junction and kill Mohammad al Maghrebi al Haushabi, a bandit leader whose most recent atrocity had been to attack the beer resupply wagon. We spent the rest of the day preparing and rehearsing; how we would group in free-fall, how to meet up on the ground after landing, and all the immediate action drills for every possible eventuality.

At one o'clock in the morning, a full moon lit the airfield and we walked out to the waiting Scout helicopters. Actually, we waddled, like penguins, weighted down with rucksacks hanging off our harnesses behind our legs. I could hardly walk and we all had to be helped into the Scouts. Most unusually for the time, the pilots had willingly agreed to break new ground, toss the book of regulations aside, and had removed the back doors. The AAC flight officer Captain 'Lowlevel' Greville Edgecombe gave us every support throughout.

I sat wedged in the back, behind the pilots' seats, helmeted and trussed like a turkey, my main parachute on my back, the reserve on my front, my rifle strapped inside my harness at my side, sitting on my bulging rucksack which was tied to the webbing under my legs, and sweating like a pig. I couldn't wait to get going. The blades started turning. We took off and climbed north over the moonlit desert hills which rose in line after line towards the high mountain range on the border with Yemen. The flight was short, maybe twenty minutes, and almost at once it seemed the pilot turned round and shouted, "Action Stations!"

Time! I struggled out of the back of the heli, let myself down to stand on the skids and held on to the floor rings as the slipstream whipped past. I looked across at the second helicopter and saw the other three clambering out, dark shapes bulky with gear from head to foot, their goggles gleaming in the moonlight. The seconds flashed past as we glided towards the invisible release point high over the target. Above were the rotor blades, the huge full moon and the stars. Below, in clear night air 10,000 feet down between my legs, was the pattern of moonlit ridges interlaced with wadis in deep shadow. And our target.

Nearly there. I glanced across at the other heli again, at the two men on the other side of my heli. Thumbs up. Ready.

I stared into the cockpit at the pilot waiting for his signal. It was vital we all left both aircraft at the same time, to fall together, to land together, or we would be scattered over the jebel one by one and terribly vulnerable on our own.

The co-pilot raised his gloved hand, twisting briefly to look at me.

I tensed, ready, adrenalin rushing.

The hand chopped down, and "Go!"

I shoved myself off backwards and fell rapidly away from the beating noise of the heli. What a relief! To be weightless in free-fall after the sweating struggle strapped up with kit like a parcel. I had a quick glimpse round at the others falling with me, face down, spreadeagled, like frogs, their helmets glistening in the moonlight. My relief was short-lived. As I reached terminal velocity after ten seconds, my rucksack began a mad gyration in the air rushing past, and I had to fight to keep stable and face down, throwing my arms from one side to the other to prevent the rucksack whipping me into a violent spin. We had no automatic BODs, barometric opening devices, to release our parachutes at a pre-set height, so it was vital I stopped the spin catching hold or I might lose consciousness. Grimly, I fought on as the seconds passed, rode it down to 4,000 feet above the ground and pulled.

The T-10 parachute had a fearsome opening shock. The harness straps ripped into my groin and I thought my eyes would pop out of their sockets into my goggles. But it opened and that is all that ever matters. I looked up. The parabolic canopy billowed satisfyingly above me, all the rigging lines in place. I relaxed a bit and checked the sky all around for the others. One, two, three, four, me, five. Fuck. One missing. There was no sign of the sixth. I guessed he had had the same problem as me, spun like a top and piled in.

The ground was coming up. Below, the track junction was exactly where we expected it. Smack on target. The AAC pilots and Green Archer radar had done an excellent job. I dropped my rucksack on its 30-foot line, put my feet and knees tight together for a good parachute-landing position and hit the earth. I dusted myself off and felt good. You never mind what happens to you in the air as long as you can land and walk away. I would remember that years later, but now I was fine. The other four landed close by, no more than 60 yards away. A quick roll call showed the missing man was Woody.

We looked up into the sky and I spotted a lone parachute, small in the distance. Several thousand feet above us, Woody was drifting off down the wadi. He appeared to be trying to drive against the wind rather than pull handfuls of brake lines down and spiral his parachute out of the sky before the upper winds could carry him away. Robin Letts was not very happy. He agreed to my suggestion that we put out a couple of two-man groups to search out from our position to try and find him. Two hours later, Woody turned up tired and drenched in sweat. Needless to say, our relief that he was still alive was tempered by more tactical concerns.

I snapped, "You seen any Arabs?"

"No."

"Where's your parachute?"

"I hid it under a bush," he admitted.

"A fucking bush!" I hissed at him. "I hope the goatherds and Arab boys don't find it."

There was no time to go looking for Woody's parachute then. Faint signs of dawn lightened the eastern sky, so we hoisted our heavy bergens and marched down into our ambush position in the wadi, where we remained for two days as planned. The euphoria of the jump quickly wore off. We were back on the ground in the same old sweltering hot routine of ops in Aden and, typically, Mohammad al Maghrebi al Haushabi failed to turn up. We wondered if drinking the spoils off the beer wagon had anything to do with it.

Of course, when it was time to pull out, we had to fly in all of 7 Troop to picket the high ground while we searched for Woody's parachute. They thought that was a huge joke and there was much talk of useless 'Sky Gods'. Needless to say, when we started looking for Woody's bush, we saw the wadi was covered in bushes, millions of them. But the quartermaster was emphatic. The T-10 parachute was American, on loan, and had to be returned.

There has been only one other British operational free-fall jump since Op Snowgoose, also by the SAS. In 1970, 24 Troop of 'G' Squadron jumped into the Musandam Peninsula in northern Muscat and they lost a man. Trooper Rip Reddy, who exited the aircraft a fraction later than the others, missed the wadi they were supposed to jump into and plummeted into the side of a mountain before his BOD height-finder operated or he could pull his parachute. He was found and his body returned to England.

I love free-fall but I have doubts about free-fall operations. How many jumps actually go right? How many are cost-effective in time, manpower and resources? So much time is needed to train the freefallers for that one method of entry when there are all the other techniques to train for in preparation for whatever they have to do when they have landed. Free-fall operations are dangerous, with the chance of injury ever-present, and they take up valuable and usually scarce aircraft resources. Good parachute equipment is not always available where you actually need it, and no one has yet produced any workable anti-radar suits – increasingly necessary when even 'Third World' countries operate quite sophisticated radar systems to spot intruder aircraft passing in or near their airspace. I wonder if the effort is really worth it. In my opinion free-fall should only ever be used for certain very specific applications.

However, there is a spin-off. All these technical difficulties, the danger and exhilaration of jumping out of an aircraft at night, have established an exclusive club of free-fallers in special forces units everywhere. But the benefit is more sexual than operational! The 'Triple Diver's Badge' is coveted more for its success in pulling women (it stands for 'Sky Diving, Skin Diving and Muff Diving') than for giving the holder any special skill in war.

Properly set up, free-fall is ideal for the covert entry of small groups, in direct contrast to static-line parachuting which delivers large groups into action, like the famous drops at Arnhem and Suez. I think I strike a note of sympathy with most Paras when I say we have all grown bored of listening to the veterans of Arnhem and Suez who retell their few seconds of glory over and over, but the fact is there are very few, if any, British soldiers left who have jumped any operational para jumps, especially onto drop zones defended by the enemy. For me, Op Snowgoose was the first of several 'hot' para jumps, but a vastly more technical and complicated way of going into action compared to operations in the jungle.

CHAPTER 3
PLANNING AND
PREPARATION

I went to Borneo in February 1965 as a young trooper in 'D' Squadron, in between trips to Aden, and I quickly learned that jungle operations are relentless hard work.

During the Malayan Emergency, military statisticians worked out that a soldier might expect to be on patrol for 1,800 man-hours of slogging through the forests before even seeing the enemy, let alone being quick enough to shoot him before he disappeared into the gloom. When you think that the British Army took twelve long years to kill 6,710 communists in Malaya, you begin to get some idea of the sheer effort (1,378 man-years!) required to defeat an insurgent terrorist enemy in the jungle.

The war in Malaya ended in 1960, though terrorism continued sporadically on the Thai border, and only two years later the British Army was committed again to fight in the jungle, this time in Borneo. At first, on December 8, 1962, insurgents tried to overthrow the Sultan of Brunei and were defeated by prompt action, flying troops from Singapore where the British still boasted a 'Headquarters Far East'.

Next, in early 1963, President Sukarno of Indonesia decided to prevent the formation of Malaysia by destabilising Borneo in a campaign of confrontation. He applied a harsh mix of international political pressure, subversion inside Borneo by the local Clandestine Communist Organisation (CCO), and cross-border attacks through the jungles using his own regular Indonesian troops or CCO terrorists trained and supported by Indonesia. His ultimate aim was nothing short of conquering Borneo.

Borneo, in complete contrast to Malaya, was very vulnerable to such attacks. Borneo has a jagged 1,500-mile-long coastline and a 1,000-mile land border with Indonesia, which runs through dense, remote mountainous jungles. Administratively, the country was divided into penny packets, five divisions in Sarawak, and four residencies in Sabah. People in the small coastal towns were poor and liable to subversion by the communists' CCO, while the aboriginal tribes in the interior lived in settlements on rivers, which are the arteries of travel in tropical forests, where they were easily terrorised and hard to visit, let alone protect. Lastly, the enemy were regular Indonesian troops, properly trained, equipped and supported by an outside power, advantages the Malayan communists never had.

There was really nothing to stop the enemy crossing the border under the jungle canopy, and it was hard to know if he was there at all. That was how the SAS fitted into the tri-service strategy used to counter Sukarno's aggression. In the first years, SAS patrols worked inside Borneo in the remote border areas, winning the hearts and minds of the tribes, establishing a network of information on enemy movements, and mapping uncharted areas of jungle.

Then, in August 1964, Sukarno made a terrible error of judgement, sending a special force of his soldiers to attack targets in Johore Province in Malaya. Thoroughly alarmed, the Malayan government in Kuala Lumpur gave their support for secret cross-border operations, initially to recce, then later to attack regular Indonesian troops in their camps inside Indonesia, eventually up to 20,000 yards across the border. The strategy behind this was similar to that employed by the South Africans when I was fighting with them in southern Angola years later. The aim was to push the enemy back from the border and harass them inside their country before they could cross and cause trouble in ours. *These* operations, codenamed 'Claret', were politically most sensitive, as Britain never actually declared war on Indonesia, and they were classified 'Secret'.

The conditions of patrolling secretly in the jungle imposed a very strict code of professionalism on the SAS, and it was here that I really learned what the SAS means by absolute attention to detail.

First, I joined a jungle training course run by 'B' Squadron, under the excellent tuition of experienced Malayan hands like Sergeant Geordie Tyndale and Corporal Anthony Hallan – who was a Malay interpreter – and Lance Corporal Chuck Hinde – who had just done a Borneo tour in Tyndale's patrol. They set up a jungle shooting range camp, a tactics camp and a recce camp, and we moved around all three in sequence, learning how to move silently through the shadowy green foliage, eat and sleep without leaving sign of our presence, and read the enemy's tracks. We spent hours moving slowly and noiselessly in four-man patrols – learning how to navigate in the forest – and tested our skills against each other – playing games of 'ducks and drakes' in our four-man groups through small areas of jungle to see who could spot the other patrol first and react quickest, to hone our contact drills against a 'live' enemy with blank ammunition. For a month, we lived in these camps as if we were actually on operations, under the harsh tactical disciplines of no noise, whispering all the time, and living on 'SAS hard routine' with no lights, no smoking and no hot food. We became very lean, very pale from not seeing the sun under the canopy, and very fit. Our numbers were made up with men from Patrol Company of 2 Para and the majority were very good men well capable of being in the SAS.

I went back to 'D' Squadron in Kuching, where Major Roger Woodiwiss was about to hand over command of the squadron to Major Glyn 'Punchy' Williams, and joined Sergeant 'Smudge' Smith's four-man patrol. The other two were Trooper Derek Gorman and Trooper Terry Falcon-Wilson, who was later killed in the Yemen working for David Stirling's civilian company called Watchguard. Smudge had learned the art of SAS soldiering in the jungle in Malaya. He was fair-haired, well built, strong and had endless patience. He had come from the Somerset and Cornwall

Light Infantry, never took shortcuts in his work, and rarely swore; indeed, I never saw him lose his temper or drink, and he was always at tremendous pains to explain things, even when he was correcting the stupidities of us young troopers. In short, he was an excellent jungle soldier, typical of the older SAS NCOs in Borneo whose professional attention to detail, no matter how tired you felt, was impeccable. I was wild and aggressive, but these formative years in the SAS showed me there is no substitute for the seven Ps: proper planning and preparation prevents piss-poor performance!

My first recce patrol was in the Sarawak First Division in the area of Sidut, where the squadron set about gathering all available intelligence on the enemy, in particular paths used by the Indonesians to infiltrate across the border. Two years before, Captain Ray England had reported that the Indonesians had established a regular army camp at Sidut for about 100 troops, and they had forced the local people to cut infiltration routes for them through the jungle.

The day before going into the trees, Smudge made us practise all our immediate action drills in case we bumped into the enemy; we test-fired our 7.62mm SLRs and packed our bergens, keeping the weight down as much as we could. We tried all sorts of tricks for this, taking low-weight foods with high-fat and high-protein content, like peanuts or sardines, to limit our bergens to a target of 35 pounds, but with fourteen days' rations to carry it was always a real struggle to bring them under 50 pounds. The heavier the bergen, the more tired, sweaty and drained you become, which gives the enemy advantages you cannot afford. At least the weight went down day by day as we ate our rations, but I still lost an average of 1 pound of body weight per day on patrol.

We came up to the border by Wessex helicopter and crossed it on foot. Smudge always worked on the principle that the enemy might just be the other side of the next tree, so we moved very cautiously, usually only 10 kilometres a day, stepping softly on the dead leaves of the jungle floor, our eyes constantly probing through the gloom for danger signs hidden in the thick undergrowth. We were conscious all the time that we were only a small recce patrol of four men while our Indonesian enemy were trained regular troops familiar with jungle conditions and usually moved in large groups of ten or more.

Although the chance of meeting the enemy was slight, the risks if we did were high. The squadron had lost Sergeant Bexton and Trooper Condon killed in a previous tour, and the new tour had started badly, with Sergeant Lillico and Trooper Thompson being seriously wounded and lucky to escape with their lives from an enemy ambush just over the border opposite Plaman Mapu, about 22 miles from Sidut. The SAS standard tactics at the time were for us to shoot and scoot, leaving a wounded man if necessary to save others being wounded or killed. None of us agreed with this at all. In fact, Smudge, Terry, Derek and I thrashed over the details of the Lillico/Thompson mishap and decided to ignore that particular SOP completely. We were determined that we would never leave anyone. We were all in the shit together and either we all came out or we all stayed, but we would never split up and never leave a wounded man.

The jungle across the border looked the same as our side, but as soon as we slipped over the border ridge, we felt vulnerable. We depended wholly on every detail of our routine to survive, to avoid being ambushed by the Indonesian patrols.

Every morning at five o'clock, in total darkness, we slipped out of our hammocks and packed up ready to go. First, I took off my dry lightweight para smock and wriggled into my clammy, freezing, wet shirt and trousers which had been hanging on the end of the hammock from the day before. This was bloody miserable, but there's no better place than the jungle for a really good night's rest on operations, because no one can see a damn thing to creep up on you in the utter dark, and it's spoiled if you have to wear clothes soaking with sweat and rain from the day's march. Then I strapped on my webbing belt kit, with the pouches which contained my water bottle, escape rations and four spare magazines of ammunition. I quickly packed away my hammock and shelter sheet in my rucksack and finally I sat down on top of it, my rifle across my knees, to wait for first light.

As soon as we could see each other in the grey shadows of dawn, we silently hoisted our rucksacks onto our backs and moved off on a bearing Smudge had selected the night before. We had to get away from our sleeping position in case the enemy had located us at last light the evening before. After an hour, he led us round in a loop and we halted for a brew and to send the morning radio message back to squadron headquarters. The loop brought us back near the track we had made coming in so that we could hear anyone tracking us, and we could either ambush them or slide away into the green shadows. By the time they could track us round the loop to our position, we would be long gone.

Exercising great caution, we continued on our day's march through the endless green gloom, every hour looping back for a five-minute break. Not being able to see more than a few yards in any direction meant we had to constantly check our position. We judged it against the map, which gave so little topographical information in this area that we recorded the ridges and streams we had crossed for future patrols, comparing the time we had spent moving against our distance travelled, and we took it in turns to count every single pace and step of the way as a final measurement of where we were. At 4 PM, we looped again and stopped for the afternoon radio schedule to send the daily SITREP and eat our main meal of the day. We worked in pairs. Two cooked the inevitable SAS 'all-in' curry, Derek Gorman, the signaller, tapped out the Morse code on his PRC 128 radio, and the fourth man lay on guard, facing down the track we had made into our position.

These stops were the only breaks during the day, but we never relaxed. Even going for a shit, we always kept our belt kit on, or right to hand. We buried the shit, as a bonus for the dung beetles, but we carried out all our other rubbish. Discarded paper and tins give away a great deal about a soldier's morale, feeding and habits.

After this main stop, we marched for another hour before looping for the last time, at last light, into our night lying-up position. I constantly wondered what the Indo soldiers from Sidut were doing, whether we might bump into them creeping through the jungle to cross the border themselves. I tried to imagine their own routine in their company base in Sidut. Did they have patrols out looking for us? The tension of never being able to see any distance, of always imagining the enemy were so close, never let up. It's a part of jungle operations. I constantly hoped we might meet the enemy, but the days passed frustratingly with nothing to change our slow, dogged, sweaty, wet, tiring routine. Day followed day of inflexible sameness, so we all knew each other so well we hardly had to whisper even to explain what we were doing, but we never saw the enemy. After fourteen days, we marched back over the border.

A few weeks later, Captain Ragman had more success in Sidut. Supported by 9 Company, 1st Battalion Scots Guards who were placed in ambush positions on the border, he combined his own and Sergeant Alex Spence's patrol to make a stronger party of eight, and they crept 5 kilometres over the border through the jungle to ambush an enemy courier in his wood and atap-leaf hut on the edge of Sidut. After a short skirmish, throwing No. 36 grenades and firing at the hut, they seized the courier and some important documents which created an impressive stir in Headquarters. The whole essence of this success was that the operation was based on precise intelligence provided to Major Woodiwiss by 'certain persons' who were probably Special Branch acting on source information. The lesson was not lost on me. It was all very well for us to patrol about the jungle in the hope of finding the enemy, but, no matter how highly trained and motivated your best troops may be, there is no substitute for good intelligence. The point has been proved in every campaign I've fought.

I met other Commonwealth troops for the first time in Borneo and saw an interesting comparison between the Australian SAS and the New Zealand SAS. Smudge was ordered to take his patrol and assist with a jungle training programme for the new arrivals. The Aussies' attitude was typically blunt: "Us being trained by a bunch of Pommy bastards? No way!" The Aussie SAS have their own jungles in the Northern Territories and they resented our training programme. However, in private some of their sergeants came round to Smudge and other senior British SAS NCOs and admitted, "You Poms 'ave been here some time and we'd like to hear what you've got to say."

The Kiwis were quite different. They were big tough men, mostly Maoris, with superb morale and they were naturals in the jungle. We got on well with them and I recall characters like Lieutenant Manuera, who later won the Military Cross in Vietnam, Trooper 'Kingie' Ihaka, and Trooper Snooks, who was built like a gorilla. He loved to carry the Bren gun and slid effortlessly through the jungle bare foot. Once I watched amazed as he and two others casually packed away forty boiled eggs between them, not for a bet like Paul Newman a year later in *Cool Hand Luke*, but just because they felt peckish!

All the same, just to bring the Kiwis back to earth, I should add that 'D' Squadron beat the Kiwi rugby team from the 1st Battalion Royal New Zealand Infantry Regiment, which astonished us as much as them, and the consequent party was a night to remember.

There was a routine in camp too. The first day back, we went through a thorough debrief with the ops officer and the squadron commander and cleaned all our kit, the second day we were allowed to drink and relax, the third day we received our orders and prepared for the next patrol, and then on the fourth we went back into the jungle.

Life was simple. We were either in the jungle or billeted in a concrete 'hotel' in a seedy part of Kuching, conveniently opposite a whore house. Living on top of each other, working and drinking together month after month, all ranks in the squadron became very close-knit. Personal character traits became exaggerated and the usual behaviour patterns of the British regular army were often stretched to a limit which would have been unacceptable in any other unit. Captain Ragman was nicknamed 'Dapper' because of his dress sense. One day he was seized after a drinking session on our day off, held down and Sergeant Alfie Tasker cut off a great lock of his blond hair which always fell over his face and irritated the men. On another, we tied Jock Thompson to the fence in the yard because we were fed up listening to his stories. And Kevin Walsh, whom I had known in 1 Para when I first joined the army, had a problem with the ArmaLite assault rifle. Kevin was short, stocky and famously ugly, a great SAS character who sadly died. With a cigarette between his teeth, as was considered cool at the time, he swaggered into our room and gazed round at the iron bedsteads, bare boards, khaki webbing, trousers and shirts hanging everywhere on strings, towels and bars of soap left out to dry – which is always a vain hope in the tropics – and piles of spare rations which accumulated from stripping out the food packs we had to carry on patrol. Some of the men were dozing and some sitting on their beds in shorts, cleaning kit for the next patrol.

His eye lit on Jake Allsop's weapon, the American 5.56mm ArmaLite which had just been issued to us.

"This looks a good bit of kit, Jake," said Kevin knowledgeably, his cigarette bobbing up and down between his teeth as he talked.

Taking his hands out of his pockets, he picked up the ArmaLite, admired it a moment – turning it over in his hands – and before Jake could move, he cocked it, aimed it through the window and pulled the trigger.

A burst of automatic fire blew out the whole window frame and a whore across the street was shot in the arse.

"Oh my God!" said Kevin, appalled, standing in shock in a slight haze of blue smoke. "What the fuck do I do now?"

Raising my voice slightly over the hysterical screaming from across the street, I said casually, "Just go downstairs, Kevin, and put yourself on a charge!"

Smudge took me on a helicopter recce along the border for our next patrol, not so much because we could see anything through the canopy, but to see the flow of the endless rolling green canopy itself. The maps were so bad, with so little contour detail, that we wanted to see the lie of the land, the ridges, the re-entrants and lines of the streams under the endless acres of green trees and arching fronds.

We were escorted up to the border this time by a ten-man patrol from 2 Para out of Gunan Gajan Company base, not far from Plaman Mapu. They chose a route near the border so that we could step off their line of march, literally stepping over low undergrowth, to prevent anyone who might be tracking us from seeing the marks of our boots where we had quit the main group.

We patrolled carefully for the rest of the day over the sharp border ridge and then, in the late afternoon, dropped off the ridge to loop round on the steep slope and come back close under the ridge again for the night. Here, there was no level ground to lie on, so no one would expect anyone to sleep there, but we could sling our hammocks between the trees, and secondly, we could hear anyone blundering along the ridge above us and have plenty of time to 'bug out' if necessary, before they tracked us round on our loop in. That evening, as the sweat dried on our backs in the warm dusk after we stopped to tie our hammocks, we heard the artillery in the Paras' company position at Gunan Gajan firing off the 105mm guns onto defensive fire positions thousands of yards over the border, on paths which the Intelligence Staff reckoned the enemy would use to approach the border. We lay in our hammocks and listened to the shells whining overhead and crashing into the greenery beyond, happy we were tucked safely under our ridge.

We moved with especial care on this patrol as this was the area where Lillico and Thompson had been wounded and the Indonesians had regular troops all along the River Sekayan which ran parallel to the border, from Jerik through Balai Karangan to Sentas further west. On our twelfth day out, we found a sign of the enemy.

I was lead scout, as usual, and peering ahead through the dripping greenery I saw the track quite clearly. It was enormous. I sank to one knee, keenly alert for enemy, and signalled with my fingers for Smudge to join me. He slipped up silently behind me. He was a good tracker, and after some time peering at the signs on the track from the shadows, he explained what he could see in a low whisper. There were several striking facts: the track, which ran parallel to the border, was very wide, maybe 10 feet, which is unusual in the jungle where everyone is trying to avoid leaving signs of their presence; it had been made by a large body of men, judging by the amount of trampled mud and crushed leaves; they were all wearing army boots, the marks of which were clear; they had only recently passed, maybe the day before, or even earlier that morning, as there was no dust or detritus lying on the leaves; and they had all been going in the same direction, from west to east, from our company area to the Plaman Mapu company area. The important question was, *Where were they now and what were they doing?*

After some while waiting and listening in case anyone was using the track, we crossed swiftly, one at a time. When we were in the safety of the dark undergrowth on the other side, Smudge whispered, "I had a good look as I crossed, and I'm sure there's a camp down the slope."

Under his guidance, I moved forward, very slowly indeed, advancing step by step, sinking to one knee to listen every few steps, alert like a wild animal, trying to make absolutely sure we would not stumble into a sentry without seeing him first. The footprints on the track had been leaving the camp but it would be quite usual military procedure for the Indonesians to leave guards.

For nearly two hours we inched forward, straining our ears and eyes through the green foliage, for giveaway human sounds ahead, which always pitched at variance with the soft natural noises of the jungle, and for signs of straight unnatural edges among the leaves immediately in front of us, like the delicate line of a tripwire which might be connected to a grenade. The Indonesians often laid these booby traps round a camp, as a defensive measure, as Captain Ragman had found before his ambush at Sidut. Finally, I spotted openness ahead through the trees. The enemy camp.

The place was enormous. For maybe 100 yards across, all the undergrowth had been cleared between the tree trunks which still provided their green canopy to prevent any sign of the camp being visible from the air. Huts were dotted round with all the signs of being recently occupied by troops: muddy paths between the huts, crushed and rotting leaves, broken poles and discarded paper, boxes and hessian sacking. It was eerily silent, though Smudge was sure he could hear 'people' sounds on the far side, out of sight down the slope.

We waited hidden in the shadows for some time, just in case the Indonesians had left a guard behind. Nothing stirred in the oppressive, damp heat except the birds and chattering insects. Smudge signalled at me to circle the camp, keeping in the jungle fringe, and we moved very cautiously indeed.

It was a very strange experience moving with an open piece of ground on one side after spending all our time enclosed in green gloom. We felt terribly vulnerable, like nocturnal animals suddenly exposed to the sun.

Suddenly, Derek Gorman, climbing a small bank while he glanced rapidly around, dislodged a stone which rolled down the hill, sounding in our ears like an avalanche.

We froze. Long minutes passed, but nothing happened, no one came, and no one murdered us with fire, exposed as we were. Smudge later said he was sure he could smell something unnatural which made him think humans were not far away, but no one reacted to the noise of Derek's stone.

Had the camp been fully occupied, we would have been in real trouble. We were lucky, but plain logic dictated that, if the enemy were not in their camp, they were somewhere else in the jungle, and we all wondered what they could be doing. After Smudge was satisfied we had seen enough, we moved very carefully indeed, but as fast as we dared back towards the border to make our report, first at the Paras' base in Gunan Gajan and then at our SAS base in Kuching.

The director of operations, whose authority was required to mount all Claret operations, decided we should cross the border again, this time in force with the Paras from Gunan Gajan under command of the Para company commander, and ambush the track we had found.

I was unable to go on this operation, as I flew back to Aden to assist 'B' Squadron there, who had no Arabic speakers. However, a few days later, Smudge flew back to Gunan Gajan by Wessex and landed there on the afternoon of April 26, for briefings and final preparations with the Paras.

The base was like all the others, an organised mess of rich mud, logs tied into walls, defensive trenches, sandbags, wriggly tin huts and atap roofs, the stumps of trees sticking up where the ground had been cleared and coils of barbed wire around the perimeter, all baking under the sun or drenched by rain. Beyond was the green wall of jungle on all sides.

On Tuesday, April 27, while Smudge waited in the crisp air of dawn for the Paras to go out on our ambush, there was a sudden stir round the company HQ trenches. A Para ran out and called, "To your positions! Stand-to!"

His patrol found an empty weapon-pit near the 105mm howitzers and regretted it at once. The gun crews ran over, jumped into the pits beside their guns, orders were shouted, shells prepared and they opened fire with a deafening salvo over Smudge's head. And followed it with another, and another, and another. The air filled with the pungent smell of cordite and a great pile of empty steaming shell cases built up behind the guns.

The guns were firing in direct support of 'B' Company at Plaman Mapu, which had been deliberately sited within the 17-kilometre range of the 105s. That same dawn, 150 first-class Indonesian troops had surged out of the jungle and tried to overwhelm the few Paras left in the base. The Indonesians had obviously been watching the base. Most of the Paras were out on patrol, leaving the hasty defence to the gunners, cooks and bottle washers who fought desperately, hand-to-hand in the weapon-pits, to drive off repeated and determined enemy attacks. During several bloody hours under the exemplary leadership of Company Sergeant Major Williams, who was later awarded the Distinguished Conduct Medal, the Paras inflicted thirty casualties to their own two dead and eight wounded, and finally drove the Indonesians back into the jungle. Though only a small battle over a company-sized position, it was vital. Indonesian success would have received huge publicity and done immense damage to the British cause.

By the time the guns next to Smudge fell silent, the pile of shiny brass empty shell cases was enormous. Gradually the story of the attack filtered back and it was plain the Indonesians on this attack had used the track we had found, and come from the camp we had seen. There is no knowing what might have happened had we found the track a week or so earlier, but we were too late. Luck is a vital ingredient in all military operations, as in life.

To counter incursions like the attack on Plaman Mapu, Claret operations became gradually more aggressive. After 'D' Squadron came home, 'A' Squadron took over under Major Peter de la Billiere and began aggressive patrolling in squadron-sized groups, sometimes with the Gurkhas, which had considerable success. In one patrol, a large Gurkha platoon killed twenty-seven Indonesians, who were clearly under the impression they were attacking the usual, small, four-man SAS recce patrol. The Indonesians began to move their camps away from the border. 'B' Squadron followed 'A' with several successful full-squadron Claret operations, but they were not replaced, because, in March 1966, there were definite signs that Indonesia was changing its policy and looking for peace. Sukarno had been placed under political house arrest in Jakarta due to the failed policies of his 'Konfrontasi', and the Indonesian Army stayed well back on their side of the border for fear of being hit, and hurt, by Claret operations. Inside

Borneo, all subversion and terrorism had been gradually reduced by the tireless work of the police and conventional troops, and the danger of invasion receded.

In the jungle, Australian and New Zealand SAS patrols continued to cross the border on recces and confirmed Indonesian troops were withdrawing from the border areas, so that by the time I returned with 'D' Squadron in July 1966, there was little to be done except 'hearts and minds' patrols to maintain British influence among the remote jungle tribes.

Finally, in August 1966, peace was declared because, quite simply, Indonesia's attempt to overwhelm Borneo had totally failed. British tri-service co-operation, relentless, meticulous and aggressive patrolling, police work against the communists in urban areas, and the pursuit of the hearts and minds of the population, had triumphed. Victory parades were held in Kuching, but now the fighting was over, politics ruled the day. No British units were allowed to take part, or even be in Kuching. We were back to the old form, so familiar to British regulars and which Kipling expressed so well: "It's Tommy this, an' Tommy that, an' 'Chuck 'im out, the brute!' but it's 'Saviour of 'is country' when the guns begin to shoot."

The Malayans wanted to trumpet the success as theirs – with a little help from the British. The Royal Malay Regiment marched in triumph through the streets and no one seemed to recall that the only other time they had been in the headlines was in 1963, after being soundly defeated at Kalabakan by Indonesian troops (who were then almost wiped out by a quick-reaction force of Gurkhas).

So, we had our own victory celebrations. I had been in trouble again, and 'Punchy' Williams banished Jock Allison and me to Santubong Island, not far from Kuching. His orders sounded serious: "Go away and fix up some boat skills training," but this was really an excuse to stop us fighting in the bars in Kuching at such a delicate political time. Sitting on a sun-baked, sandy island with nothing about except coconut palms quickly concentrates the mind on the comforts of life, and, following Punchy's orders to the letter, we arranged a clandestine boat patrol in our Rigid Raider, in the dead of night, cruising noiselessly into Kuching docks for a secret rendezvous with Tony Woodhouse. His boat was filled to the gunwales with Merrydown Cider, which we lovingly cross-decked to our craft and took back to Santubong Island. I don't think any of us 'deportees' to Santubong were too surprised one morning to see a flotilla of assorted craft chug round the headland with the remainder of the squadron sent, so Punchy claimed, for "more boat training". I have no idea what he understands by 'beach recce parties', but we had some great barbecues with fresh meat rations, Merrydown Cider and a number of fit, though not very beautiful, young whores covertly boated out from Kuching, who enthusiastically dragged various members of the squadron into the sea for underwater sex and a dose of the clap.

I regret not having contacted the enemy in Borneo, but very few people did. Thinking of the hours we all spent in the jungle, all patrols were hard work and most were uneventful, but they formed the basis of my soldiering. Smudge, and the other sergeants I worked with on different occasions – like Alec Spence, 'Tanky' Smith, Mo Copeland and Geordie Lillico – were meticulous, and what I learned from their example made the hard discipline of jungle operations seem worthwhile. Close bonds were formed between us in our fourman patrol and, though it was frustrating because I never saw the enemy, these were intensely satisfying times.

My last impressions of Borneo are of the Bungo Mountains, a remote border area south-west of Kuching where the true value of our work was clear as day. Captains Willy Firrel and Keith Farnes, Staff Sergeant Bob Crichton, Corporal Roger Tatersall and I came upriver by boat and then walked into the steep hills beyond. We patrolled as carefully as before. The politicians might have been talking of ceasefire and victory, but in the jungle all men are equal, except those who cut corners. A few incursions of diehard communists still occurred from time to time and we had no intention of being caught out. We saw no enemy, but we were royally entertained by the tribespeople.

These small brown-skinned Indians with their blue tattoos were the real essence of Borneo. Their kindness repaid several times over the efforts of our medic with his sutures, crepe and drugs. They honoured us in their long-houses with their food, music and dance, their betel-stained teeth yellow in the flickering light of their fires, and we felt good to have helped protect their way of life. This was their land. A dramatic green jungle land of steep mountain ridges joined by fantastic swaying bridges of rattan and bamboo over raging torrents of water far below, an exotic land of birds, animals, trees, flowers, and spirits which inhabit all things visible and invisible. Communism is not their way.

CHAPTER 4
RETIREMENT

Camp life with no prospect of operations is always difficult for soldiers. By 1967, Aden and Borneo were over and nothing else loomed. This is an age-old problem for army commanders, and for special forces in particular. How do you occupy young men constantly at a peak of training and fitness when there are no operations?

I found it hard.

The regiment did its best to keep us busy but exercises were basic compared to today. We thought a regimental escape and evasion or a two-week NATO exercise in Germany was impressive. We were content as long as we had plenty of para-cord and masking tape, while nowadays SAS soldiers demand a vast array of weapons, air and naval assets and high-tech equipment, of which exotica like SATCOM, LTDs and MAGELLAN are but small examples.

We were sent off on skills training. In September that year, Tony Ball and I went to the French Parachute School at Pau in south-west France on their free-fall course. The French are full of the reputation of the Foreign Legion and 'les Paras', and they have a robust attitude to parachuting. They make sure that Paras and aircrew keep up to date with their jumps and consequently their parachute jump instructors (PJIs) are so well practised you could set a metronome by the way they despatch men from their big Transall transport planes. They fill the aircraft to capacity for every drop and empty the whole lot on one run-in over the DZ, in simultaneous 'sticks' out of both doors, left, right, left, like clockwork. In Britain, the RAF keeps adding restrictions little by little to the point where big 'sun-stick' drops are now rare and proficiency must be in question.

Tony and I enjoyed the free-fall, so when Arnhem Day came on September 17, we repaired to a café in Pau to celebrate. After a while, four French Paras drinking in the same bar kept saying Arnhem was a disaster. True or not, we couldn't take this from the French, so we beat them all up. Horrified by this melee of brawling soldiers, the café owner phoned the police, and we legged it into the night across a stinking cabbage field with the wailing siren of a French police car and headlights swinging wildly through the darkness on the roads behind us. The gendarmes called in reinforcements and we were cornered by a group of furious French military police, who held us at gunpoint with their MAT machine guns. This was too much. Inflamed with the spirit of Arnhem and several litres of French ale, I grabbed a MAT off one MP and proceeded to batter him with it till all the others leapt on us and we were both jailed. French English relations were patched up and we were sent home. Tony and I were marched in to see the commanding officer, who listened to our story and gave me a 'final warning', but nothing worse. I suppose beating up cocky French Paras is a permitted SAS pastime.

During the Sixties and Seventies, the regiment was inundated with requests for training teams abroad. Foreign leaders, often in the Commonwealth, had seen the successes of Malaya and Borneo, and wanted their own special troops and bodyguards trained too. It was not just a question of prestige. These 'praetorian guards' were set up to keep them in power. This rush of jobs quite suited the regiment at a time when it had little else to occupy itself, and it kept us out of trouble. 'Team jobs', as they were called, were not operations, but they were better than hanging about in camp and getting into trouble in Hereford. Plus, they offered the chance of lucrative local overseas allowances and lovely claims to supplement the miserable pay in those days, long before SAS Special Pay.

In October, shortly after returning from France, I was picked by Lawrence Smith to join his training team in Guyana. He was an excellent soldier who had been awarded a Military Cross in Borneo while still a sergeant major and was then commissioned. Our job was to train the Guyanese Defence Force (GDF) and I found myself thoroughly absorbed. We were in the jungle again and we worked hard. There were some light moments in our rest house, or 'pub' – which we found in the middle of the jungle – and we coped with the inevitable dramas: such as having to casevac Brummy Stokes with a repulsive infestation of huge maggots in his back, quite a different problem from those he faced climbing to the summit of Everest. We spent six weeks there and achieved a great deal with the GDF soldiers. I enjoyed the challenge, lifting the GDF standards from nothing to a quite reasonable level, and the experience of training local black troops came in very useful later in Africa.

However, back in Hereford I found my reputation set me up as a target again. One night after we had all had too much to drink, Danny Oldham, a big man in 'A' Squadron, took me on in the 'Grapes' pub, a famous SAS haunt. We repaired to a piece of rough ground behind the Welsh Club, badly lit by distant street lights, and set to. He was a very powerful man with immense hands and I went down straightaway under a haymaker. He stood on my head and kicked me for a count of seventeen, but I refused to give up. I managed to grab his foot, hit him and turn the tables. I am afraid Major Bill Dodds, my squadron commander who had taken over from Punchy Williams, did not see this incident in the epic David and Goliath light I have described. I was locked up at once, marched in front of the commanding officer, Lieutenant Colonel Viscount John Slim, and banished from the regiment back to the Paras.

Once again, I offer no excuses. I grew up in a place where it was common to carry knives and use them. We weren't so 'advanced' as some hooligans now who fix two blades separated by a penny into a Stanley carpet knife so the double cut is very hard to suture, but we were rough enough and I suppose my character attracted those who wanted to fight me. That's how I was before I took a long look at myself and settled down somewhat, later. However, it takes two to fight and the army 'system' made little effort to find out the other side of the story. None of the officers realised that I was often picked on by men who wanted to make a name for themselves, like the first time I was in trouble in America, and the other fellow got away scot free.

Of course, there is another aspect that I can see more clearly now, which I do not think the junior ranks saw then and which many people reading this now might find hard to believe. The SAS was trying to make sure it was not disbanded, as it had been after the Second World War. After Malaya, Aden and Borneo provided new opportunities and the SAS did well, but their reputation was not established enough for complete impregnability at Army Board level, and many senior servicemen could not see a role for the SAS in peacetime. The anti-terrorist role, which rightly guarantees the SAS a job now, was not developed then, and nor was the conventional special forces role now so widely accepted after the wars in the Falklands and the Gulf. So, while SAS officers beavered away trying to influence army politics to the regiment's lasting advantage, the last thing they wanted was adverse publicity.

However, don't think I was the only one who enjoyed fighting. I had a higher profile, if you like, than some of the others, but fighting was very much a 'work ethic' of the SAS in Hereford at that time. A lot of us came from tough backgrounds and we used Hereford as a big playground; the women we knew enjoyed the excitement and the police were marvellously long-suffering in the days before drink driving became a serious issue. The rivalry between us was stirred up in the pubs, where we all drank and then usually settled with a fight. None of it meant very much to us, because we knew that a good scrap never made any difference to our work. Only some of the officers took a dim view.

There was one thing that always rankled. One or two officers told me I was a useless soldier. They said, "Since you patently can't keep yourself under control in camp, you would be useless on operations, in action." What can I say? They were probably right to come down hard on the fighting in Hereford, and I was high profile, as I say, but those officers were in no position to judge. They were never on operations with me. I don't think I ever let down anyone on operations, I don't think anyone ever criticised my performance in action, and I do not recall ever failing to perform my duty in camp because of drink.

Anyway, I was back to 1 Para again, in Aldershot. Fortunately, the banishment was deemed sufficient punishment and I stayed a corporal. However, I was marched in 'on orders' to see the battalion second-in-command, whom I had never met before, and to my disgust he took the same view as the officers in Hereford. I stood to attention in front of him in his office and his actual words were, "Corporal McAleese, you won't make a fucking biscuit storeman in this battalion."

Annoyed, I replied at once, "If that's the way you think, sir, I'd like to revert to private."

"Request denied."

I was puzzled by this, but I replied at once, "Then I would like to go on the next Brecon course, please, sir!"

I meant the British Army's Senior NCOs Tactics Course in Brecon, under the Beacons where I had done selection for the SAS. This course provides the core of the British Army's sergeants, the best NCOs in the world.

The second-in-command snorted at that. He said, "Don't be a cunt, Corporal McAleese. They want highly skilled men at Brecon and you certainly won't fit. However, I'll send you on the course and if you pass, I'll eat my fucking hat."

He probably just wanted to get me out of the way, but British Army logic is legendary and he later promoted me to sergeant to go there!

I was delighted. Brecon is not far from Hereford and I had married a local Hereford girl, Marlene Good, in 1966. She refused to leave Hereford when I was posted back to Aldershot, so, in Brecon, I was close to Hereford and we continued to try to make our rocky marriage work. She was a small, attractive, dark auburn-haired girl and I sup pose she did her best. She was six years older than me and wanted to be married to a fit young SAS soldier, settle down and have children. However, I must admit, my way of life did not give us much hope. I was only 25 years old and wild. I alternated violently between an obsessive commitment to my work and an equally intense commitment to play. It was not that I drank any more than anyone else, just that I could not resist a scrap. Marlene enjoyed nights out too, but never understood the aggression. Her aspirations were different to mine and I don't blame her for the subsequent failure of our marriage.

Contrary to the presumptions of these aforementioned officers, I did well at Brecon. I was given a 'B' grade, but when Brigadier Findlater – who was in command of the Tactics School – read my report, he said the results deserved better than the grade given and put it up to 'A'. I don't know what sort of results soldiers get now at Brecon, but in those days, the course was a very important part of infantry soldiering and heavily emphasized by the Parachute Regiment in particular. 'A' grades were rare and I was pleased with mine.

What was more important, I learned a great deal about good solid infantry tactics and soldiering. The course taught us everything, starting with the fundamentals – the 'Principles of War'. The age-old cynicism of the British 'Other Ranks', often useful in the hardship of war, had provided me with four short, unofficial principles so far:

1. Slaughter of the fighting men
2. Search for the scapegoats
3. Punishment of the innocent
4. Decoration of the non-participants

Now, I learned the official Ministry of Defence 'Principles' and I will repeat them here, as I have found them invaluable since.

There are nine, as follows:

1. Maintenance of morale
2. Offensive action
3. Surprise
4. Security
5. Concentration of force
6. Economy of effort
7. Flexibility
8. Co-operation
9. Administration

Above these is the master principle: the selection and maintenance of the aim.

All these struck a chord in me, though I suppose I still had to learn about co-operation. In the army, co-operation means knowing who is in support and knowing when to give a little for the common good, but I have always been someone who prefers to get stuck in and do the job myself, rather than waste time asking other people to do it.

These principles struck a parallel to my own life. I liked accepting a challenge and being determined to win. I have learned since that they are fundamental to success at any level, and must be instinctive in any situation. I think I applied them before, but certainly tried to do so from then on.

Unfortunately, when I left Brecon after this course and went back to Aldershot, I was unable to present the second-in-command with a hat for eating as he had been posted elsewhere. Pity too, because I was flattered and pleased when the staff at Brecon invited me to return there as an instructor. At least someone thought I had something to offer. Also, I was promoted to staff sergeant instructor. Not bad after eight years in the army, plus a bit of up and down in the process.

So began a very enjoyable period of my soldiering. Actually, it was my last job in the British Army. I was helped immensely by the officers running the course, Major Hopton and Captain HarringtonSpear. Harrington-Spear was an officer in the Royal Anglian Regiment and, after watching me grappling intensely with the techniques of instruction, giving lectures, and taking practical periods on the course, he called me over one day and said, "Staff McAleese, ever thought of taking up a hobby?"

"What d'you mean, sir?" I replied, more puzzled than astonished. The army was my life. There was no time nor place for anything else.

"Er, a hobby," said Harrington-Spear vaguely. He looked at me standing to attention in front of him, my hair cut down to the wood, my uniform immaculately pressed, my DMS boots gleaming, and he could think of nothing at all with which to take my mind off the military. Instead, he said lamely, "There are other things than the army, you know."

"Yes, sir?"

"Forget it," he said after a moment.

But I liked him. Only later did his words make any impression, but I still don't have a hobby.

Major Hopton, a typical Royal Green Jacket officer, tall, dark-haired and fit, was also a help. He watched me giving a lecture once, on first aid, which I had spent hours preparing in my room, laboriously rehearsing it over and over in front of a mirror. My delivery was strictly by the book and when I had finished, he said, "That was fine, but a bit stilted."

"Sir?"

"You are an extreme man, Staff Sergeant McAleese. Take a look at yourself in the mirror. Your obsessions are your strong points, your qualities and your character. Stop trying to suppress them inside the manual of the School of Infantry. You have experience, you have drive. Use it."

I took his advice and allowed my enthusiasm for the subjects to determine the method of instruction, rather than the other way around, and I think it worked. I think the students thought so too. Even when they were soaking wet and cold, marching up and down the Welsh hills or digging filthy dirty trenches and filling them in again, I was always there with them, encouraging them and trying to pass on my own enthusiasm for soldiering.

I felt that these infantry officers gave the Brecon school depth and balanced the sometimes-selfish obsessions of the airborne, the Parachute Regiment and the Special Air Service. In fact, I grew to respect the line infantry regiments, particularly the Argylls, though maybe my old grandfather's ghost affected my judgement. Overall, I saw that although Mr Average Top Student was a Para, so was Mr Average Bottom Student, while most line infantry NCOs produced steady 'C' grades. They were good men with experience to offer and I enjoyed working with them.

My nine years were up. I enjoyed working in Brecon very much, but officers in the Parachute Regiment had told me that I had reached my ceiling in the British Army, that my record stood against me and I could expect no further promotion. This was a bitter pill to swallow at 26 years old. I knew nothing else and I wasn't interested in anything else.

So, on March 1, 1969, I retired from the British Army.

I owe my early training and experience to the Parachute Regiment and to 22 SAS, but, as it turned out, I did far more soldiering elsewhere.

CHAPTER 5
CATHARSIS

As a Catholic, but not a very good one, I might say that the next few years were my time in the wilderness. Sister Loyola and Father Brett would rise up at the sacrilege, but the analogy ends there, abruptly. I did not go to a high place, to observe the world. I saw the world from a very different place altogether, among the low-life. I was tempted and gave in.

This was a grim part of my life and for a time I debated whether to include it in this book at all. I'm not proud of what happened, but then I thought I should present the whole picture. If I did not mention it, some smart-arse would bring it up and accuse me of trying to hide it. I don't want that. I have said before, I'm not making excuses for myself. My life has been about soldiering, I'm proud of my achievements as a soldier, I've served in three different regular armies and I won't allow myself to act like so many Walter Mitty 'mercenaries' who prefer not to talk about their past. Besides, I learned a lot in these next miserable years.

I did not make a very good civvy. In fact, the first thing I did was to try and join the United States Army, to go to Vietnam. A wellknown SAS officer, nicknamed 'the Rat', who left the army as a colonel, took a considerable risk for me by pulling strings with friends of his in RAF Air Movement and fixed a flight for me from RAF Lyneham to La Guardia airport in New York. I equipped myself with a passport and visa and arrived in the USA with no money to find bureaucracy was alive and well. The US Army recruiting officer was perfectly happy to sign me on, but first he wanted me to obtain a US work permit. When I went to the Labor Office, they were quite happy to give me a work permit, once I was in the US Army.

Depressed and penniless, I was walking down Broadway and bumped into Tim Holt, who had won the Military Medal in the SAS in Malaya. He was broke too. We both went to a hospital and gave blood for a few dollars. Finally, I succeeded in phoning my contact in RAF Air Movements and he wangled my flight home. On reflection, I think it was probably a good thing I was prevented from joining the US Army. I don't think I would have survived Vietnam.

I got a job laying gas pipelines. This was the golden period of British North Sea Gas development and I worked up and down the country on sites in Fife, and at Braintree and Hereford.

However, I was still married to Marlene and when I came back to Hereford between jobs, I saw all my old mates who were still in the army. I found this hard to bear. I was on the outside – looking in to where I wanted to be – my marriage was in tatters and I was having to come to terms with civilian life at a time when the student revolution, peace marches, long hair, beads and bells were in full swing. It sounds odd now, but I used to go to work on the gas pipeline in Hereford dressed in my Parachute Regiment smock. I changed into work clothes on site, and at the end of a day's work, I put the smock on again. I resented being a civilian and I was disgusted by the 'you-owe-us-aliving' attitude in people at that time. I wanted to show I felt differently. I suppose I took it out on Marlene; we argued constantly and it made no difference when, in 1969, she produced my first child, a son we called Jason.

The Hilton Assignment was a brief light in the darkness. In June 1970, I received a telegram from Lawrence Smith, who had finished his time with the army, saying there was a lucrative prospect of earning £5,000 for one night's work. I was immediately interested, especially as I felt confident in the nature or ethics of the job if it came via someone like Lawrence Smith. I was even more encouraged when the follow-up contact was made by John Ragman, who had been in Borneo with me and had been awarded the Military Cross.

If there had been any lingering doubts, they were dispelled

when he took me to a black and white block of flats in Kensington in London, where the offices of Watchguard were located and where I met David Stirling, founder of the SAS in the Second World War in 1941. With him were a roomful of people. Of the twenty-plus men there, I knew more than half, as they were all ex-SAS soldiers. Stirling and Ragman did not give us all the details, to preserve secrecy on a 'need-to-know' basis, but they gave us a sketchy outline in which they explained that we were being asked to commit ourselves to an operation in Libya and must be on stand-by till called forward.

Given the people running this job, and that we knew the revolutionary new leader Muammar Gaddafi had seized power from the Libyan King Idris in Tripoli only the previous September, none of us had any qualms about agreeing to join in. We were each given an envelope containing £200 and left to await 'call-up'.

Nothing happened. Eventually, I was amazed to read about the whole operation in the national newspapers. I found out that we had been planning to raid the prison in Tripoli, ironically called 'the Hilton', and release all the anti-Gaddafi political prisoners before withdrawing via the beach nearby. The British government found out what was intended, decided it was all too embarrassing and aggressive, and the operation was blown by British MI6.

This was a real let-down, and, about the same time, Irene, a girl I had met in Hereford that winter, reported me to the police for kicking in her door and back-handing her. She was a slim girl with brown hair and long legs perfectly shaped for the mini-skirt which was all the rage then. Our relationship rocked violently from extreme excitement to violent brooding rows. She became a dangerous obsession for me. Our affair finally shattered my tottering marriage with Marlene, who divorced me in February 1970, and was the cause of my spending nearly two years in Gloucester Jail.

To start with, in June 1970, I was given three months for kicking in Irene's door, and three months for smacking her, both suspended sentences. When I did it all over again in November, the Justice of the Peace implemented both previous sentences, added some more and sent me away for nine months.

HMP Gloucester was old-fashioned – slopping out, army-style food, full of low-lives – and it never bothered me at all. Remember Lethamhill Road. I spent as much time as I could in the gym working the weights, a number of my friends in the SAS visited me regularly, and I got on well with the prison officers who were mostly ex-army or police. I was put in charge of the detail in the mailbag sewing shop, which may sound grand but in prison is the bottom of the pile, and immediately ran into trouble with the prison's subculture.

Sewing mailbags may sound a straightforward job to you, but since a few pence was given for each finished bag, rules had been invented by the prisoners to benefit the old lags. The understanding was that new men had to do a week's 'training' at the end of which they received only a nominal sum, which might just cover them for a bit of tobacco, irrespective of the number of bags they had sewed. The old lags took the rest, booking in all the new men's bags as their own. I decided this was unjust. Everyone should get paid for the number of bags they sewed.

That day, a thin-faced man who had spent years in and out of prison, an old lag though he had never done a real lagging (i.e. served a sentence longer than three years), took a finished bag off one of the new younger men, gave it to me and said, "Book this in for me, mate."

"That's not yours," I said. As the prisoner in charge of the shop, I recorded the number of bags each man had sewn.

He sneered unpleasantly. "You better book it as mine, mate; I've got ways of fixing guys like you."

To which I replied, "And I've got ways of fixing guys like you." And I hit him.

He staggered back, and suddenly ducked out of the bag shop to find the nearest prison officer who was on duty outside the door.

Standing right beside the screw, he screamed at me, "Come

on, you bastard. Take me if you can!" He waved his hands at me, encouraging me to attack him again. He knew perfectly well that, whereas he did not care how long he was in prison, I risked an extra sentence for fighting.

I glanced at the prison officer and, to my delight, he tactfully turned away. So, I hit the guy again, as hard as I could. Needless to say, the prison officer saw nothing and I had no further trouble.

On May 28, 1971, I tore off another sheet from the missionary calendar hanging on my cell wall and read, "Be ye transformed by the renewing of your mind". An hour later, I walked through the gates into Gloucester city to the bus station, free again. I have never forgotten the text, though the timing wasn't quite right. Irene was waiting for me in Hereford and we swung back into another violent roller-coaster chapter of our relationship, which ended predictably with me standing before the Bench again. I was back inside Gloucester nick by August 16 for a second six-month stint of weight training.

I read a good deal and had plenty of time to think about my life. The prison was not rich in reading material, and it was common to find that prisoners had deliberately torn out the last page, but the *Reader's Digest* was allowed and I found much to read there which mirrored my reflective mood. Maybe that's why they allow the *Reader's Digest*. The majority of the prisoners were thieves and conmen who spent their whole lives in and out of jail, and I don't think many of them bothered too much with articles like 'Unless You Deny Yourself' by A.J. Cronin, or 'The Joy of Doing Good on the Sly' by the Reverend Gordon Powell, which I found in the *Reader's Digest* and still have, or Patience Strong's 'Quiet Corner' in the *Sunday Mail*.

Believe it or not, I began learning calligraphy during my weeks inside. As you might imagine, I had plenty of time and got rather good, laboriously copying out Patience Strong or 'Quotable Quotes' like, 'Only a life lived for others is a life worthwhile' (Albert Einstein). Of course, the other prisoners watched me doing this and I ended up writing letters home for them, to wives and girlfriends. One man who could hardly write his own name dictated to me as follows, "My dearest darling beautiful wife, I miss you very much, my dearest darling beautiful wife, and I am looking forward to seeing you again, my dearest darling beautiful ..."

A crowd of others round us nodded appreciatively.

"For fuck's sake," I interrupted, pen poised over the inkwell. "She's got the message! What d'you want to say?"

He looked hurt at that, paused and continued, "My dearest darling beautiful wife, you will notice my handwriting is different. This is because I'm taking calligraphy lessons here ..."

There is no shame in prison.

Out again, I worked with Bob Varey over the winter of '71–'72, but by June I was in trouble with Irene once more. She just got to me, drawing me in like a bee to a honeypot. We got wildly involved with each other, and then as always ended up arguing like cat and dog. One night I went from the pub to her house to find she was out, but the rest of her family were waiting for me. Her mother and brothers began a screaming argument, a fight ensued and the beak sent me back to Gloucester, this time for another six months.

This time, the weight training, my calligraphy (which was quite good by now) and a good deal of reading sorted me out. I never felt that I was a villain, a thief or a criminal. I was in prison over a woman, for drinking and fighting – which seem to me clean enough reasons – but this time I had resolved the whole thing in my mind. I had done my time and that was that. I had reached the bottom, seen the low-lives there and wanted no part of it. It was time for a change.

The doors of Gloucester Prison closed for the last time behind me in December 1972 and I went back to Hereford. Irene found me near the old stone bridge over the River Wye and began her usual game. She strutted along the pavement ahead of me, tantalising in her mini skirt like some houri in a Hollywood movie. This time I stopped, turned around and walked away.

CHAPTER 6
LNTERESTING WORK ABROAD

In 1975, I was 32 years old, working on the oil rigs, like so many ex-soldiers, and bored. So, when Bob Varey said he had seen an ad in the *Daily Mirror* offering 'interesting work abroad', I thought, *Why not?* This turned out to be the 'Rhodesian Contract' and it was very bizarre indeed.

On Saturday, May 27, three of us – Bob Varey, Norman Duggan and I – went up to the Centre Court Hotel in London from Hereford. We sat about in a large conference room, curtains drawn, facing a stage, and around 7 PM the drama began, just like something from a movie. All the lights gradually dimmed, leaving us in darkness till suddenly spots lit the stage curtains, which began to slide open. A small man with a long face appeared neatly dressed in a suit, and he began to give us an extraordinary briefing. This was my first sight of John Banks, who had been dishonourably discharged from 2 Para in 1969 for driving a car without insurance. Behind him, a big sheet hung on the wall on which someone had drawn a free-hand map of Africa, with 'Kariba Dam' written on it in free-hand in huge letters right in the middle. Banks had taken

trouble with the stage effects. He dropped his voice and began to speak softly and penetratingly, using maximum amplification on the public address system.

"All those who have scruples about fighting for the black man against the whites should leave the room now," he boomed.

Only a few got up and left the room. I reckoned there was

plenty of time to leave later if necessary and we were all too intrigued to walk out at once. However, there was a lot of muttering and John Banks, sensitive to this, wisely announced there would be a tea break. I took the opportunity to talk to some of the others and quickly realised that the place was full of dreamers, Walter Mitties and space cadets with no military experience whatsoever. One very younglooking 18-year-old, wearing a natty combination of suit and baseball boots, tried to convince me he had been fighting with Frelimo against the Portuguese. Another, called Les Aspin, certainly came from Norfolk, East Anglia, in the UK, but insisted he was from Norfolk, Virginia, in the USA and swore blind he had been working for the CIA. He later claimed in a book to have single-handedly killed four Palestinians who burst into his hotel bedroom to assassinate him, and even said he threw one out of the window to his death on the pavement below.

The briefing continued. Banks told us there was a big op coming off in Rhodesia. He had been asked to find two squadrons of mercenaries to fight out of bases in Zambia against the Rhodesians and we were to use medium machine guns. He meant the Vickers watercooled MMG, which is good value in defence but massively cumbersome and particularly unsuitable for mobile operations (we never used it in the Rhodesian SAS). The three of us from Hereford listened intently, but Banks's briefing matched the whole atmosphere and was entirely phoney.

Once he found we had been in the SAS, he was at pains to impress us and revealed that he had some secret backers for this operation. All ears, we listened fascinated as he told us he had been to a covert meeting with Paddy Mayne in a caravan on Hankley Common. Colonel Paddy Mayne was one of the first great SAS names, who was awarded three DSOs for bravery in the Western Desert and commanded I SAS. He left the army after the war and went back to Northern Ireland. Banks's claim was certainly impressive, as Paddy Mayne had died in a motor accident two decades before in 1955. I later found out that the story had more to do with deceiving his wife, as Banks was screwing a bird in this caravan on Hankley Common at the time, the dull bit of sandy ground near Aldershot which the Paras use for balloon jumps.

Banks had to come up with something about 'backers' because he had to explain how he was paying for the hire of the conference room, how he paid for expenses and how he managed to afford handouts. He gave £200 to Bob, Norman and me, probably because he wanted ex-SAS soldiers to balance the young dreamers we had met during the tea break. Banks had been a Para and if he stopped to think, he was able to tell the difference between a complete bluffer and someone who had a little soldiering experience. Later, as Banks's own delusions developed and the cash rolled in, he didn't bother to try.

However, most journalists who follow mercenary stories cannot tell the difference. Or, more likely, they prefer not to. Most opted for Banks's story, leaked by him to the press, that he had gone to another secret rendezvous with a "tall, aristocratic officer" in a "large country house near Hereford". The implication was clear. Colonel David Stirling, who had certainly been behind the Hilton Assignment, was the backer, 'the Paymaster', with other decorated SAS officers in the shadows. The press also implied these men enjoyed the tacit approval of the British government. This scenario was supported at long distance by the fact that Britain was far from cosy with Ian Smith of Rhodesia. Such speculation made more suggestive and exciting copy than the sordid truth, and suited the arrogant imaginations of Banks and journalists alike.

The Rhodesian operation was plainly fantasy and we went back to Hereford, where I was living with my sister Molly. I thought no more about it till Banks phoned and asked me to help him with security at another weekend meeting, this time in the Post House Hotel in Heathrow. On Saturday morning, I arrived in a suit and tie to find a group of nineteen people who looked even stranger than the first lot – in jeans and loud-coloured T-shirts, swinging heavily tattooed arms like muscled gorillas – and, needless to say, they spent most of their time sitting at the bar telling each other war stories.

"I was in the Foreign Legion," said one. "I flew a glider into Dien Bien Phu in '58." Apart from the fact that he would have been four years too late, the Legionnaires parachuted in, and this fellow didn't look old enough to have been a pilot then, sixteen years before.

"I was a marine," replied his co-drinker. "A commando."

"Yeah? I work for Mossad," said a man with a mass of fair curly hair.

"We came across your guys when I was in the CIA ..." said a fourth.

"I was a navy diver for seven years," said another, enormously fat man on a bar stool. "I was with the Special Boat Service then. A medic. Of course, that was after my radar technician's course, which I did to supplement my sonar skills." He twisted with difficulty on the stool to bring me into the circle at the bar and said, "So, you were in the SAS as well?"

How could a man do so much in so short a time? Amazed, I nodded, actually catching myself considering his question seriously, when I knew perfectly well I had never seen nor heard of him in Hereford before in my life.

He smirked knowledgeably and shouted, "Barman! Another round of beers!"

All the drink went on expenses, which Banks paid for, and I began to wonder what the point of it all was until a reporter for the *Daily Mirror* turned up, on Banks's tip-off. The meeting was supposed to be a continuation of the Rhodesian operation, but that was all nonsense. Fantasies are no good without publicity, which the journalists willingly gave him.

After the press left, I decided to find the man with the mass of curly hair. He had been behaving very strangely, peering at me from behind pillars and ducking out of sight round corners. I decided it was time to have a talk. In his room, he backed away from me, saying, "John Banks told us if we did anything wrong, you'd kill us!"

I was speechless. The man was completely bald and on the

dressing table stood a hairdresser's polystyrene head supporting a fine, curly blond wig.

More bemused than angry, I went downstairs to find Banks, who airily said, "Oh, that? I told them you were the adjutant of the SAS and that you would not hesitate to kill them if they misbehaved."

Such was the 'mercenary' world of John Banks, but the casual attitude to 'killing' had a more sinister impact the following year.

By the Sunday, Banks decided the farce had gone far enough and told me to dismiss the men. Of course, as soon as they heard the jamboree was over, the hotel telephone lines were jammed with orders for bottles of gin and whisky on room service, and some could hardly lift their suitcases as they staggered out of the hotel. The final bill must have been vast. The price of fantasy and, do not forget, publicity. Banks drove me to his house, where he thanked me when I left and said cryptically, "Peter, there could be pressure on. You'd better take this." And he pressed a loaded .32 calibre pistol into my hand. Puzzled, I returned to Hereford and handed it straightaway to one of the officers in camp.

Of course, Banks phoned again. A few days later, he offered me £60 a week to act as his security advisor, which was reasonable money at the time, so I agreed. He wanted to sell commercial security, and needed my enthusiasm and experience to make presentations to clients, to persuade them we could train their bodyguards and security officers. This was a genuinely good idea. At this time, the so-called 'security industry' was only just starting and there was money to be made by new firms in the market. If Banks had not allowed his penchant for intrigue and fantasy to ruin the plan, we might have made it work. Management for the Churchill chain of hotels came to us and Banks spoiled a good presentation of mine by calling himself Patterson. Of course, the hotel's security boys immediately recognised him as Banks. His face had been all over the newspapers as the central figure of the mercenary stories only weeks before. Once again, it was clear nothing would come of these plans and I went back to Hereford.

This, then, is the shady world of mercenaries and security men, where it is difficult to separate fact from fiction, where most of what you hear is lies, but – beyond that grey twilight zone of nonsense – there was (and still is) a hard core of reality.

Ironically, it was directly because of the publicity generated in these first ludicrous fantasies that John Banks was contacted to supply men for an actual contract. On Saturday, January 17, 1976, I was sitting in Molly's house in Hereford when the phone rang and a voice said, "Is there a Major Nick Hall there?"

I had never met a 'Major Nick Hall' but the line went dead before I could ask for an explanation.

This happened again, so the third time I was surprised when the caller announced, "This is Major Nick Hall speaking." He explained that he was a colleague of John Banks and there was a job abroad, with pay and expenses. "If you're interested, come to the Tower Hotel tomorrow and we'll tell you what it's all about."

The following morning, I packed some overnight things, dressed in a suit and took the train to London. I was under no illusions about John Banks and the rest, but the chance of excitement abroad made scaffolding in Hereford look pretty dull.

We sat about in one of the bedrooms and I recognised a number of the same faces I had seen before in the Post House Hotel. The door opened and a tall, upright man of medium build with a pasty face came in and introduced himself as Major Nick Hall. He said he was an officer of the FNLA (the National Front for the Liberation of Angola). He seemed about 25 years old and I don't think any of us were convinced by his rank. I did not at that time know that he had been a private in I Para, where he had been courtmartialled for selling weapons to the Ulster Volunteer Force and sent to prison for two years with a dishonourable discharge from the army. With him were John Banks and Les Aspin, the man who had singlehandedly taken on the PLO in a hotel room, and they briefed us about the war in Angola.

On November 10, 1975, the Portuguese had announced that they were leaving Angola after five centuries of colonialist rule.

On the same day, two liberation movements fought a battle 25 miles outside the capital Luanda. Holden Roberto, the leader of the FNLA, was defeated by the Marxist troops of the MPLA (the People's Movement for the Liberation of Angola), so it was the communist-backed MPLA who celebrated Independence Day in Luanda while Holden Roberto retired to Ambrizete, 100 miles up the coast. In the south was another group, UNITA (the National Union for the Total Independence of Angola), led by the bearded Jonas Savimbi, who was supported by South Africa.

The USSR then stepped in to bolster the MPLA. First, 'advisors' arrived in Luanda, then Fidel Castro willingly supplied thousands of his black Cuban soldiers (who were a nuisance to him in Cuba at that time). Most importantly, Russia supplied plenty of arms: the ubiquitous AK assault rifle, T-34 and T-54 tanks, and artillery in the form of mobile 122mm Katyushka rockets, known in the Second World War as 'Stalin's Organs' because of the horrendous noise they made when fired.

This massive effort shocked the United States, which tried to solve the problem with money. President Ford agreed to allocate $32 million to prevent Angola becoming communist. The fund was managed by a handful of CIA agents but the reaction was too late and produced too little on the ground where it mattered, facing the reality of the MPLA.

African armies usually fight each other with the minimum martial contact, moving from town to town terrifying the opposition with propaganda. When the attack comes, resistance is slight because most of the enemy have run away. Those foolish enough to stay can expect little mercy. The hallmark of African wars is slaughter. I have heard it said that 90 per cent of casualties are the result of massacres and atrocities, mostly after the fighting is over, and after my time in Africa I can believe it. This was certainly true in Angola.

However, the MPLA was a serious proposition. Stiffened by Cubans, Soviet arms and the Soviet doctrine of concentrating the maximum forces in the attack, the MPLA had a distinct military advantage over its rival FNLA. Like a juggernaut, the MPLA advanced along the long, thin roads through the bush, taking town after town, and by the time we all met in the Tower Hotel, Holden Roberto's FNLA was confined to the north of the country.

Nick Hall tried his hardest to paint the FNLA in the best light, but those were the facts. He finished off with, "We need tough guys. Men who aren't afraid of fighting tanks and overwhelming odds."

I can't say what the others thought about that, but there were about twenty-four people in the room and they all accepted a first payment of £200 each and a Sabena airline ticket to Kinshasa. Flight 614 left Heathrow that evening for an overnight stop in Brussels, before carrying on to Kinshasa. Passports were no problem. Banks and Hall had foreseen that most if not all would come without their passports and spent hours on the telephone and visiting the Belgian Embassy. Presumably, the British made no connection between John Banks of recent mercenary fame and the new manager of a group called the Manchester Sporting Club, nor thought it odd that such a club should carry identity cards in the name of the SAS! (Banks called his company Security Advisory Services.) Anyway, they had no objections to letting us out of the UK, the Belgians were happy to clear us through Brussels, and plainly we would have no difficulty flying into Kinshasa. Holden Roberto was related by marriage to President Mobutu of Zaire, who allowed him a base in Kinshasa.

Banks and Hall had been given very little time to supply men for the FNLA and keep up the momentum. Without delay, we drove to Heathrow. I went with Hall in a car, while the rest piled into a coach. By the time they reached the airport, the total number had shrunk to twenty. Four or five had gone absent with their cash, stepping out of the coach at traffic lights on the way, preferring to spend their money at the bar rather than face 'tanks and overwhelming odds'.

I left Britain as a trooper in the new mercenary army and I was one of the few who did not get paralytically drunk that night in the hotel in Brussels. I enjoy my drink but not when I am working, and however Mickey Mouse this adventure seemed to be, I stayed off the booze. Perhaps 'Major' Nick Hall was impressed with this, as by the time our Sabena DC-10 landed at N'Délé airport in Kinshasa, he had promoted me to captain.

We waited on the plane for all the other passengers to leave and then emerged into the hot, sweet-smelling, oppressive African night to find Holden Roberto standing by a car on the tarmac. He was a tall, stocky black man dressed in a grey, light-weight suit and wearing dark sunglasses, which he was never without. He appeared tired, like a man with his back against the wall, but he shook us all by the hand as we filed past and we all replied as instructed by Nick Hall, saying, "How d'you do, Mr President."

Feeling dirty and exhausted after our long flight, we walked across the tarmac to a gate beside the terminal buildings. The whole place was in a terrible state of repair, with paint peeling, concrete walls chipped or broken, and windows smashed. Outside the perimeter fence, we boarded an old American-style bus and followed Holden Roberto's car at speed over bumpy, potholed streets littered with garbage to a restaurant where he gave us all dinner. He was surrounded by a crowd of black bodyguards wearing slacks and bright calypso shirts bulging with under-arm pistols. With them was a white man, called 'Canada' Newby. After dinner, 'Major' Nick Hall, Holden Roberto and Newby climbed into the big limousine and led us on another mad drive through the sprawling city to the other side of Kinshasa. Holden Roberto lived in a large white villa called 'Kirkuzi', which was set in spacious grounds on a rise overlooking the city and was guarded by patrols of well-armed black men.

It was late, we were exhausted, but we were immediately issued with Belgian-style army uniforms, belt webbing and pouches, boots, M2 carbines with several 30-round clips and one magazine. Four of us picked FNs, the Belgian equivalent of the British .762mm SLR. Contrary to what has been said, these M2 carbines were in good order, though there was some grousing because some of the men with British Army experience were more familiar with the SLR. Nowadays, everyone wants the M16 (Armalite), a smaller calibre still at .556mm, but everyone recognises that the ammo is lighter and you can carry much more of it. Important, that, in a firefight.

Perhaps because we now looked more like soldiers, 'Major' Hall announced we were to drive to Angola that same night. He told me to inform the rest and we set off again in the old Yank bus. No more than fifteen minutes later, still inside Kinshasa, we were stopped by a Zairean Army roadblock. In the manner of black African troops, all the Zairean soldiers pointed their rifles at us and expressed ignorance of our 'mission'. To my horror, everyone in the bus started leaping to their feet and cocking their carbines, and there were shouts of, "Let's shoot our way out!" I saw our expedition ending in a bloodbath before we even reached Angola and took command. I calmed them down, made them all unload their carbines till we reached Angola, and put an armed man in each corner of the bus. Outside, an interpreter sorted out the roadblock and we set out again. Others have written up this incident as taking place inside Angola, where it doubtless looks better on their combat CV, being in the war zone.

The bus rumbled on towards Angola, 200 miles away. We spent a very uncomfortable night lying across the seats, trying to sleep as the bus jolted along through the darkness. The roads were so bad that we occasionally had to detour into the bush to avoid potholes, and everywhere I saw the remains of cars and trucks which had simply broken down or crashed and been left for want of spare parts.

At dawn, we reached Luvo, a largish place on the Zaire-Angolan border, and crossed into Angola for the final leg of our journey to São Salvador, another 50 miles south-east, where Holden Roberto had his headquarters. This town was no different from any of the others we had passed through, with peeling, whitewashed, flat-roofed houses, a few with Cape Dutch fronts, very colonial and very Portuguese, with loose-limbed black Africans wandering about in desultory groups in streets filled with rubbish, or just sitting and waiting, with small black children playing in the dust, and the warm air was rich with the rank smell of decay. Our bus passed the police station, which was patriotically flying the red, yellow and white FNLA flag from a pole on the roof, and then turned off the main street through big, wrought-iron gates into the grounds of a large, colonial-style palace. This was the FNLA HQ and it looked very run-down indeed.

Stiff from our journey, we stumped out of the coach and stood about. I was wondering at a number of black soldiers slouching about under trees wearing only their underpants when a tall English officer appeared suddenly at the top of the steps of the house. He was neatly dressed in uniform, in the rank of captain, with a moustache and a military bearing. He surveyed us from the steps, waved his hand at us – his face a picture of disgust – and demanded succinctly, "Who the fuck are these fuckers?"

This was Mick Wainhouse. He stood there like an officer on parade and cursed everyone for having long hair, and being dirty, unshaven and overweight. I must admit I felt rather guilty, as I was not as trim as I had been, but I was reasonably fit and saw no point in lambasting men who had been travelling solidly for thirty-six hours. I sent the others off to find billets upstairs in the palace and had a quiet talk to Wainhouse on my own and cooled him down.

Once again, we were not to know that 'Captain Mick', as he was called, had been a private in the Parachute Regiment and, like Hall, Callan, Copeland and others, had been dishonourably discharged from the army. Maybe his court-martial proceedings had given him an extra opportunity for studying officers' behaviour, because he aped their mannerisms rather well.

Callan appeared and I must say he made me apprehensive. He was of medium height, tanned, well-muscled and looked very fit. It takes a lot to make me frightened, but I could see he was a complete maniac by the arrogant way he strutted and threatened.

All around us, everyone was running about standing to attention in front of everyone else, stamping feet, saluting, shouting orders and generally carrying discipline to extremes which would have been ludicrous in any regular army.

There was something dangerous about this man, which warned me to act very carefully with him. I was on his ground and vulnerable. Remember, all these people had been brought together at short notice, with no allegiance to each other. A pecking order had to be established out of nothing, with no army system, traditions, ethics nor any body of law to help – in other words, by force of personality alone. Callan, Wainhouse, Charley Christodoulou and Nick Hall were the first in, and Callan had taken charge. It was as simple as that. We were the men hired to help Holden Roberto and his FNLA, and the whole lot of us were being funded by the CIA, just as the Cubans fighting with the MPLA were the vicarious agents of the Russians. While the enemy were about 70 miles down the road at Damba, this was the absurd situation at the headquarters in São Salvador. It was the mercenary jungle at its worst.

Callan asked me for my background, which I told him. He did not volunteer his own. His real name was Costas Georgiou, born in Cyprus in 1951, and he had been a moderately successful Para until he foolishly decided to rob a post office in Belfast, with Wainhouse. The postmaster observed them driving up and down the road and telephoned the nearest army base.

"I'm going to be robbed!" he said to the Para officer on duty.

"How do you know?"

"Take it from me," replied the man. He had been robbed numerous times before by the IRA and recognised the signs.

"All right. Can you give me a description of the men?"

"To be sure, I can," said the postmaster drily. "They're in uniforms of the British Army, wearing Para berets and driving one of your Land Rovers!"

This ill-conceived frontal assault was typical of Callan, as was his threat to kill the officer prosecuting him at his court martial. He was given five years and dishonourably discharged.

While I was talking to him, a black soldier came up to us. Callan snapped at him, "What is it?"

"Sir, we've got no food or ammunition."

"Fuck off!" Colonel Callan shouted in the man's face, and drove him off with a few broken words of Portuguese. Then he turned to me and said indignantly, "Can't you see the trouble I have with these people?"

We went in his Land Rover for a drive round the town. The streets seemed filled with more black soldiers carrying rifles and wearing nothing more than their underpants. I remarked on this and Callan said absently, "I requisitioned their uniforms."

He pointed out the salient features of São Salvador as we bumped along the potholed streets. The place was dominated by the palace and the police station. Outside the town on the edge of the bush, he showed me the airstrip and I saw how São Salvador was built on an escarpment overlooking a lovely valley full of green trees and grazing land.

Back in the HQ, everyone was trying to get their hands on weapons of their choice, mainly AKs, which they simply grabbed off passing black soldiers. Typically, no one was satisfied with what there was available. Callan told me there was a shortage of weapons, but I made my own tour of inspection and found a large shed behind the palace full of weapons and ammunition. There were no FNs but there were plenty of Soviet AK assault rifles, boxes of brand new M2 carbines complete with magazine pouches and rifle cleaning kits, a couple of American 106mm M40A1 recoilless rifles, a couple of Soviet anti-tank RPG rockets, American 66mm light anti-tank weapons, the free-flight disposable missile launchers called LAWs, and plenty of ammunition. In brief, it is complete nonsense for anyone to say we had a shortage.

There was an atmosphere of chaos in São Salvador and especially round Callan, which had nothing to do with the proximity of the MPLA. He was totally unpredictable, manic and brutal. He called over a black man who was wearing a new camo' hat he wanted to give to one of the new arrivals. When the black

FNLA soldier didn't spring to attention quickly enough, Callan screamed at Sammy Copeland and another mercenary to take the black man away and beat him up, which they did. For an hour, we could hear the man screaming inside the palace before they flung him out into the dusty yard covered in blood. It was common to see, or hear, black soldiers being beaten up and it added to the atmosphere of anarchy. I don't know what Holden Roberto thought of this, or even if he knew, but hitting people who can't hit back, black or white, has never been my style. Unfortunately, beatings were only just the beginning.

By evening, the new arrivals were in some shape. At least, the pecking order was established – with a trail of broken sunglasses, the budding mercenary's indispensable accessory, trampled underfoot to prove who had lost out in various petty little arguments – and we went to bed. Callan gave me a room on my own and I slept deeply until a terrified mercenary burst in during the middle of the night shouting, "We're being shelled! We're under attack!"

I rolled out of bed and ran to the window. Great flashes of light lit the brooding African night, with cracks of distant thunder, but this was an electrical storm, not the MPLA gunners. Some of the men had never been out of England and they were very inexperienced.

A couple of days later, I was pleased when Callan let me pick six of the men for a separate detachment. I took Brummy Barker, Mike Johnson, Doug Saunders, Stuart McPherson, Mick Rennie and John Tilsey. We drove to the airstrip and flew to San Antonio di Zaire, about 150 miles from São Salvador, in a Fokker Friendship piloted by a Portuguese and paid for by the CIA.

This was my first real look at the battleground and the first time I appreciated the sheer size and emptiness of Africa. Dry, grassy hills rolled out in all directions beneath us, covered in scrub bushes, spreading baobabs and low, thorny trees, and I could see the importance of the roads which linked each lonely town in the bush. San Antonio di Zaire was in the north-west corner of

Angola, rather out on a limb, but it was an important coastal town on the south bank of the Congo River estuary. Zaire lay several kilometres away across the water, on the north bank.

From the air, I could see that San Antonio was ideally placed to defend. The town was built on the side of a flattish, triangular peninsula sticking out into the mouth of the river. The road to the town from the south passed a crossroads at Corpus Christi, which I identified as a crucial defensive position. However, it was clear that the only escape once the attackers blocked the land approaches would be by boat from the quayside where the local fishing boats were tied up.

Holden Roberto met me in the town and explained he believed it was vital to hold out against the MPLA, who had already taken Ambrizete, the next important town south about 100 miles down the coast. If the MPLA took San Antonio, he said, there would be nothing left to the FNLA except an enclave in the north, pressed against the border with Zaire.

I installed my small group in a villa in the town and drove out to Corpus Christi, where some nuns lived in a convent. It was perfectly obvious that an immense effort was required to organise any sort of defence. The whole town was in a state of virtual anarchy. FNLA soldiers ambled round without any sort of discipline, and the black civilians had given up, complaining that the FNLA stole food and anything else they wanted. There was talk of rape, and the two hospitals we went to, one civilian and one military, were utterly disgusting, fly-blown, stinking of gangrenous rot, with dirty bandages lying about and no drugs.

Angry more than depressed by all this, I asked Holden Roberto to gather the people together on a worn basketball pitch in the centre of the town. We looked out on the sea of black faces, and Holden Roberto adjusted his black sunglasses and shouted, "This man is the new commander of San Antonio di Zaire." He turned elegantly from side to side, looking over their heads, trying to muster enthusiasm, a spare, somewhat menacing figure. "You will come to him if there is anything you need. He is in charge of everything!"

No one seemed very impressed, but then, the place stank, they were starving and I don't think anyone cared who ruled them: the FNLA, the MPLA or UNITA. Come to that, for me, the state of the place rather took away the pleasure of being called the supremo of a whole town for the first (and only) time in my life.

There seemed little point defending somewhere as close to total anarchy as San Antonio was when we arrived. There was no sign of law and order, even African-style, among the civilians, while even the rudiments of military discipline were lacking among the soldiers. I decided the place needed ripping into shape at once. I detailed my small group to make various assessments – of the weapons available, of the numbers and state of the FNLA troops, of the rations and ammunition – and went myself on a closer look round the town. On the outskirts, near the river, I discovered a large warehouse belonging to Petrango, Angola's petrol company. Amazingly, while the country starved for lack of organisation amid the ravages of civil war, it was full of food.

"All this," I told the black man in charge, waving my arm at the boxes of rations stacked inside, "is requisitioned."

He was terrified, not by me, but for his stores, and reminded me of a certain type of British quartermaster sergeant who likes his stores neatly arranged on the shelves out of use, and, therefore, out of harm's way.

I tore off a sheet from my notebook and wrote him out a receipt for the entire shed, signed simply, 'Peter.' He was happy with that, and I have often wondered if he got into trouble, if he survived. Back in our villa HQ, I ordered the food to be issued to the 800 black FNLA soldiers so they had no excuse to steal from the civilians.

These soldiers were useless. They had been trained by the Chinese at Kinkusu in Zaire and spent their time learning stupid communist slogans rather than training with their weapons, which they had hardly fired, even on the range. Their commander was a large black man called Commandante Lima, who had been a policeman in the United States. He passed his time working on a boat down on the river, presumably preparing for his escape. I can't say whether it was his American police experience or the way of life in Africa, but when two smugglers were brought in with boxes of cigarettes they had tried to smuggle into Angola, Commandante Lima generously waived the usual death penalty, took charge of the cigarettes for his own profit and let the smugglers go.

On the second day, I got Commandante Lima to gather all the townspeople on the basketball pitch again. I don't know what they all thought of a stocky white man getting all fired up on their behalf, as none of their own leaders seemed to have bothered, but they really perked up when I announced that the riverside fish market was out of bounds to all soldiers. I even saw a few smiles when I explained the soldiers now had their own food and would not be allowed to loot any more. I told the people that if there were any infringements, complaints must be made direct to me in our little villa in the middle of town. I finished by saying I had found a vast quantity of flour, in Petrango's warehouse, and I wanted bakers to bake bread. There was no shortage of volunteers. For the time being, we had solved the food and discipline problems. There was enough food and flour to feed both civilians and soldiers. Thanks are overdue to Petrango.

We got several generators going and restored electricity to the main buildings, including the two hospitals which we cleaned up. I used the FNLA military police for this job, huge men wearing armbands and dirty white helmets with 'MP' painted on the front. For some reason I could not explain, one particularly massive MP sported a single sinister black glove. Certainly, human rights were not top of the agenda in San Antonio di Zaire, or anywhere else in Angola. Out on a limb in San Antonio, none of us knew just how true this was in Maquela, where the self-styled 'RSM' Sammy Copeland had moved to conduct operations. In between mad dashes at the advancing enemy, these two were conducting their own grown-up version of *Lord of the Flies*, torturing and murdering blacks without compunction.

I was worried at the lack of intelligence in my sector from Holden Roberto, who had not returned to see me, from Commandante Lima or anyone else. So, when I heard a light plane overhead, I drove out to the airstrip outside the town in one of the Land Rovers I had commandeered from another Petrango shed. I found a Portuguese pilot standing by a Cessna. He said he worked "for the Americans" without elaborating on that and agreed to take me on an air recce. I needed to find out how close the enemy had come. Rumour said the enemy might be approaching through the bush to our south. We took off and flew towards Ambrizete, following the coastal road, which I could see was in an appalling state of repair. We saw nothing, so we turned east towards Tomboco, along the road. Again, there was no sign of enemy forces, nor did there seem to be many locals about.

On my return to San Antonio, I sat down in our small house and wrote a 'Military Appreciation', depending heavily on the training I had received at the Brecon Tactics School, for I had never had to do an appreciation like this in the SAS. I looked at all the factors I thought were important, such as the available intelligence, our own forces (seven of us whites and 800 black FNLA), the enemy forces we could expect to attack us, and detailed weapons, ammunition, food and the tactical possibilities. Then I finished off with my principal conclusion, which was that there were various local skirmishes going on in the Maquela and São Salvador districts, but there was no overall control or staff work of any kind to coordinate the anticommunist FNLA effort. My conclusion was that trained staff and logistical support was needed at once. I gave the report to the Portuguese pilot, who promised to give it to the 'Americans' he worked for.

One morning some days later, the precarious rule of law and order I had instituted fell apart. A group of black civilians appeared at our house all jabbering at once. The FNLA soldiers were busy looting from the fish market, in direct contravention of my orders. "Let's go!" I said to my group of six and, with the huge black MPs, we grabbed some rubber batons we thought might come in handy, piled into Land Rovers and drove down to the river to the fish market, where we set about the looting FNLA soldiers. The locals were amazed and delighted, and shouted their encouragement. "Viva O Peter!" they cheered. "Viva O Comandante!"

Feeling rather pleased with this Robin Hood-style action, we put the men we had arrested in the Land Rovers, dumped them at the town jail and went back to our house for a brew.

This may sound rather rough justice, but you should understand that at the same time, acting on Callan's order, two mercenaries murdered a black FNLA officer called Zeferino for stealing tins of pilchards. By the end of January, Callan and the other lunatics with him had probably murdered more than 200 blacks, excluding those they claimed to have killed in action. They were murdered on 'suspicion' of being spies, for infringements of discipline, for fun and to see what certain weapons did to the human body. Callan decided to test Charley's shotgun for himself. He called over a FNLA soldier, stuck the barrel in the terrified black man's mouth and blew off the top of his head, leaving nothing but his jawbone. Many were shot behind a mud hut in Maquela by Callan and others, while Sammy Copeland preferred to take the blacks to Quiende bridge, where he shot them, pushed them over the parapet and watched them fall in the river.

We knew nothing of this, having left Callan in São Salvador, and our problems were mild by comparison. Two days after the episode in the fish market, the MPs came to say that the jail was full. I went down and found the place looking like the Black Hole of Calcutta! With all the men we had arrested, the prisoners were standing shoulder to shoulder, hardly able to move and suffocating in the intense heat. I ordered, "Let them out!" meaning release only the soldiers we had arrested at the fish market, who had suffered enough after a beating and a couple of days in the overcrowded jail. However, the MPs – always men to overdo an order – promptly opened the doors of the jail and the whole lot streamed away to freedom, including all the long-term prisoners and a murderer who promptly vanished into the bush.

Meanwhile, we had been organising the defences. I saw that there was a terrible mix of weapons among the FNLA troops. Some had Russian Kalashnikovs, firing Soviet 7.62mm by 39mm medium rounds; some had FNs, firing standard NATO 7.62mm by 51mm; and others sported Spanish CETMEs, also firing 7.62mm NATO rounds but using different magazines which would not fit in the FNs. This would be a nightmare for ammunition resupply in combat. I told Commandante Lima we would divide the FNLA into four groups and redistribute all the weapons. There was considerable resistance to this, because some liked their AKs, or FNs, but I insisted, and we ended up with four companies which at least carried the same weapons: one with all the AKs, a second with all the FNs, a third with the Spanish CETMEs and a fourth carrying an impossible mix of carbines, Sten guns, shotguns and all sorts, which I kept in reserve in the middle of town.

I placed my few men with the FNLA companies and we started training. I had seen little enemy activity from my air recce, but the word was that they were loafing about in Ambrizete only an hour or so down the road, and there was not much time to knock our soldiers into shape. The six men with me were excellent, no matter what they were accused of later, and worked hard and enthusiastically, encouraging the black soldiers. I made out a simple training programme, starting with the basics: weapon handling, some range work, section battle drills and fire control. While this was going on, I selected defensive positions round the town and in between training periods we set the FNLA soldiers to digging trenches and filling sandbags.

A day or two after giving my report to the Portuguese pilot, he flew back in with two Americans – John, who was very fat, and Stuart. They wore civilian shirts and slacks, and offered no explanation of who they were. Stuart said he had been a US Marine and I assumed they were CIA. They said, "Who wrote this Situation Report?"

I owned up and John said, "Can you take Ambrizete?"

"Yes," I replied. "But I'll need trucks to get there, ammunition to take the place and resupply to hold it." I was beginning to see the frustrations of staff work. The lack of communications was probably the most acute problem. I had no idea Callan and the others had moved down to Maquela more than a week before; that Jamie McCandless, an enormous ex-SAS soldier, had been the first white mercenary to be killed in an ambush; or that John Banks, now elevated to the rank of 'Major' in the FNLA, had flown into Kinshasa with another draft of new mercenaries on January 29.

I said, "I need radios."

The Americans promptly flew in radios, American-style PRCs common in the US Army in Vietnam. With this new source of supplies came the suggestion that we should begin to prepare for a breakout south towards Ambrizete. They also said they would give me some Rigid Raiders. These were steel boats with outboard motors which took a section of ten men – like those I had used on the rivers in Borneo – and which I could use to loop south along the coast during the advance on Ambrizete.

I managed to raise the FNLA Headquarters base station in the palace in São Salvador, where I heard 'Fuzz' Hussey's voice. He was an ex-SAS soldier and he was manning the radio shack. He told me, "Some boys have come out, but I'm on my own and I can't say any more." I knew his voice and he sounded very odd. I felt something was badly wrong.

I used the Cessna again to make another air recce. This time I landed on the rough airstrip at Tomboco, where I spoke to a Portuguese mercenary called Colonel Bento. He said his men were starving for lack of food and he was delighted when I told him about the Petrango food stores. By this time, I had found another store in San Antonio, an even bigger aircraft hangar full of food which I 'liberated' with a second promissory note that full payment would be made after the war: I arranged with Colonel Bento for the FNLA Fokker Friendship to fly in rations to the Portuguese at Tomboco, and also to our people at São Salvador.

However, all our efforts in San Antonio di Zaire, and come to that, the FNLA effort everywhere, was destroyed by what happened next. There is no doubt in my mind that the Americans I had met were keen to help. They had the finances and were already supplying me, but it all fell apart.

On the morning of Thursday, February 5, I was making another appreciation of the town's defences with Brummy Barker, whom I had known in Support Company in 1 Para when I joined up. He was in the anti-tank platoon and his advice was crucial in defending the town against the MPLA tanks. He agreed that the key to defending San Antonio was at Corpus Christi, just outside the town, where there was a small river and marshy ground which was impassable to tanks. My plan was to move a company of FNLA out to the area and we were talking about this as we drove back to town, when we saw the Fokker Friendship coming in to land. Holden Roberto and his bodyguards turned up at our villa HQ in town with an attractive blonde American reporter called Robin Wright. Holden Roberto looked worn out. I asked one of the others to show Robin around, just to get her out of the way, and we sat down to talk.

Speaking in broken English, he said, "Terrible things have been going on with the others. At Maquela."

"What?"

"Killings," he said, with an air of defeat.

Sensing disaster, I grimly wondered if he meant blacks or whites, but he added, "Many men, also white men. Colonel Callan gone for three days. I want you to take over. Please? You go now, with me, to Maquela, to sort it out?"

Even through his bad English, the picture was not encouraging, and it struck me as typical of my luck that all of a sudden, I was expected to be the saviour of this mess. I must say the idea did not appeal. Roberto had heard these stories by word of mouth and he had no more idea than me what we were likely to find on landing in Maquela. Other mercenaries, who had 'deserted' from Maquela by jumping into the Fokker during an admin run to

Maquela the day before, said they expected Callan to return any moment with a large number of mercenaries. I knew very well that Callan completely controlled his henchmen who would do, and seemingly had done, anything he ordered. There was no news of Callan's other two associates, Sam Copeland and Shotgun Charley. All in all, it did not sound as if Callan and the rest would take much notice of me or Holden Roberto's new decision to put me in charge.

However, we took off in the Fokker for Maquela, Holden Roberto, his bodyguards and I with Mick Rennie as my support. I left Brummy Barker in charge of San Antonio in my absence and refused to allow Robin Wright to come with us. God knows what Callan would have done with a woman as pretty as her when he found out she was a journalist. Few soldiers get on with reporters, especially not British Army soldiers with experience of media reporting in Northern Ireland.

We left San Antonio under a brilliant blue sky, but as we approached Maquela we faced a great tower of black cloud. Visibility was limited, the rain lashed across the windows and the pilot refused to land. Instead, we flew to Kinshasa, only about thirty minutes' flying time north.

In Kinshasa, we drove from the airport to Holden Roberto's secretariat in the big villa on the hill above the city, and I tried to piece together the threads. I found one man called Dave Tomkins in Mama Yemo Hospital (a very grisly, dirty place in a similar condition to the hospitals we had found at San Antonio). One of his grenade booby traps had exploded and wounded him in the arse, but he knew very little of the massacres except hearsay. Others added their evidence but it was sketchy. 'Major' Nick Hall had spent most of the time in Kinshasa, sleeping all day and living it up all night in the Kinshasa nightclubs, such as the Intercontinental Hotel, where he terrorised black girls and swaggered about for the benefit of reporters like the BBC's John Simpson.

I discovered we had collected another selfstyled captain, an American called Tom Oates, who was an ex-Los Angeles policeman. He wore a ridiculous, bright scarlet beret, squeaky-clean camouflage fatigues and polished boots, and he had hardly crossed the border into Angola. Barry Freeman, a mercenary who had allegedly been involved in the killing, understandably played down his involvement, but the story which emerged was truly appalling.

Briefly, on Sunday, February 1, only a day after arriving, a number of the new intake of mercenaries had been ordered to set a night ambush on the south side of Maquela facing the expected approach of the enemy. In the dark, they had seen a vehicle approaching, not recognised it as one of theirs, and fired a rocket at it. In fact, the Land Rover contained Dempster, Freeman, Boddy and a new man called Max, who were severely shaken but, amazingly, unharmed.

Believing themselves to be under attack by a large enemy force, the ambush party then piled into trucks and drove pell-mell for the border, where they were turned back by an ex-SAS soldier called Terry Wilson. Then they drove to São Salvador but met Callan on the road. After some confusion, Callan brought everyone back to Maquela, found the town had not been taken by the enemy and called a parade in the square, lining up the new intake, unarmed, and surrounding them with his henchmen and a number of Portuguese mercenaries.

When the Land Rover party explained how they had been 'treacherously' attacked, Callan lost his temper and ranted about the 'cowards' who had tried to 'murder' his men. He screamed at the accused men to strip to their underpants, which older hands recognised as his usual prelude to a killing, demanded to know who had fired the rocket and, when a young man called Phil Davies owned up, Callan shot him on the spot, firing three bullets into his head. Then, casually, he ordered several of the Portuguese to throw the body over a low wall, and told Sammy Copeland, "Take them away, Sergeant Major. You know what to do with them."

Copeland, it seems, was delighted. Perhaps he was bored

of killing blacks. He took Barry Freeman, Tony Boddy, Chris Dempster, Paul Aves, Andy McKenzie and a Portuguese called Uzio; they shoved the eleven condemned men into the back of a Dodge troop carrier, drove them outside the town and gunned them all down.

I was badly shocked, more so when I heard that my cousin, Tom McAleese, was reported to be among the dead. Thankfully, the report later turned out to be false.

The images of this disaster would not go away and I never stopped moving that day. I thought of the shame it brought on British soldiers, of my cousin, and the senseless criminality. As the story of the massacre leaked out, an atmosphere of doom settled on Holden Roberto's camp in Kinshasa. It deepened on hearing that Tomboco had fallen. I met the Portuguese Colonel Bento, who said he had blown a bridge on the Ambrizete-Tomboco road but that the enemy tanks had found a way round. In spite of my suggestion, he had no intention of going back. To add to our troubles, it seemed the enemy had ended their rest and refit, and they were on the move.

The military situation was grave. My initial assessment of the dangers of not having any staff support, command or control was confirmed. The Americans had poured in millions of dollars to support the FNLA. A good deal was creamed off en route, but there was plenty left to buy the tools of war. Contrary to what anyone may say, there were enough radios, arms, ammunition and all the necessary supplies. The problem was that no one was coordinating any of it. All this equipment was hidden in sheds and lock-ups all over Kinshasa, but no one was in command to pull it all together. Holden Roberto was not strong enough to control his undisciplined, murdering mercenaries, he spoke little English, and this was Africa, where chaos and corruption were endemic in peace and rampant in war.

The massacre at Maquela had to be pursued. Holden Roberto had realised that when he had flown to see me at San Antonio. The political consequences were nothing short of disastrous for him. When the news burst on the world, his support would vanish. In mitigation, the very least he could do was show he was doing something to clear up the mess. He asked me, "Will you go to Maquela? To arrest Callan and the others?"

I felt inadequate for the task. This was an adjutant's job, a policeman's job. I don't think like a policeman, I was not trained as a policeman, but I knew I had to go. Besides, no one else wanted the responsibility.

That night, I spent some time working out exactly what I was going to do once the Fokker landed in Maquela. It was easy for Holden Roberto to say, "Arrest them!" but I could not see these maniacs coming quietly.

I was up very early the following morning, February 7, and dressed with a bulletproof vest hidden under my combat shirt. A small group gathered at the villa. Holden Roberto was wearing his dark glasses as usual, but the sharp edge of his control had gone, which gave away how worried he was. With him from the nightclubs came 'Major' Nick Hall, wearing uniform for a change, and Mick Wainhouse, who may have been intrigued to see what his old colleague Callan had been up to. Tom Oates, complete with pressed fatigues and scarlet beret, made up the group. We drove to the airport and the Fokker took off at 6 AM.

There was little talk during the flight. As we approached Maquela, I looked down and saw a Land Rover with a mounted Soviet Degtyarev 12.7mm machine gun driving onto the airstrip. I frankly admit I was worried. Callan plus his cronies and firepower like that would be hard to beat on my own.

But I was committed. The Fokker lost height, landed and taxied to a halt. All the others immediately hopped out and doubled away from the aircraft to the sides of the airstrip, pretending to take up defensive positions, ready to see who came out on top, I suppose. A blasphemous thought flashed through my mind. At that moment, I understood what Jesus had felt like in the Garden of Gethsemane, abandoned by everyone and facing the Roman soldiers.

The Land Rover drew closer and I saw just Copeland and Shotgun Charley aboard.

"How're things, lads?" I asked cheerfully. According to my plan, I produced a map, spread it out on the ground, and asked them to give me a brief.

Copeland knelt down in the dust, putting his AK beside him on the ground, and began to talk. I watched him carefully as Shotgun Charley wandered over to look at the map. As he came within reach, I kicked Copeland's AK out of the way with my foot and began setting about Charley with the butt of my M2 carbine.

He dropped his shotgun and screamed, "I wasn't even fucking there!" He knew perfectly well why we had come.

I stepped back, covered them with my carbine and told them I would kill them at once if they moved a muscle. Copeland just stared.

Of course, the rest came running over as soon as they saw it was safe, and we all drove into Maquela, with Hall and Oates shouting for blood.

Maquela was a shambles. The usual whitewashed colonial flat-roofed houses were wrecked, garbage littered the deserted streets and there was a terrible smell of death.

We found the rest of the mercenaries sitting dejectedly in a burned out building in a state of shock. They were utterly drained with the constant fear of being shot too. I galvanised them out of the house and lined them up in three ranks. Once they saw we had arrested Copeland and Charley, the accusations flooded out and we learned that nothing had been heard of Callan since he had disappeared into the bush on a vehicle patrol with a handful of other 'meres' to recce the Cuban positions.

Hall set up a 'court martial' which was nothing more than a kangaroo court. Holden Roberto, the Commander of the FNLA in Angola, was present and provided some semblance of authority. He certainly knew exactly what was happening, even though he spoke little English, but the fast pace was dictated by 'Major' Hall. Naturally, there was no shortage of witnesses. They queued up. They had all seen Callan murder Phil Davies on parade, an extraordinary and quite inexcusably callous act, though no one present had actually seen the other mercenaries killed outside the town. Sammy Copeland was found guilty of 'mass murder' but Charley 'Shotgun' Christodoulou was acquitted.

Copeland was sentenced to death by firing squad without further ado.

Once the sentence was declared, numerous people began jostling to take part in the firing squad.

"I'll do the bloody thing myself," shouted an enthusiastic young man who was in 21 SAS, a Territorial Army SAS regiment which recruits in the south of England.

The firing squad was selected and took its place, still arguing. Copeland stood a little way off. Holden Roberto made a last-ditch attempt to salvage something of justice. He spoke up and suggested that Copeland should be handed over to the British authorities in Kinshasa, and that the matter should be dealt with by the British police.

Nick Hall shouted him down, bellowing, "This man must be executed!"

Suddenly, taking advantage of the momentary confusion, Copeland made a break for it. He jinxed left and right through the bush and nearly escaped, but the firing squad was so keen to shoot him that all they had to do was lift their carbines and a fusillade brought him down wounded. Casually, one of the mercenaries walked over, drew his pistol and shot him dead.

In retrospect, I blame myself for what happened. I weakened and let matters get out of hand, allowing men like Hall to force their will against my better judgement. We were all working for Holden Roberto and those two were his so-called 'officers', but I should have taken control. Killing like that is not my scene, though Copeland was certainly guilty and none of us felt any sympathy for him.

Further, I want to clear up an accusation made against me, typically by people who have never taken the trouble to speak to me about it, and make it absolutely clear that while I was unhappy with what was going on, I made no suggestion whatsoever to exchange Copeland for Christodoulou. There were some terrible things done by various people under Callan, but it is certain that Copeland was guilty. He had led the execution party which had cold-bloodedly murdered eleven British citizens. He rightly suffered a rough justice.

In fact, I will never forget the scene of the killings. Some of the men thought they knew where to look and we drove out of town in a Land Rover. A detailed search was not needed. We drove up a hill and I could smell the place at once. I stopped the Land Rover. I got out slowly, unbelieving, feeling a great sense of emptiness as I stared out over the gentle slope of the valley which dropped away from the road. Copeland had told them to try to escape and the eleven corpses lay scattered in ones and twos among the dry green grass and thorny bushes, tapering away to the last victim who lay some 100 metres away. The bodies were bloated, fly-blown and rotting horribly, but I could see they had been grotesquely murdered, shot to bits, limbs torn off with bursts of automatic fire. For fun. One body was kneeling, his head hanging over a bush, the twigs covered with his brains, which had been blown out of the front of his face.

I whispered to myself, "God in heaven! How can men do this to their own?" I have never seen anything like it. I felt ashamed to be a British soldier.

But the war went on. The enemy were closing in on the FNLA São Salvador-Maquela pocket in a classic Soviet pincer strike on the two roads through the bush from the west through Tomboco and from the south through Damba. Advance elements were only a few kilometres from Maquela.

Morale among the guys was understandably low. I gathered the shattered group of mercenaries together and I told them the situation as far as I knew it. There was no disguising the facts. The communist MPLA and Cubans had the upper hand and looked like seizing the whole of Angola, and I bluntly finished with, "Those who want to go, leave now. The rest will carry on."

Armchair critics had their opinions of this war, but always with the benefit of hindsight. You should know that when I asked that question, all the 'big men' from the Foreign Legion, the TA SAS and the Paras, in that order, fucked off. Oh yes, they all had various plausible excuses, which were all bullshit. The men who volunteered to stay were old hands, many from the Second World War, and young men with no previous military experience, who were bloody impressive.

As a result of the depletion of our numbers, we were in no position to hold Maquela and I gave the order for the men who stayed to fall back on São Salvador. While in Maquela, I had also found that our supplies were very low, particularly diesel fuel. There was nothing left except what was in the vehicles already, probably a lot less than 60 gallons. Since I was now cast in the role of 'Commander', whether I liked it or not, and since there was no one else ready or capable of doing it, I flew the half-hour back to Kinshasa to find the Americans again and ask for more of their support.

When I got there, I found more bad news. San Antonio di Zaire had fallen. On Saturday, February 8, Brummy Barker had chosen to go hunting rather than move the 106mm M40 recoilless anti-tank rifle and a company of FNLA into the position we had agreed at Corpus Christi. When he got back, he found the tanks rolling up the road without means of stopping them, and in minutes the port was in communist hands. The FNLA soldiers had evaporated into the bush and the rest of the men, with Robin Wright, were chased through the town to the river, where they only just made good their escape in a boat. I expect they were pleased that I had had the foresight to put it there. They reached the Americans in Zaire on the northern bank of the Congo, who arranged for them to fly to Kinshasa. Barker quite literally missed the boat and was taken prisoner.

At the same time, the news of the massacres at Maquela leaked out to the world's press.

"Mercenaries in Big Killing!" screamed the headlines, and

Murray Davies of the London *Daily Mirror* reported later, "I saw the horror of Massacre Valley!"

When I met the two Americans, John the fat one and Stuart the ex-Marine, they were glum. They agreed to supply fuel but they said, "These killings have really screwed us. The State Department has ordered us to hold off completely, and if we get caught helping you, we're in the goddamn shit."

The communist enemy were making the most of the massacre, using the propaganda coup to show the capitalist system 'at its worst'. Costas Giorgiou, the 'Colonel Callan' for whom, astonishingly, various journalists had a 'sneaking admiration', did more damage by his criminal killings of blacks and his massacre of white mercenaries than the enemy were doing with their army. After the news broke, the United States dissociated itself as fast as possible from Holden Roberto's mercenaries and the FNLA. At that point, if not much earlier, Angola was lost to communism, and there was nothing the few mercenaries left could do about it.

However, we all tried for the next week or more, while the FNLA soldiers deserted over the São Salvador palace wall in large numbers every night.

In São Salvador, those who remained went out on patrols to identify enemy positions and intentions, and prepare the Quiende bridge for demolition.

In Kinshasa, the two Americans had reluctantly agreed to give me the five vital airlifts of diesel fuel we needed to move our vehicles, and also radios for the lads, under cover of supplying the Zairean Army. Nick Hall was supposed to have coordinated the delivery so that it came to us (after all, he was Holden Roberto's 'senior staff officer'). I arrived at Kinshasa's N'Délé airport at 5 AM and the Fokker landed two hours later with only three drums of diesel. Someone had ripped us off. I was furious. At eleven o'clock, Nick Hall, in civvies, finally arrived at the airport with Holden Roberto and I lost my temper with them both. Hall had done nothing, having spent all night drinking and chasing women in the Kinshasa Casino, while Roberto put on his 'shy and timid' act. It didn't wash. By that time, I knew that during wars in Africa, especially in Angola, a lot of innocent people die off the battlefield. Holden Roberto's FNLA was as bloody as all the rest, and he knew it.

The corruption in Kinshasa was getting me nowhere. I decided I must get back to São Salvador, but the Fokker was grounded for repair (in retrospect, perhaps deliberately so, because the Americans had been funding this plane all along). I looked round for something else with Dave Bufkin, a recently arrived American pilot, and found the Cessna parked up near a hangar. It was guarded, but a few weeks in Africa had added a worldly-wise cynicism to the refreshingly clean violence of my youth. I bribed the black guard for the keys and stole the Cessna. Dave Bufkin, who had been a crop-duster in the States, flew us both the half-hour south to Angola using a Shell road map.

As we approached São Salvador, I was suspicious at once. Whereas normally there were crowds of people milling about in the town, this time the place was deserted. I told Dave to land anyway, but none of our men in the town could explain why the locals had vanished. They said there were a couple of Land Rover patrols out along the roads south-west, towards Quiende and Tomboco, looking for the enemy, but since there were no radios, there was no intelligence reported back from them. I told Dave to turn around and we took off again for a look around.

Dave flew towards Maquela and after about fifteen minutes I saw them. What a sight! It was the biggest armoured column I have ever seen, stretching for a mile or more back along the dusty road through the bush. Bufkin had an awful lot of guts. He flew us low right along the column so I could see everything. The Russian BRDMs and T-54 tanks were easy to spot in their typical harsh green Russian paint, though they had made a good effort to conceal some vehicles from the air by removing the canvas canopies and tying bushes to the metal frames. This neatly broke up the shape of the vehicles and made them impossible to see among the scrub trees of the African bush till we actually flew right over them. At the back of the column were endless trucks filled with Cuban and Angolan soldiers, and plenty of massive 40-ton articulated supply vehicles filled with combat resupplies. I counted about seventy fighting vehicles and estimated some 2,000 troops. It was a hell of a sight.

"We've seen enough! Let's go back," I said drily to Dave, and we returned to São Salvador, where I got all the guys together by the steps in front of the palace.

"Who feels like fighting?" I asked sarcastically, and when several put up their hands, I told them flatly, "Well, you're going to have all the fighting you ever wanted." And I told them what we had seen from the air.

While I was talking to the men, two small Chinese tanks appeared in the road and turned into the palace grounds. Horrified, we watched one smash through the wrought-iron gates and lumber towards us. It flattened the flag pole and headed straight for the steps of the palace where we were standing. We scattered. Engine roaring and tracks grinding, it lurched halfway up the steps and halted. A big black tank commander emerged from the turret.

"Great!" I shouted, as I watched him extricate himself from his tank. "We've got tanks!" They were certainly not the last word in tank design but better than nothing against the armoured vehicles I had seen from the air.

I had no sooner spoken than this gallant black tank commander jumped down from his tank and legged it, disappearing through the broken gates into the town.

Dave Tomkins appeared in the second tank. Though not fully recovered from his wounds, he had discharged himself from Mama Yemo Hospital in Kinshasa and volunteered to lead these small Chinese tanks and a vehicle with anti-tank missiles down to Angola. He was obviously in pain and I advised him to go back to Kinshasa. I found a Portuguese bulldozer driver who could drive the tank instead.

We made preparations to defend the town but I don't

think anyone expected to be able to repel the Cuban/Angolan armoured columns. There was some excuse for wondering why we were bothering at all. The town was deserted, people had gone to ground in their houses or disappeared into the bush to await developments, all but one hundred of Holden Roberto's FNLA soldiers had also deserted São Salvador, we could expect no resupplies from the Americans after Callan's behaviour and our leader, Holden Roberto, was in Kinshasa with his FNLA 'senior officer' Nick Hall.

The air of doom worsened when one of the black soldiers with me fired off his rifle by mistake. At once they all started shooting wildly and several more took the opportunity to desert. A group jumped on the anti-tank truck mounted with our 106mm M40 to drive it off. I lost my temper and leapt aboard after them, physically throwing them off it again.

Mick Rennie came back from his recce along the São Salvador-Tomboco road and reported the enemy in that direction too. Again, I felt suspicious. We were out on a limb in São Salvador; we only had a small corner of Angola left in our control, two enemy columns were closing in like the horns of a bull on the only two 'main' roads and a glance at the map showed me that our retreat would soon be blocked off when the armoured column I had seen from the air reached the Lucosso-Cuimbata crossroads.

Again, I saw no realistic option but to withdraw. The decision to retreat is easy to talk about, bloody hard to take on the ground. Also, I knew there were several mercenaries (including Callan) still unaccounted for on patrols of which nothing had been heard for some days. However, there was nothing our little group of thirty mercenaries could have done at that time to alter the course of the Angolan war. The march of world politics had stridden past us and we were now a lost cause. I gave the order to pull out of São Salvador.

Not far outside the town, one of the little tanks threw a track. We left it and drove on towards the Lucosso-Cuimbata crossroads, passing great crowds of black refugees. Among them we saw deserting FNLA soldiers carrying all the loot they had stolen. Everyone was trying to reach Zaire.

We dug defensive trenches at the crossroads just in case we were surprised there by an advance party, and I sent Dan Aitken with some others to blow the river bridge just beyond, on the Maquela road. They returned two hours later and confirmed my gut feeling that we had been sitting in a trap in São Salvador. The Cubans had been only 1 kilometre away when they had blown the charges and dropped the bridge.

If I had not pulled the men out of São Salvador when I did, the Cubans would have taken the crossroads, cut us off and many more mercenaries would have been captured. At that stage, the outcome of the war was not in question.

While we were at the crossroads, I got a message to drive to the border, where I found the two Americans again. They confirmed it was hopeless. For the last time, I returned to the lads at the crossroads and gave the order for complete withdrawal to Kinshasa. The last mercenaries left Angola to the communists on February 17.

Holden Roberto and Nick Hall were furious.

"Why have you deserted your positions?" shouted Nick Hall, back in uniform again. "You should be court-martialled!"

Bluntly, I told them the situation. Angola was lost. Had been for a long time. As I spoke, I wondered if these two could have done more than just spend their time lurking in Kinshasa, but I kept this to myself. My position was weak. I was in uniform as a white mercenary working for a defeated black Angolan leader in someone else's country.

The Americans had also heard that two patrols were missing and let me take the Cessna (which they had recovered from São Salvador) out the following day to see if I could find them. I took a radio in the hope of raising someone on the ground. There was nothing. Only one young boy, from Barry Island in Wales, emerged from the bush in Kinshasa a week later after a two-week escape and evasion from deep inside Angola. He had been with Mick Rennie's patrol driving towards Tomboco when they had been ambushed and split up. Without any training, he had instinctively done all the right things to avoid capture: not stayed in huts, hidden at night under bushes, stolen food and crossed the border to safety. It was an impressive performance.

Thirteen mercenaries were missing (some for days) before being captured. Later, after a show trial in Luanda, four were executed: Callan and Andy McKenzie, Brummy Barker (who always complained of his bad luck) and Daniel Gearheart – who was executed for being an American. The rest were imprisoned. They are all out now.

The others who did appalling things while pretending to be soldiers got away scot-free. The British police investigated the massacre, but not any of the other killings of blacks. When Freeman and Dempster returned to England, they were arrested and questioned about their part in the massacre of the British mercenaries at Maquela, but a decision to let the whole matter drop was taken by Prime Minister Harold Wilson at Cabinet level.

I hung on in Kinshasa till the end of March, staying in a shed at the back of Holden Roberto's villa. I went to see the two Americans who were slumming it in the Hotel Intercontinental and we discussed their plans to mount a guerrilla campaign. Of course, after the terrible publicity of the mercenary killings, this was nothing more than a CIA pipe-dream.

The British Embassy was very stuffy and formal, but I helped them put together a list of as many names as I could, so they could inform relatives in Britain. There was still doubt about exactly who had been murdered at Maquela and killed in action, as no one had kept a record of the latest arrivals or could remember their names.

Holden Roberto withdrew into his villa and refused to speak to any of us. Finally, when only Dan Aitken and I were left, I persuaded the two Americans to talk to Roberto. We had no money at all and they told him to pay for plane tickets for us to return home. On March 23, after two months in Africa, I flew back from Kinshasa on Alitalia flight AZ827 to Rome and then by AZ282 to Heathrow, London, with very little money in my pocket.

CHAPTER 7
TRIGGER TIME

"What are you going to do next?"

"I'm going to join the French Foreign Legion," I replied morosely, staring into space over a pint in a pub in London. After another intense but disastrous affair with a girl in Hereford, I was bored and depressed, and had taken the train to London with just enough money to reach Paris.

Murray Davies was a reporter I had met in Kinshasa when he was covering the end of the Angolan war for the *Daily Mirror*. He laughed and said, "Forget it. If you like fighting, why don't you go to Rhodesia? They've got a hell of a fight on their hands."

I nodded. An ex-SAS friend in Hereford, Chuck Hinde, had told me the Rhodesians were really going for it, on their own, and surrounded on all sides by hostile black 'Front Line States'. Chuck was then in the Central Intelligence Organisation (CIO), on leave in England, so he said, but I found out later he was doing 'passport work'. This meant leaving Rhodesia by scheduled air flights and travelling to another country – usually in Europe – from which he could innocently enter one of the black Front Line States as just another tourist on a scheduled flight. Once inside the target country, he could carry out a covert attack or assassination. ZANLA's Operations Chief, Herbert Chitepo, was blown up in his Volkswagen car by Rhodesians on 'passport work'.

I said, "Sounds great, Murray, but all good ideas need finance and I've hardly got enough to cross the Channel, let alone reach Rhodesia."

"I'll give you the money."

What a stroke of luck! He generously gave me £200, which

was a lot in those days, and I flew to Rhodesia on New Year's Eve of 1976.

In Salisbury, the luck ran out.

I went straight to the Army Recruiting Office, and while my papers were processed, I stayed in the Ambassador Hotel and kept out of trouble. By contrast, several other new arrivals – including a man about my build and appearance – spent their time drinking and whoring on the town.

Some days later, to my horror, the Recruiting Office told me I was unsuitable and refused to sign me on. I was really at a loss and deep depression loomed, but fortunately I ran into Chuck Hinde again. He listened sympathetically and set up a meeting for me with Jack Berry, who was in Rhodesian intelligence.

"I see you like the dusky maidens," said Berry cryptically in his office.

"What're you talking about?" I replied, puzzled.

"Don't give me that shit," he said bluntly. "We check out all foreigners who come here to join up, and you didn't pass the test." He told me a covert surveillance team had followed "a short stocky man with a moustache and intense blue eyes" who spent all his time out on the town screwing black women. They had mistaken me for the other man in the hotel who looked like me! I confirmed this with my room number.

On January 4, 1977, once satisfied my intentions were good, I was allowed to sign on in the Rhodesian Army as Number 728200. I immediately volunteered for the Rhodesian SAS, went to Cranbourne Barracks in Salisbury, and began one of the most serious periods of my soldiering career.

I had joined the army of a country which was locked in a fight to the death. Mr Ian Smith, the Rhodesian Premier, headed a government which refused to countenance a change to the system where 240,000 whites held total political and economic power over 6 million blacks. He had steered the country through the Unilateral Declaration of Independence from Great Britain and by 1976 Rhodesia had no international friends. Even South Africa was under tremendous pressure from Dr Henry Kissinger, the American Secretary of State, to stop supporting Rhodesia with arms and oil. It is not surprising that the atmosphere amongst all Rhodesians, not just those in the army, was tense.

A glance at the map shows that the country was surrounded by antagonistic communist black states. My experience in Angola the previous year came in useful, because the Portuguese withdrawal from Angola on one side and Mozambique on the other – both countries which succumbed to post-colonial communist regimes – placed great pressure on Rhodesia. Both communist governments, along with Zambia and Botswana, gave safe haven for the black communist organisations outlawed in Rhodesia. The two principal groups were Joshua Nkomo's ZIPRA and Robert Mugabe's ZANLA. Communist bloc countries, headed by Russia and China, were applying a sort of domino theory to South Africa, as countries became communist one after the other, with the object of making the whole of South Africa communist.

The Rhodesian communists were Joshua Nkomo's ZIPRA (the Zimbabwe People's Revolutionary Army) and its political wing ZAPU (the Zimbabwe African People's Union), which recruited from the Matabele tribal lands. The fighters lived in camps in Zambia and infiltrated Rhodesia from the north and west, from Zambia looping south through Botswana to cross Rhodesia's long western border with Botswana.

The second and largest communist group was Robert Mugabe's and the Reverend Ndabaningi Sithole's ZANLA (the Zimbabwe African Nationalist Liberation Army) and its political wing ZANU (the Zimbabwe African National Union), which recruited from the other side of Rhodesia, from Mashonaland, where the Shona tribe was the country's biggest population group. They had built some very large training camps in Mozambique, which was seriously Marxist and gave ZANLA great support, and they infiltrated Rhodesia through the thick bush along her 1,300-kilometre eastern border.

In fact, only 220 kilometres of Rhodesia's 3,000 kilometres of border was with a friendly nation; South Africa was friendly, but international pressure was increasing to stop the South Africans assisting the white regime in Rhodesia. Quite simply, the Rhodesians had their backs against the wall, and there was no time for frivolity. Even the drinking and womanising was pursued with intensity.

By the time I arrived, both Robert Mugabe's ZANLA and Nkomo's ZIPRA had carried out countless attacks on the local population, and the blacks suffered more than the whites. The newspapers were filled with a constant stream of accounts of brutalities from all over the country. One of the most grotesque incidents occurred just before I arrived, in December 1976, when ZANLA terrorists attacked the African families on a British tea estate in the Eastern Highlands because the blacks refused to stop working for the owners. The ZANLA 'fighters' lined up and machine-gunned twenty-seven of the men, then kidnapped the women and children by walking them back to their training camps in Mozambique. By Christmas that year, 716 civilians had been murdered by terrorists but only sixty-one were white.

Training for the Rhodesian SAS matched the mood of the country. I buckled down to it, enjoying the feeling of being back in a regular army with a clear purpose, but I found the first phase frustrating. Recruit training is fine when you are 17 years old and fresh to uniform, but I was 34 and for me the first six weeks were absurd. There was the usual regular army obsession with imposed discipline, endless drilling, digging holes, filling them in again, and making up 'bed blocks' in the billet, which meant folding the bedding to an exact measurement, which we checked with carefully measured lengths of string before the inspection. In fact, it was easier to make up the bed block, leave it on the bed and sleep on the floor. Colour sergeants Kruger and Paul Fisher were insanely keen on a vast number of star jumps for every infringement. I have always thought that punishment should be relevant to the training. Leaping up and down like frogs in drill uniform certainly improved our fitness, but it did nothing to better our drill.

I discovered Rhodesians just love wearing shorts. Sometimes we just wore PT vests and at other times we wore camouflage shirts with a full complement of webbing pouches on our belts, but we always seemed to wear shorts. One awed American said, "The Rhodesians are the only goddamn soldiers who have ever been to war in their underwear!"

Most of the men with me were little more than half my age, called up for their National Service, straight from civilian life. The SAS had to select men like this because the only alternative source, the Rhodesian Light Infantry (RLI), was fed up with having its best men poached by the SAS. In fact, the Rhodesian SAS found the system worked reasonably well, as they were able to teach men to think in SAS terms from the start, rather than having to reteach them. Of course, Rhodesia was at war, so they had the great advantage of testing everyone very quickly on active service as soon as they finished selection and training. Furthermore, by direct contrast with most recent British Army campaigns, soldiers in Rhodesia expected to see and fight the enemy every time they went out on patrol.

This was probably why the weapon training and live firing exercises were realistic and good. For another six weeks, we lived in the Gwaii River Mine area, north of Wankie National Park, and worked through an intensive programme. I loved it. There is no better feeling than being really fit and totally involved in training when you know it will lead to action. Every morning, we started with PT before a busy schedule of speed marches in full equipment and live firing with a wide variety of Western and Soviet weapons.

The Rhodesian Army depended on captured equipment so, apart from the usual FNs and GPMGs, it was essential to train with Soviet arms, AKs, RPDs, the anti-aircraft KPV 14.5mm super-heavy MG, the US 106mm M40 recoilless rifles and so on, many of which I had seen and used before. In addition, we did plenty of realistic mortar shoots and imaginative battle drill exercises in the bush, always firing live ammunition, of which there never seemed to be any shortage.

We became very brown, lean and hard, and we were ready for the SAS selection course. Mere walking over the hills meant nothing to us anymore, so the instructors deliberately set about trying to wear us down. This was called the 'Rev' and it was twenty-four hours of continuous PT. We were made to do anything the instructors could think of as long as it was really tiring and pointless. Anyone not totally mentally committed to passing would give up. In the UK, we had something similar during British SAS selection and called it the 'sickener', but it did not last twenty-four hours solid and we were allowed to eat normally in the cookhouse. In Rhodesia, we were given only water, glucose and salt tablets.

They gave each of us a brick. We had to carry the bloody thing everywhere.

"These bricks are your loved ones!" shouted the instructors. "You know what women are like. They hate being left alone. At *any* time." All the bricks had girls' names. Mine was called Elsie. One man had a brick called Arthur because he was an ex-Royal Marine. For hours they made us climb in and out of an empty swimming pool, stark naked, time after time sliding down the water-less slide till our arses were raw, and finally, around 4 AM, I dropped Elsie on the concrete pool floor, breaking her in two.

"Stop everyone!" screamed the instructors, horrified. "McAleese has dropped his loved one!" In the dark, naked, we formed up in three ranks and gave Elsie a solemn funeral with full military honours. Then she was unceremoniously dumped in a dustbin. I was given another brick which, to underline my promiscuity, was called Elsie too.

The actual selection course was short after all this and took place in the Matapos mountain area, where Cecil Rhodes is buried. We turned out in complete patrol gear, with full ammunition scales, our FN rifles – loaded, of course – and an 80-pound rucksack.

And they gave us a log. One huge telegraph pole between twelve of us.

"This log likes travel!" they shouted. "Take this log to Bambata Cave!"

Rhodesian SAS selection is much more emotional and more group-orientated than British SAS selection. In the UK, if you don't want to carry on, no one cares. You're just invited to 'have a brew and jump in the truck'. In Rhodesia, it was a bit more like the Parachute Regiment 'A' Company in Aldershot.

"Keep a tight arse and you'll pass!" we shouted together, as we sweated to carry our log for 15 kilometres up and down hills which looked like Ayers Rock in Australia, and which were just as hot and quite as horrendous. We never reached Bambata Cave with our log. No one does. They don't even give you the right map. The point is the effort of trying.

We finished off with a week of walking about 30 kilometres a day in full kit and rucksack. At night we slept on the ground rather than waste time returning to barracks, and the medics injected our blistered feet with yellow proflavine to harden the skin. Out of the hundred who started, I was one of about twenty-five who finally passed.

During this selection period, on leave in Salisbury, I met Jane Crist. She was, and still is, a tall, gentle person, with soft dark hair and big brown eyes, a beautiful person, but with a certain inner calm which I think comes from her quiet conviction in her beliefs as a Roman Catholic. She was born in America of an Irish American father who achieved fame as the role model for J.P. Donleavy's drunken fictional character Sebastian Dangerfield in his book *The Ginger Man*. She had been living in Salisbury for two years, where she was training to be a nurse at the Andrew Fleming Hospital, and we met at Sharon Hinde's apartment some while after Chuck Hinde had been killed on operations. Jane was there with her little 4-year-old daughter Emelda and as soon as I saw her, I knew I wanted her.

Determined not to let her go, I played the 'sweetie trick' on

Emelda. While Jane was talking to Sharon Hinde, I chatted up Emelda, gave her a couple of sweets and casually asked her where she lived with her mummy.

"Alderbury Court," the little girl replied, her eyes fixed on the sweets in my hand.

When I had a couple of days' leave, I just happened to be walking through Alderbury Court and rang the bell.

You will probably wonder what on earth such a person could find in one of the 'brutal and licentious' such as me, and I can't really say, but we fell in love. I admit that I may not have put it quite like that to her, but that's how I felt. I always get hopelessly absorbed with the women in my life, but I can honestly say this was true of no one more than Jane. Almost, perhaps, my feelings amounted to obsession. Maybe she loved me for that, as we were happy then in Salisbury, in spite of the risks to life and limb in Rhodesia at that time.

My continuation training finished with more Soviet weaponry, laying and lifting mines, parachute jumps from Dakotas, and Klepper canoe training on Lake McIlwaine. Then I joined 'A' Troop, in Kenyemba in northern Rhodesia on the Mozambique border, and we went on operations.

My first operation with the Rhodesians was against FRELIMO, the Mozambique black national army. This was not the first time the Rhodesian SAS had fought FRELIMO inside Mozambique. In 1973 and '74, the Rhodesian SAS had been obliged to cross the border and conduct specific operations against FRELIMO from a place called Macombe, as the conscript Portuguese troops spent their time sitting about in their bases afraid to go out. Portuguese troops were never very good at controlling FRELIMO guerrillas but at least they had kept the lid on the problem. So, Portugal's withdrawal from Mozambique in 1976 was a strategic disaster. It fuelled black African communism and nationalism, and exposed Rhodesia's whole eastern flank. Rhodesia was left with no support at all along the 1,300-kilometre border.

By the time I arrived, FRELIMO was in power, supplied

with shiploads of arms by the Chinese and giving full support to Robert Mugabe's ZANLA, allowing them to open more routes of infiltration into Rhodesia through the excellent cover of the low, scrubby trees which covered the broken, hilly country of Mashonaland. These routes were also used to bring new black recruits out of Rhodesia to the huge ZANLA training camps in Mozambique, where Cuban, Russian, Chinese and East German 'advisors' came to train and indoctrinate them. Once trained, the guerrillas – or 'terrs' as we called them – slid back through the bush into Rhodesia to cause mayhem and death.

Our job was to ambush one of these routes. Dave Berry, who had been in the SAS in Macombe and who ultimately received the Bronze Cross of Rhodesia for bravery, was our patrol commander and set the four of us in position among one of the typical outcrops of rock in the bush, hidden in the scrub grass and low, thorny trees overlooking the track which passed in front of us in a slight dip in the ground. For three days no one came along and we sat silently in the heat, concealed among the low bushes and rocks, taking it in turns to sleep, basking in the soft warmth of the day and grateful for the slight cool of the night.

Then, Dave received radio permission to pull out. (We used Morse as it is less subject to static and interference over long distances than voice.) We began to creep away from the track between the boulders, trying to minimise the crunching noise of the dried, fern-like grass which covered the ground, when suddenly the last man out turned round to check his back and spotted four of the enemy in distinctive green FRELIMO uniforms, moving along the track. Swiftly and noiselessly, we slid back unseen into our ambush positions and opened fire. We killed all four. I remember that Baz Joliffe was most effective with his RPD, the Soviet light section machine gun. We searched them for documents and left the bodies for other guerrillas using the infiltration route to find.

Operations with the troop continued relentlessly. In contrast to my experience in Aden and Borneo, when contacts with the enemy were rare, we expected to meet them every time we went on patrol. And win. Our morale was high, though the pace of operations never seemed to let up and reflected the continuously worsening political situation.

We ambushed another track and this time the guerrilla was riding a bicycle. They often used bikes; they could move faster along the winding trails through the trees, give nothing away from footprints and no one would normally be suspicious of bike tracks. I often wondered if the idea came from the Vietnamese through the Viet Cong's use of bikes.

We also did vehicle operations, on one of which Corporal Ian Suttil, an ex-British Royal Marine, was blown off his truck by a land mine. He sailed past me and broke his arm on landing. Ian never had any luck. He was wounded four times, once passing an apparently dead terrorist who then leapt up and shot him, and was finally killed on a cross-border raid into Mozambique with the SADF. On another, led by Lieutenant 'Dangerous' Darrel Watt, we did a vehicle sweep looking for terrorists in an area where the thorn bushes were so bad that we carried three spare tyres per vehicle. We spent the day stopping to change burst tyres and all night sitting up repairing the tubes in our overnight laager position. We had a tremendous mix of weapons on that patrol, which was typical of the campaign, as there was so much communist bloc weaponry supplied to the ZANLA and ZIPRA guerrillas. We even had a Russian 12.7mm and a Browning .50 cal on the same truck.

After six weeks of operations, we came back to our camp in Salisbury for ten days to clean up, refit and prepare for the next six weeks. During these ten days, we were allowed leave. The Americans in Vietnam went for 'R and R', or Rest and Recuperation, but the Rhodesians, with an endless routine of non-stop ops for six weeks with ten days off before the next six weeks, headed for the bars and the women with the same passion with which they pursued the enemy, and rechristened their leave 'I and I', or Intoxication and Intercourse.

I went to see Jane and we found her daily shifts at the hospital were really awkward. She worked from 7 to 12 AM, and the matron was very strict indeed with the student nurses. I used to wait impatiently for her in her apartment and listen out for her coming back along the avenue outside, which was lined with lovely big jacaranda trees. The purple blossoms fell all over the place and I could hear them crackle under her feet as she walked back up the road to see me. Sometimes, if she knew exactly when I had my week off, she would work a seven-day stretch of night duty – which entitled her to nearly four days off – and we would spend the whole time together.

Back on operations, our enemy was ZIPRA. We were to attack Namumba Farm, just inside Zambia on the northern bank of the Zambezi River. Special Branch sources and recce patrols confirmed that ZIPRA guerrillas used the old farm as a staging point when crossing in and out of Rhodesia. ZIPRA were trained by the Russians along more conventional military lines than ZANLA, as Nkomo's eventual plan was to invade Rhodesia in full strength. They usually crossed into Rhodesia in uniform, looking like soldiers, but their terrorism was as brutal as anything ZANLA could offer.

"Missionaries are the enemy of the people," said ZIPRA leader Albert Ncube, trotting out the awful tired phrases which the communist guerrillas used to excuse their killings, and he fearlessly murdered the Right Reverend Adolf Schmitt, the former Roman Catholic Bishop of Bulawayo, another priest and a nun. The bishop's successor, the Right Reverend Henry Karlen, sent a message to Joshua Nkomo and Robert Mugabe saying, "Is this the reward for our work for the Africans?"

Captain Bob McKenzie, who was awarded the Silver Cross of Rhodesia for his continued leadership and bravery, gave us a very thorough briefing on the Namumba Farm attack. Then we piled into a big 3-ton Mercedes, which had been sold to the Rhodesians by the departing Portuguese, and drove north from Salisbury through Makuti, then on to the edge of the Kariba basin. Everything seems big in Africa but the view from the escarpment overlooking the Kariba valley is truly impressive, across the trees and grasslands of the huge valley floor, with the immense Lake Kariba lying in heat haze far off in the distance.

We zigzagged down the steep escarpment onto the floor of the valley and set up our forward base. After a further recce, McKenzie, whose preparation and planning were always excellent, gave his detailed attack orders. After final rehearsals, we assembled the canoes and carried them through the grass and trees to the Zambezi, arriving at about 2 AM.

The Klepper was the only collapsible canoe rugged enough for special forces infiltration (and is still used by the British SAS) but it is designed to carry only two men. The river was about 100 metres across at this point and McKenzie decided we would carry a third man: one paddling at the back and two lying over the front. On command, we all slipped down the muddy bank into the slow-moving dark brown water and paddled silently to the black line of the opposite shore. As Mo Taylor was on his way back to the Rhodesian bank by himself, a great shape suddenly burst through the glistening surface under his canoe and dumped him in the water. A vast hippo opened its mouth angrily, wallowed and disappeared. Mo Taylor was gone, and there was no time to look for him in the dark.

The riverbank was green and lush, like England, but as soon as we moved off, we were back in typical African bush: thorn trees, boulders and long, tough grass. As planned, we divided into two groups, the assault group and cut-off group, and – as the first streaks lightened the eastern sky above the spreading baobab trees – we attacked the farm.

I, and every second man, carried an RPD to give the maximum firepower and we opened up with everything. I was close to McKenzie in the assault party, next to Imre Baka, an American serving with the Rhodesians, and we cut down two ZIPRA who ran across our front, clearly disorientated by the firepower, noise and suddenness of the attack. The enemy dropped two or three 60mm mortar rounds on us without effect and we swept through the farm, driving most of them out towards the cut-off groups beyond.

One enemy held us up for a couple of minutes, firing from thick bushes, which all three of us peppered till he stopped. As we went past, I thought of Ian Suttil and looked back, saw the enemy soldier lying down and noticed he was still very much alive, holding his AK ready to use. I fired a burst into his chest. The black terrorists often feigned death like this. Overall, Namumba Farm was a success. We killed seventeen enemy, most as they tried to escape through the cut-off groups by the river, and Mo Taylor turned up the next day, half-drowned but in one piece. Unfortunately, Mo was killed later on another operation.

These operations were high in effort and low in results. Killing four or even seventeen terrorists may sound rather horrid, but it simply was not enough. The number of ZIPRA and ZANLA guerrillas operating inside Rhodesia was increasing year by year, from 400 in 1974, to 700 in 1976, and to 2,500 by the time we did that patrol in early 1977. They were losing some, captured and killed, but seemed to have no trouble recruiting, even forcing young blacks from their villages to join the big guerrilla training camps in Zambia and Mozambique. There, safely beyond the reach of the Rhodesians, morale was high, numbers swelled and if a few were killed when they went back to fight for their cause inside Rhodesia, the number of successful terrorist attacks continued to increase.

ZANLA were hard to detect. They went back to their villages, mingled with the people and struck out in civilian disguise, though some were assigned to different areas so they could 'lean on' the locals if necessary. On February 7, another seven white missionaries were massacred at St Paul's Catholic Mission, near Mrewa, only 37 miles from Salisbury. Rhodesia was losing the war and many officers wanted to hit back.

We got the chance at Chimoio.

Before I describe the battle, I must explain the background, because this operation was no ordinary one-troop cross-border op.

The decision to go ahead was taken by Prime Minister Ian Smith himself, advised by his Combined Operations HQ. It is amazing, in retrospect, that the Rhodesians only formed a ComOps HQ in March 1977. Before that, the war had been regarded as a police operation with military support, rather like the British experience in Malaya, Aden and lately Northern Ireland. The new structure put the emphasis the other way around. It brought together under one political-military control all the essential elements needed to win the battle against the terrorists. But it was too late.

On Monday, November 21, we were ordered to drive all the way back to Salisbury from an operation in Mabalauta, right down in the south-east. When we arrived back, tired and dirty, we were told to prepare our personal equipment for a para jump. OPSEC (operations secrecy, or the need-to-know principle) was very good. We thought we might be attacking a FRELIMO/ZANLA base at Malvernia, a small town on the border in Mozambique on the end of the Maputo-Malvernia railway near the border with South Africa. We were way off the mark. We arrived at Salisbury military airfield at midday and were told to wait outside a hangar. A PJI (Parachute Jump Instructor) came out with a look of amazement on his face. He had caught a glimpse of the briefing map inside and said, "You want to see what you're going to fucking attack this time!"

When we were called in, I saw at once what he meant. We were on for Chimoio, the biggest ZANLA training area in Mozambique. It was massive. We sat down with the RLI on bleacher benches, which had been specially set up in the hangar, and listened to the intelligence briefing from the SAS I.O., Captain Scotty McCormack.

For more than a year, intelligence had been collated from the Rhodesian Special Branch and the Selous Scouts, who had sent two man patrols 90 kilometres deep into Mozambique to reconnoitre the whole target. The Chimoio camps occupied an area of 28 square kilometres near the town of Chimoio, formerly known as Jilla Peria, on the railway line which led from Beira on the Mozambique coast to the Rhodesian border in the south-east.

No less than thirteen separate ZANLA training camps had been identified in this area, containing between 9,000 and 11,000 guerrillas. Air photograph interpreters had counted 700 on one rifle range alone. They were being trained by Cuban, Chinese and East German advisors, and supplied with all the weapons and equipment they needed on the railway straight from the docks at Beira. The Rhodesians wanted to attack with only ninety-seven SAS and eighty-eight RLI troops.

SAS Commander, Major Brian Robinson, gave the finest ops brief I have ever heard. He was a small, wiry terrier of a man who was the driving force behind the success of the Rhodesian SAS, and he really wound us up for the fight that day. He, McCormack, and the air force commander, Group Captain Norman Walsh, had pushed ComOps for over a year for this operation, arguing against staff reluctance to commit so many valuable resources of men and planes to a project in which the odds were so heavily stacked against us. Failure would be a serious blow to morale, deeply embarrassing, and any lost assets would be impossible to replace.

Senior officers were also constrained by the limit to cross-border operations, which had always been till then only 45 kilometres. However, by November 1977, the government realised they had to do something dramatic about the guerrilla camps, to hit back at ZANLA, and they gave the go-ahead. Major Robinson used a superb model made up on the hangar floor, with aerial photographs stuck together to give a realistic impression of the whole target area. He explained we could only attack five of the thirteen camps, because the place was too vast an area and we lacked numbers and assets.

We were to parachute onto the five parts of Chimoio in groups around the target area to form a tight airborne envelopment box to hem in the enemy and prevent him escaping, like lines of guns at a shoot. Then, Vampire and Hunter ground aircraft, Canberra bombers, and Alouette helicopter gunships flying out of temporary admin bases, were going to bomb and strafe the camps for several hours. Finally, we were to fight through to the centre of the camps, attack the secretariat and cause as much devastation as possible before withdrawing by chopper to Rhodesia. If successful, it would shake ZANLA to the core.

"We've totally stripped everything from everywhere for this op," said General Walls in his introduction. Walls had been in 'C' Squadron SAS with the British in Malaya and was now the Supreme Commander of the Rhodesian Armed Services. Major Robinson listed our attached resources: aged Vampires, Hunters, six Dakotas, thirty-two Alouette III choppers, a DC-8, Canberra bombers and a mountain of logistical support, most critically AVGAS to refuel the short-range choppers.

The logistical problems were immense but had all been solved. We hoped. The Daks could reach out from Salisbury to Chimoio and back, but the Alouettes could not, so the Rhodesians had invented the Forward Admin Base concept, setting up a self-protected base inside enemy territory which the Alouettes could use to refuel quickly and not leave their station over the fighting area for too long.

Using the model, Robinson explained how the Vampire and Hunter ground-attack planes would be used. Six Dakotas would drop the ninety-seven SAS and forty-eight of the RLI, twenty-four per aircraft, and some of the Alouettes would put down the other forty RLI troops, four per chopper, who would be flown from Grand Reef airbase via Lake Alexander nearer the border, where the command-and-control chopper with Group Captain Walsh would orbit high over the camp to direct operations. The Alouette 'K-car' gunships, mounted with twin .30 cal machine guns, would circle and give fire support.

As he talked, he pointed out different-coloured circles and points suspended above the big model, which indicated the flying patterns of each aircraft, cleverly building up a clear picture of what was going to happen all around us as we fought through on the ground. As we listened, I began to feel the tension of the impending operation building up inside me, thinking of the para jump onto a hot DZ, the distance inside enemy territory, the sheer excitement.

At the end of the briefing, Major Robinson said to the RLI, "Good luck!" Then he turned to his own SAS and added, "Professionals don't need it!" Which really pissed off the RLI, but it was done with the dry, hard humour typical of soldiers before action. We all knew Chimoio was going to be dangerous work. That evening, we were billeted in an isolation compound on the airfield, to preserve OPSEC, and we spent the time preparing our kit. I remember a tremendous atmosphere of commitment.

On Wednesday, November 23, we were up at 3 AM, everyone mucking in and helping each other. That evening, they gave us free soft drinks and beers, which really told us how seriously they were taking it, but no one drank more than a couple before lying down on the floor to catch some sleep. We drew and fitted our parachutes in the hangar. There was an atmosphere of intense concentration and common purpose, everyone helping each other in pairs to pull up the straps and adjust the webbing. I was paired with Steve Kluzniak, a tall, fair-haired, rather studious-looking man who was an excellent soldier. I tucked my Russian RPK behind my shoulder straps and Steve checked my harness.

We heard the Daks' engines roaring outside and walked out to the aircraft. Six Daks did not make a second Arnhem, but it was a stirring sight and I don't think any man among us could deny the thrilling rush of adrenalin and excitement when we took off and headed into the brightening south-eastern sky, one aircraft droning purposefully after another. Shortly after seven o'clock, we reached Lake Alexander near the border in clear morning light. Perfect weather for parachuting. We circled round over the placid waters of the lake before our final run to target and I shall never forget the sight, all the Dakotas circling round, the dozens of Alouette choppers stacked up waiting to go, and further out the fast-moving Vampires and Hunters holding off like eagles before their ground-attack run-ins.

Meanwhile, out over the enemy camp, a lone civilian DC-8

piloted by Jack Malloch flew low over the muster parade taking place and thousands of ZANLA dived for cover thinking they were under attack. The DC-8 passed and minutes later, when nothing further had happened, all the guerrillas emerged from their foxholes and formed up to continue their parade in the belief they were safe after all. The deception plan had worked.

Sixty miles away over Lake Alexander, Group Captain Walsh and Major Robinson in their command chopper gave the order to attack. The Dak pilots streamed out for the run-in.

Inside our Dak, everyone appeared to be asleep. All around me were men festooned with para harnesses, weapons and ammunition, leaning against each other, their heads lolling. These Paras were apparently so cool they could sleep as we roared through the sky on our way to the drop zone. However, adrenalin was racing inside them all. Their eyes were closed but no one could actually sleep. Somehow, they all sensed the despatchers moving decisively to the door, and when they heard their orders in their headphones, every man snapped fully alert.

"Prepare for action!" yelled our despatchers over the noise of the plane. We all stood up and automatically started our parachute drills, mentally running through the sing-song mnemonic, "Top hook and pin! Bottom hook and pin! Static line cleared to left-hand side! Stowage of the static line and centre pack tie!"

"Tell off for equipment check!" yelled the despatchers. We shouted our response, one after the other, and shuffled towards the open door, a long line of helmeted, grim-faced men strapped with gear and weapons. I was near enough to the door to see the tops of the yellow green trees flashing beneath us, the big baobabs and scattered African huts only 400 feet below. Operational jump height. We were close now. A Vampire jet roared past rocketing a cluster of farm buildings on the target area, and I could see an Alouette K-car swinging round with its .30 cannon blazing at the ground. I checked my watch. It was thirteen minutes to eight, exactly as Major Robinson had ordered.

"Red on!" I felt wild, aching to get out, to get started.

"Green on, GO!" I shuffled and stamped towards the open door, and leapt into the rushing air.

My chute had hardly opened before I was nearing the ground, and I plunged straight into a tree, my parachute hopelessly tangled in the branches. All around there was the racket of the K-car choppers firing, the explosion of the jets, rockets but, more personally, I could hear bullets cracking past my head. I looked around desperately for the bastard firing at me. I couldn't see him and there was no time to climb out of my harness completely, so I yanked the capewell attachment clips at my shoulders to release myself from the chute, fell out of the tree onto the side of an anthill – still in the harness – and struggled to release my RPK from under the straps. I fired off a couple of shots at where I thought the guerrilla was hiding but then Big Steve Kluzniak shouted he could see the man behind a tree. Steve had landed more easily and was out of his harness. He fired at the terrorist, despite the fact that the butt of his RPK had broken on landing, while I scrambled out of my harness completely.

Staying in cover behind my anthill, I watched Steve to see where he was aiming and located the man's fire position not far away. I opened up on him too. Suddenly, I heard quite distinctly the tell-tale click of the hammer on an empty chamber, which told me he had run out of bullets. To my amazement, the bastard tried to surrender, having seconds before been happily trying to kill me as I hung from my 'chute. What a fucking neck! I shot him, and Steve and I moved to join the others.

Bob McKenzie formed us into a stop line, hiding spread out in the dry grass and scrubby trees, waiting for ZANLA troops running away from the K-cars which were wheeling and circling overhead. From time to time, terrorists in ones and twos would break cover, trying to get through our line, and we shot them as they came.

At midday, we began to move forward, careful to keep our line straight so we would not shoot each other by mistake, and we began to flush them out, one group after the other. Really, it was a slaughter. We took a lot of fire but it was mostly inaccurate. They had been poorly trained by the North Koreans and Chinese, who had fired them up with political slogans and taught them lots of dialectical non sense about the masses, but very little about individual fire and movement and battle skills. They simply had no chance against trained, disciplined troops.

A lot tried to change out of their uniforms into civvies or just run away naked, but we looked for the tell-tale piles of coloured clothes and found them all the same, hiding in bushes, under tree trunks and in holes in the banks of a dried river bed we swept through. Some fought, some pretended to be dead, and some who I thought were feigning death were actually dead, the victims of the tiny steel flechettes from the anti-personnel bombs dropped by the jets that morning. You couldn't always see the tiny rips caused by the flechettes in their rumpled uniforms, but I always fired a couple of rounds at the chest, just to make sure.

By three o'clock, we halted the line on the edge of the centre of the camp, where we could see the communist secretariat and admin huts – large wooden huts with corrugated steel roofs, African grass huts and white army bell tents scattered among the trees. Our tracer set fire to some of the grass-thatched roofs and fires blazed all around us, creating a scene of utter mayhem, with the smell of burning, the constant noise of gunfire and the explosions of rockets and bombs further away. One hut turned out to be an ammunition store and suddenly exploded, sending a firework display of RPG rockets zooming off in all directions across the camp.

We reached our objective, the communist party secretariat, and found mountains of documents and ledgers stacked inside big trunks, called trummels in Rhodesia. It was a gold mine of information for the Special Branch. I had never realised what detail the communists keep about every single party member. They even had information on who was sleeping with whom.

So far, we had only had two wounded in 'A' Squadron, but I heard on the radio that Frans Nel, a young, fair-haired Rhodesian in 'B' Squadron, had been shot dead by a female terrorist. Let no one tell you the women were innocent bystanders. They trained and fought just like the men, sometimes better.

A group of us in 'A' Squadron found a little white Peugeot truck near the secretariat which had been stolen from North Side Service Station in Salisbury and had a service sticker inside showing it had been in Salisbury only three weeks before. With Steve Kluzniak, Rob Rodell, Chunky Chesterman, Dick Borman and I, Bob McKenzie used it to drive round the camp picking up the enemy's 14.5mm anti-aircraft guns and other equipment we wanted to take back to Rhodesia. He said, "We felt like Kelly's Heroes all tooled up and perched on the bonnet and roof of this pickup."

Shooting and explosions went on all day all around us, as other SAS and the RLI worked through other target areas further away. I spent the rest of the day blowing up large enemy trucks and other vehicles, doing the maximum amount of damage that I could. I can honestly say it was a lot of fun.

The attack had been a complete surprise and success so far. Our commanders assessed that the enemy were disorganised and incapable of counter-attack, so they decided to leave us on the ground overnight. When darkness fell, we hid among the trees in an ambush position of all-round defence and waited lying on the dusty baked earth under the trees, keeping all our gear on. I took it in turns with Steve Kluzniak, who was nearby, to sleep or doze, our weapons cradled ready to use. The shooting died away at dusk and in the silence ZANLA must have thought we had pulled out, because they began to emerge in small groups out of the darkness, calling out, "Comrade! Comrade?" We imitated them, calling back, "Comrade! Yes!" When they came on towards us, we shot them. All through the night there were bursts of fire from all around.

All this may sound very hard-nosed, but these were people who were being trained to go back to Rhodesia and kill. They weren't innocent civilians, like the people who they killed in buses and schools, like the eight men and women and four children of the Elim Pentecostal Church Mission they killed at Vumba in the Eastern Highlands. These victims were raped and bayoneted to death by twenty of Robert Mugabe's ZANLA. The ZANLA guerrillas were not, in the words of ZANLA commander Josiah Tongogara to US Ambassador Andrew Young, "moving round villages conducting political seminars and singing songs'".

They killed their own as well. The following morning, we went back to the farm building which had been attacked by the Vampire as we parachuted in and found four prisoners who had been murdered by ZANLA just as our para drop started. Their hands were tied behind their backs and they had been shot in the head.

An odd thing happened near here, where we found a store of rations with Portuguese markings. One of our guys called Andy Johnson found a can of milk, drank it and immediately went into shock. He collapsed on the ground, kicking and spluttering and shouting in panic. He was casevaced at once in an Alouette and recovered, and I later discovered that our own very dear and lovely Special Branch had somehow got access to these rations during Portugal's withdrawal from Africa and poisoned them. "What swine!" you may say, but the fact is that the gloves are off in guerrilla warfare and enemy soldiers are fair game. On both sides. For example, in Malaya during the Emergency, the British Special Branch poisoned enemy food caches or put ground glass in rice rather than lift or destroy the food. A wounded or sick enemy is better than a dead one: he is an embarrassment for them to look after and is not likely to give the others much confidence. A dead guerrilla is a martyr who can be buried with stirring songs and forgotten. Come to think of it, that's rather the same for soldiers anywhere.

The Chimoio plan had been well thought through. We spent some time that morning burying homing beacons in the ground so that, after we had gone when the camp was reoccupied and bulging with guerrillas again, our bombers could home in and bomb them to bits.

We made a second sweep of the camp, working through the area where we had found so many of their trucks and Land Rovers. The sheer size of Chimoio, an area of 28 square kilometres, meant there were masses of places to hide. We had landed in the middle, scattering the guerrillas in panic, and we wanted to find as many as we could before we had to leave. Forty were hiding in a ravine and a brief firefight flared up till their resistance collapsed under the discipline of our fire orders. They broke and ran and we killed them all.

As you may imagine, the overall 'body count' was huge. This antiseptic American phrase is the newest label for an ages-old military obsession to quantify the outcome of a battle. At Chimoio, the count was enormous, well into the hundreds, and duly trumpeted by soldiers and press alike. I accounted for thirteen, excluding those 'shared' when one enemy is hit by several people at once. As may be imagined, the impact of this attack on ZANLA was immense. Before we left later that afternoon, we set fire to everything. I lit a gas cooker and ran along the thatch roofs of the party huts, lighting the dried straw, which caught immediately. In minutes, the fires spread rapidly through the trees and great clouds of black smoke billowed into the sky. To add to the chaos, bullets they had hidden in the thatch cooked off in the heat, exploding everywhere.

Finally, that afternoon of the second day, Alouettes lifted us out back to Rhodesia, to Lake Alexander and then to Grand Reef airbase at Umtali. We flew back to Salisbury feeling dirty, hot and tired but pretty pleased with ourselves, and we were all looking forward to a good session in the bar to tell our war stories. Instead, as soon as we landed, we were put straight back in isolation on the airfield again. ComOps were so delighted with Chimoio, they wanted us to attack another big ZANLA training camp without delay.

This was Tembue. It was estimated there were 6,000 guerrillas there, so it was not as large as Chimoio, but it was more than twice as far away, right the other side of the Cabora Bassa dam lake, north-east of Salisbury, 200 kilometres inside Mozambique near the Zambian border. If things went wrong there, there was no easy way out. However, ComOps reckoned that we could use the same trick twice in quick succession, on the basis that the enemy were still in utter disarray at Chimoio and would have no time to learn the lessons, let alone pass them along to Tembue. That, at any rate, was staff theory.

We caught up with a bit of sleep on Friday, but only after we had reorganised and prepared our equipment, and received another set of excellent orders from Major Robinson. The plan was exactly the same, but the Forward Air Fields (FAFs) were even more important because of the greater distance from the Mozambique border to Tembue. ComOps set up two admin points inside Rhodesia, at Mount Darwin and near the northern border at Chizweti, one big one inside Mozambique called the Train because it was so busy feeding helicopters in and out, and one very close to Tembue, some 200 kilometres inside enemy territory.

The next morning, early on Saturday, November 26, we jumped in on Tembue at precisely the same time as at Chimoio. ZANLA never knew what had hit them. All day, we fought through the bush in our sweep line, with endless small contacts and firefights, walking slowly and methodically through the bushes and low trees as we flushed them out. At a riverbank, Steve Kluzniak and I were working together, one covering the other, when we suddenly came under fire from thick bush, near a series of trenches protecting a sloping approach to the Tembue River. We sprayed the bushes but we couldn't see our man and both ended up diving for cover into the same trench, a long, narrow slit in the hard red earth.

"You seen him?" I asked furiously, as we sat at the bottom of our trench.

Steve shook his head. But our man knew where we were. We watched his bullets smacking into the red earth at the top of our trench. After a few minutes, sitting helpless, afraid to move, I lost my temper. I dived out of the trench and zigzagged fast to another one. Bullets cracked past. As I ran, I tried to see where the bastard was hiding, looking for smoke or flash from his muzzle, or giveaway movement in the bushes. Nothing. I dived for cover into another trench and still he kept sniping at us. I shouted at Kluzniak. He had seen nothing.

I tried again, determined to find the bastard. I rolled out of my trench and doubled forward again, leaving Kluzniak behind to cover me in the dense smoke billowing from burning huts not far away. No sign. Still the ZANLA soldier stayed hidden and, as our sweep line moved slowly on, we had to leave him. His accuracy was nothing to shout about, but he was one man who certainly knew his field craft. We moved from action to action, once flushing out nearly fifty ZANLA from a ravine in a tremendous rattle of automatic firing.

On all sides, unseen actions took place, punctuated by the bigger .20 cannon of the K-cars, which we could call in with our ground-to-air A76 radios. We heard on the radios that the others in the SAS and the RLI were having as much success as us. The ZANLA command were totally disorientated and their fighters desperate to run away or hide. That night, we lay up in ambush just as we had at Chimoio, and several of the guys near me fired at terrorists trying to come through us in the night. We got little sleep but though we were 200 kilometres inside Mozambique, we felt very confident. We had taken no casualties and inflicted terrible losses on the enemy. I never realised how many till the tally was made the following day, a Sunday. They stopped counting at 1,200 dead but the total at Chimoio and Tembue was well over 2,000. In two days, ZANLA had taken a terrible beating.

As we pulled out ready for the choppers to fly us back to Rhodesia, I heard that another troop had rescued a black Selous Scout who had been held captive in a deep pit. The Selous Scouts were excellent troops, a combination of white and black Rhodesians, who moved in groups of ten to ape the communist section strength, or just in twos, so they could move swiftly and unnoticed. They had courageously recce'd both the Chimoio and Tembue camps, to confirm Special Branch information, and supplied us with the detail we needed to make the attack. This man, a black NCO, was lucky we turned up. The black guerrillas always treated black Rhodesian soldiers far worse than they treated whites, torturing them mercilessly before they were allowed to die.

These were remarkable operations. I wonder how many countries anywhere in the world would have had the political guts to mount these attacks, let alone be able to muster the necessary military skills. Political and military commitment was required from the very top, from Ian Smith and General Walls, to the very bottom, from the likes of me and Steve Kluzniak. In the middle, officers like Robinson and Walsh faced immense staff coordination and logistic problems, posed by distance, refuelling of the short-range Alouettes, ammunition resupply, and the bringing out of captured enemy documents and equipment. All this meant working out new solutions never before tried and putting them into effect for the first time on high-risk operations which, had they gone wrong, would have caused unthinkable losses to men and aircraft and dealt a terrible blow to Rhodesian morale. Impressive, no matter what you thought about the Rhodesian regime. They committed a mere 185 troops against fantastic odds.

Jane was well aware of the dangers but she never tried to stop my work in the SAS. As a nurse, she was constantly faced with the realities of the war when the wounded were brought in to the hospital. She never said anything, but she must have been very worried before the Chimoio and Tembue raids. Later, she told me the hospital had been cleared for action to receive the expected high casualties. Whole wards had been emptied waiting for us to fill the beds. None of the hospital staff knew what the operation was to be, but it was going to be big, and Jane knew very well that I would be on it.

The pressure did not let up but it varied. On one cross-border operation in trucks, we had to come back through the Rhodesian Cordon Sanitaire minefield, the CORSAN. This was an extraordinary feat of military engineering, a barrier of mines 830 miles long to prevent guerrilla infiltration across the Mozambique border. This was the second-biggest minefield in the world after the US minefields across the DMZ between North and South Vietnam, and I believe it was a total waste of time and money. It ignored the fundamental military principle that an obstacle is not an obstacle unless covered by fire, i.e., unless observed by soldiers, which the CORSAN wasn't. The enemy just ignored it or removed it. Quite often, rains washed away the fragile African soil and revealed the mines standing out of the ground.

On this occasion, we waited on the Mozambique side and watched as Rhodesian Army Engineers cleared a breach for our trucks. They had just finished when I saw a major walk forward. His foot hardly brushed the ground at the side of the breach, but the earth erupted and his leg disappeared in a cloud of red spray. He was casevaced to Jane's hospital in Salisbury, where the poor man's daughter was a nurse. Everyone was involved in the war in Rhodesia and maybe this serious atmosphere of commitment intensified our affair.

The whole country was embattled. Everyone in the country knew someone in the services which fielded year by year an average of 25,000, equivalent to more or less a tenth of the white population, and everyone suffered from United Nations sanctions. There was no chocolate or real Scotch whisky, for example, though the Rhodesians proudly made a whisky substitute which was truly revolting. The attitude among the whites was that Rhodesia had an answer for everything the international community could do to her.

However, all this pressure was too much for some people, who quit on what was called 'the chicken run' or 'taking the gap', rather than hang on to witness inevitable defeat. They took what little they could, but no financial stocks or funds were allowed to be transferred out of the country so their apartments stayed empty and unsold. With the market at rock bottom, Jane and I moved to an excellent place in Beveridge Court in the Avondale district.

There was no end of work for the SAS. The commitment to

killing the enemy, which we did with practiced efficiency on more or less every patrol, was intense, and life in the Rhodesian SAS was rather humourless, in contrast to my experience in the British Army. Maybe this was due to the fact that Rhodesia was certainly under great pressure, maybe it was a difference in national characteristics, or maybe it was because the Rhodesians in the Rhodesian SAS were under pressure from foreigners like me.

There were men from Britain, America, France, South Africa, Canada, Zambia, Spain, Germany, New Zealand, Denmark and elsewhere. Some called us the 'Rhodesian Foreign Legion'. In 'A' Squadron, twenty-eight of the thirty-three were foreigners, excluding numbers of the Rhodesian Territorial SAS who unquestioningly served their six months a year call-up, factoring out other times when they were called on for various big operations. The Rhodesians called us foreigners 'Nanny knockers' (those who slept with black nannies in white households because they could not break into the clique of local white women) and argued we were just visiting for the 'fun', whereas they lived there and were fighting for their livelihoods and future. This may have been the crux of it, for the future was bleak and, in retrospect, defeat inevitable, but whatever the reasons, the Rhodesians in the SAS certainly retained a distinct consciousness of themselves. From the outside, anyway, they grouped together in a subtle, but none-the-less real, self-protective clique.

Of course, Rhodesian officers and senior NCOs knew each other very well, often having been to the same schools, such as Plumtree, Peterhouse and Churchill, of which they were very proud. Indeed, I remember one officer who always wore his Plumtree School socks in uniform, and, when he was unfortunately blown to pieces on a mine, the joke among us foreign swine was that the only way they had identified him was by finding one of his legs stuck in the branches of a tree still encased in his Plumtree School colours.

The 'Russian Front', as we called the south-east, was hard work. The terrain was difficult, flat and sandy, so it was nearly impossible to cover our tracks, the local Africans were very antagonistic everywhere and Mozambique's FRELIMO troops actively searched for us in this area more than any other. Knowing we were usually about twenty strong, they gave themselves the advantages of working in groups of fifty or more and used 82mm mortars and HMGs. Our target was often the single road and rail link from Maputo to the coast and, as soon as the locals spotted a Dakota overhead, they reported it and FRELIMO would sweep the area. They moved along beside the railway line, checking the sand for our spoor behind the clumps of sandalwood typical of the area which provided the only cover to hide in. Then they would attack with everything they had, confident the Rhodesians did not normally use air cover inside Mozambique.

So, the Rhodesians adapted their tactics. Eighteen of us in 'A' Squadron jumped into Mozambique and deliberately left tracks which we hoped would be seen by the locals to draw the FRELIMO troops onto us. The idea was to use air strikes to give them a bloody nose, which might dissuade them from tangling with us in the future. We were moderately successful. We dug in in slit trenches, walked to a nearby road and fired some rounds to attract attention. While waiting for a reaction, we placed a parachute in a tree, but to our dismay had no response. I never saw any enemy, but Captain Colin Willis in 'B' Squadron called in an air strike on a column of seventeen trucks and destroyed them.

On another occasion, near a small town called Maxalia, we walked in covertly, to avoid the locals spotting a para entry. On all external ops we wore plain green uniforms, and no dog tags, so that if any members were killed, they could be written off by the authorities. All callsign members were assigned book numbers so they could be identified over the radio if killed or wounded, and all had their blood group written on the chest of their plain green shirts in felt-tip pen. Callsign members were also 'blacked-up' so they could be mistaken for FRELIMO at close quarters, and camouflage cream – sometimes mixed with mosquito repellant – was constantly reapplied at each halt, which for us was normally ten minutes in every hour.

On this occasion we were even 'disguised' in ZANLA uniforms, but we took casualties. As we moved through the bush, we surprised a group of ZANLA who reacted fiercely, firing RPG-7 rockets through the scrub trees and opening up with heavy small-arms fire. The man next to me, called Tony Nesbitt, went down under a fearsome burst of AK, shot in the leg. The enemy drove off and I gave him first aid. To reassure him, I joked about his wound. "It's a really beautiful neat hole, Tony. No exit wound, but the entry hole is really lovely!" He groaned with laughter till the pain got to him.

On the same op, on December 6, 1977, Dick Borman was not so lucky. If people are honest, they would admit Dick was not much liked, maybe because he was always harping on about his religion or maybe because he had no sense of humour. One day, Steve Kluzniak, Frank Tunney and I were sitting in cover in the scrub having a brew when we heard shooting some way off. Some of our lads had gone that way and, between sips of tea, I remarked with deliberate black humour, "With a bit of luck Dick Borman's got it." Everyone laughed, but when they brought a body back, it was Borman. What's worse, the poor man died in a clash between two of our SAS patrols, and, to cap it all, there were no choppers available to bring out his body. We had to wedge him in the branches of a tree, where he stayed for two days to prevent animals, rats and so on from eating him.

I must admit, the incident made us wonder if the choppers would have come for us if there had been a wounded man waiting for casevac. The modern soldier assumes he can depend on 'hot extraction' with choppers, but the fact is that there are sometimes shortages and other priorities. I think here they rightly decided a Selous Scouts live operation was more important than a dead SAS soldier.

There were numerous camp attacks after Chimoio and Tembue, but never quite so big. Muroro, a ZANLA camp, was one which sticks in my memory for what happened to me there. We jumped at a quarter to eight in the morning as usual, and found that the camp was deserted. The Selous Scouts had reported 1,800 there but we swept through the scrub and the body count tallied only twenty-five. This was an unremarkable score for the Rhodesian SAS, which was called 'a bad day's jousting', but Muroro was a turning point for me.

As we stalked through the trees and long dry grass, sweeping the area, a black guerrilla suddenly leapt up right in front of me, no more than a couple of paces away, like a rabbit flushed from the long grass. Stupidly, he turned to run. Instinctively, I lifted my FN to my shoulder and straightaway shot him in the back of the head. To my horror, his skull literally exploded. His brains blew all over my face and chest webbing. His body collapsed on the hard ground. What shook me was the unthinking speed with which I had reacted to shoot another person, so efficiently and without compunction. The lesson stayed with me all day. There was no time to clean up. I had to stay in our sweep line carrying that man's brains all day, stuck to my webbing, stinking and fly-blown.

Maybe it was justice, but I nearly got left behind at Muroro, deep inside Mozambique. Four of us were tidying up the parachutes we had hidden on jumping in. This was a nerve-wracking job, as the terrorists would sometimes try to find the cache to ambush us when we came back, or booby-trap the parachutes. We were absorbed checking the sandy ground for signs that the enemy had been around and the last Alouette took off without us. I looked up, somehow warned that something was wrong, and ran in front of it, waving my arms to flag it down. Thankfully, the pilot saw me. All the way back to Salisbury, I could not shake off the memory of killing that man at Muroro. The images of his exploding head would not go away. They never have. I was 35 years old and his death marked the moment in my life when I began to see myself as others saw me.

My old ladies in Salisbury would have certainly declared this a sign of weakness. They were a group of charming old women who 'adopted' the foreign soldiers serving in Rhodesia. They took this duty very seriously, in loco parentis as it were, and they were more committed to the fight than the soldiers. They always showed a keen interest in our successes and regularly sent me and the other foreigners nice letters, knitted socks or balaclavas, which the Rhodesian Army faithfully brought out to us on helicopter resupplies while we were in the bush on operations. During our leave, these old ladies insisted we visit them, and, because they were all part of the white war effort, the Army Staff made precise appointments for us.

Suitably scrubbed and wearing a clean shirt, tie and slacks, I often went to tea in their apartments in a smart district of Salisbury. God knows what they thought of me, a short, stocky Glaswegian with staring blue eyes, a bent nose and hair cropped to my scalp, but for an hour or more, I would perch on a chintz armchair, delicately sip tea from a cup, and face a circle of white-haired Rhodesian ladies who pressed me for stories of the fighting. While they fed me with an endless supply of dainty cakes and toasted comestibles, they genteelly plugged me for the most gruesome details. The war was our only topic of conversation. For all their apparent frailty and charm, they delighted in hearing about raids like Chimoio and Tembue. When I said goodbye at the door and thanked them for their kindness, one old lady would always whisper, "Remember now, Peter. You'll get one for me, won't you?"

The Kavalamanja camp attack began in the bar. At least, it was late Friday afternoon when the SAS were put on standby and not many were left in camp. Captain Bob McKenzie ordered me to round up the men, so I hopped in a Land Rover and toured the old haunts of Salisbury, dragging them out of places like the Coq d'Or, the Golden Dragon, the Sahara Bar and the Monomotapa Hotel, where they were drinking beer and talking shop. They came willingly, as camp attacks were popular. There is no more thrilling way of going into a fight than by parachute. We were all psyched up with the jump before we even touched the ground, we were confident of winning the firefight and there was always the chance of a bit of loot. Russian weapons were popular souvenirs, like the Tokarev automatic pistol or the new AKM assault rifle.

Kavalamanja was one of Joshua Nkomo's ZIPRA camps, just inside Zambia and not far from the Mozambique border, where the Mkariva River joined the Zambezi River. The Zambezi was enormous at this point, flowing in a great slow bend 800 yards wide between a mountain on the Rhodesian side and low, marshy ground by Kavalamanja on the Zambian side.

Two Selous Scouts, Captain Chris Schollenberg (one of only two holders of Rhodesia's highest bravery award, the Grand Cross of Valour) and Sergeant Chibanda (later awarded the Silver Cross of Rhodesia), had watched the ZIPRA camp from high ground inside Zambia, cut off from help by the river. They lay hidden on high ground for days to produce a report of enemy activity and estimated there were some 175 terrorists armed with AKs, anti-aircraft 14.5mms and 12.7mm HMGs, and they identified numerous items of East German equipment. Heavy summer rains delayed the operation, but they stayed in their hideout with a box-office view to watch us jump in.

On Sunday, March 6, our usual 0800 hours P-Hour was put back due to bad weather, and our Para-Dakota skimmed 100 feet over Nyamvuru Mountain at ten o'clock on the run-in. We jumped with a company of the Rhodesian African Rifles (RAR). The RAR began their sweep through the camp while we formed stop groups on the north-east side. Our task was to contain the terrorists and hold off the regular Zambian Army if it turned up on the dirt track from Feira (now Luangua), not far away. It didn't. In fact, Rhodesian radio intercepts picked up frantic calls from the Zambian Army at Feira to the capital Lusaka asking for reinforcements in case Feira itself was next on our list!

In the camp, the weather and dense jessie bush slowed progress badly and most of the enemy escaped under cover of darkness that night across unguarded flooded ground on the banks of the Zambezi. The final score was forty dead ZIPRA. I had little fighting to do but we found a large quantity of Russian CE compact plastic explosives and I had a great time setting the demolitions to blow a large cache of Russian weapons we could not bring back. The explosion was deafening and blew the leaves off the trees for hundreds of yards around.

ZIPRA were conventionally trained by the Russians and better troops than ZANLA, who were nothing more than terrorists in civvy clothes. We noticed the difference when we attacked 'DK-1', a ZIPRA camp near Lake Kariba.

We drove in Unimogs from Salisbury to Binga for briefings and final rehearsals, and at last light Alouette choppers flew us low level to a drop-off point 16 kilometres from the camp. Our march in was certainly covert. The weather was foul, it poured with rain all night and we could hardly see the man in front, so no one else could possibly have seen us. We carried rucksacks with spare ammo, a day's food, sleeping bags and ponchos (before the era of Gore-Tex!). The packs weren't heavy, but we were sodden, muddy and tired when we reached the forming-up point for the camp attack just before first light. As the soaking trees emerged from darkness in the miserable grey dawn, we crawled forward through the mud in the cover of bushes and long grass to the start line.

As soon as we stood up to begin the attack, they opened fire and a furious battle started, ripping leaves from the trees all around. We put down a heavy weight of fire and moved forward step by step, bound by bound, in controlled fire and movement through the bushes and long wet grass. We pushed them back onto a slight rise, but they were as fired up as we were, shouting, "White bastards! Come on, white bastards!"

Almost the whole squadron cornered a group of ten in some bushes, where we taught them the important difference between 'cover from view' and 'cover from fire' by blasting the patch of bushes and killing the lot. Driven by us and the .30 cannon from an Alouette K-car above, the rest withdrew towards a river bed, where they ran up against one of our stop groups. Boxed in, they fell back from the stop group towards us again and we gradually crushed them, once again moving our patrols together in 'bounds', firing from the kneeling position to see over the low scrub and grass, and moving short distances in a low crouching dash to fire again.

The fighting pressed them into the riverbank, where we finished them off with A8 grenades. These were South African grenades, crude but effective: a wodge of PE inside a column of metal washers. The final score was seventyeight dead ZIPRA and two wounded SAS, but they fought hard. ZIPRA were well armed with brand-new Kalashnikovs and trained as conventional soldiers by their Russian advisors, so they stayed and fought. ZANLA were terrorists who mostly seemed to work on the principle of, "If you kill and run away, you live to kill another day."

The Rhodesians were keen on mines. The CORSAN must have absorbed a lot of staff work, time and money, and the Rhodesian SAS used mines extensively too, for similar reasons – to dissuade the guerrillas from using infiltration routes.

The Rhodesian SAS did a lot of mine operations on the north side of Lake Kariba, in Zambia. One dark night, a Kiwi captain who had been in 3 Para and I were taken over the lake in a civilian motorboat to a point some 200 metres off the Zambian shoreline. We lowered two Klepper canoes over the side and carefully stowed our heavy kit.

Two more men were going to paddle us in and bring the canoes back to the mother craft, so two of us lay over each canoe and they paddled us to the dark shoreline. We waded off with our gear and the canoes slipped away into the darkness. Ashore, we hoisted our rucksacks – which were heavy with one 12-pound anti-tank mine each – and set off into the thick scrub. Our target was a road along the lake which was used all the time by the Zambian Army supplying the ZIPRA camps to the south-west. This MSR (Main Supply Route) was about 30 kilometres from the lake but the high savannah bush was dense – like dry jungle – and allowed us to walk during the day without fear of being found. We stopped at about eleven o'clock when the sun was too hot and lay up till three before going on again. We reached the road just before dusk of the second night, watched it carefully for a couple of hours and then moved in to lay our mines.

The road was not surfaced and our plan was to lay one mine every kilometre or so. The South African anti-tank mines we carried were nicknamed 'Chocolate Cakes' because they were round and painted a chocolate colour. They contained a charge of about 13 pounds of Amatol and detonated under a pressure of only 11.8 kilograms or 26 pounds. We also carried a special mining kit which consisted of digging tools, poncho, para-cord, a brush, and carpet overboots to avoid leaving the spoor of the soles of our combat boots.

Having selected where on the road the mine was to go, I laid out the poncho on the ground beside it, scraped back all the topsoil and put it carefully on one side of the poncho. At that time, a lot of people made the mistake of digging the 6- or 8-inch hole straight down, but most of us always dug holes with sloping sides, because if it had straight sides, sometimes the vehicle would catch just the corner of the mine (which was approx. 10 inches in diameter), press it out and not set it off, or the wheel would jump across the straight sides and not touch the mine at all. Chamfered sides caught the wheel no matter what.

While a couple of the others kept watch up and down the road, I put the mine at the bottom of the hole. The most dangerous part of the process was arming the mine. The Rhodesian SAS lost eleven men killed laying mines, so I always took care. Before back-filling with soil, I very gently tied a length of para-cord to the mouth of the pin holding the MV-5 pressure-release switch. This fitted into the 75 MD-2 detonating cartridge which stuck out of the side of the mine. Then, after back-filling, I warned the others and retired to safety at the side of the road. Wrapping the string round my fingers, I gently pulled out the pin with my para-cord. Nothing happened. After a short wait in case of a delayed fuse, I went back to tidy up the topsoil and very gently brushed it over to conceal the hand marks in the dust.

Once all our mines had been laid, we marched back through the savannah to the lake, where we took a bearing on Binga from the water's edge. The lights of Binga could always be seen from the Zambian side, and once we had radioed the bearing to base, they knew where to send the boat to pick us up under cover of darkness. However, we took care to retire into the savannah till it was time for our rendezvous with the Kleppers, just in case the Zambians patrolled the lake edge and found us. A friend of mine was killed like that, because he was too ill with malaria to walk back inland and wait in safe cover.

I was never convinced by the use of mines in Zambia. After that patrol across the lake, I recall sitting at Binga and listening to the controlled explosions of the Zambian regular army engineers blowing the mines we had just laid. The noise carried quite clearly across the surface of the lake. It is rumoured that on one occasion a brave and foolhardy black Zambian engineer spent twelve hours defusing a mine fitted with anti-lift, anti-prod and anti-light devices.

By contrast, the mines we laid on the important enemy roads in Mozambique in the south-eastern area, which we called the Russian Front, seemed worthwhile. The road and rail system there was used extensively by ZANLA moving into Rhodesia. However, compared to our employment attacking enemy camps, I don't think the success of mines justified the loss of eleven men from the Rhodesian SAS, out of a total of forty-two who gave their lives for Rhodesia.

CHAPTER 8
THE DEATH OF A COUNTRY

I was promoted to full corporal in the Rhodesian SAS in two consecutive promotion conferences, but at the start of 1979, I realised I would go no higher. I do not think I fitted in. The bald fact is that I was 37 years old, fifteen years older than men of the same rank, and where the younger guys held their senior ranks in awe, as men older than them and more experienced, I did not. Of course, I respected the senior ranks as fine professional soldiers, men like Sergeant Major Lutz, Captain Bob McKenzie, Major Brian Robinson and Major Graham Wilson. They did a good job, but they were not gods. However, there is a pecking order in the relationships between ranks and personalities in any army unit, especially one as tight-knit and specialised as the SAS, and I attracted criticism because I was not a standard fit in the mould.

Rumour travels fast and my reputation had followed me from England, by damaging gossip rather than by an official report, and coloured any name I might have had for good operational soldiering. I was actively trying to shake off my earlier image and I only had one fight in the Rhodesian SAS, with a Sergeant Andy Langley during my recruit training. Langley was a big ex-US Marine with Vietnam experience from 'B' Squadron who was not much liked. He enjoyed terrorising the recruits and he made it clear to other instructors that he was going to 'fix' me and see I went to jail. I suppose he singled me out as the oldest recruit. He walked in to our billet one evening while we were watching *Wizterloo* on video, switched it off and shouted, "I'm Langley, I don't like you. I've got the highest pain tolerance in the goddamn world and I can't feel your pain!" His problem was that I was by then a very keen and fit recruit, and I gave him very little to criticise.

Finally, he ordered me to call out the time as we marched along on the parade ground, and I shouted, "Left right, left right, LEFT!" always finishing with a 'left!' in the British Army style.

He objected, shouting, "You're fucking wrong, boy!" Which I did not much like. "You have to say both 'left' *and* 'right' and finish with a 'RIGHT!'"

You will probably find this rather a petty distinction, but Langley drew himself to his full height over me, worked himself into a fury and screamed with triumph, "I'm going to put you in jail for this!" Being me, I thought I might as well go to jail for something important, so I punched him and laid him out. So much for his pain tolerance. I think most people were secretly rather pleased at Langley's fate, but Major Robinson rightly gave me twenty-eight days in jail for battering one of his sergeants. I never regretted it for a moment. After all, Langley deserved it.

Or maybe I made myself unpopular for refusing to obey an order to kill a black man. I was on a four-man patrol led by a Rhodesian corporal and we were lying hidden in bushes ready to ambush a track used as an infiltration route by ZANLA terrorists. However, an elderly black civilian wandered down the track dressed in a dirty shirt and worn trousers and, by sheer bad luck, he spotted the Claymore anti-personnel mine one of the other guys had put out to hit the enemy and which was poorly concealed. I watched the old man's expression change as he became aware of the small rectangular mine standing on little metal legs in the undergrowth. Horrified, he gazed into the bushes where he suddenly realised there were armed soldiers hiding, looking at him.

"Shoot him!" whispered the corporal beside me.

I refused, and whispered back, "If you want to kill him, do it yourself!"

He insisted, "I'm giving you an order!"

I took aim and fired deliberately wide. The black man vanished, showing a surprising turn of speed for an old timer.

The corporal lost his temper, blaming me for compromising the whole operation, and we had a furious argument out there in the bush. It was true we had to leave the ambush, but the shot, wherever it had been aimed, would have compromised us, let alone having to deal with the dead body, the spoor of dragging him away and the blood on the track. Besides, there were a group of children close by who would have heard, and seen, what we had done.

The plain fact was that few prisoners were taken by the Rhodesian SAS. I am keen to be the first to kill the enemy, especially in the heat of a firefight, but I have never been someone who could kill a civilian in cold blood. I will leave that to men like Callan, Copeland and other mercenaries in Angola. I was unpopular because of this incident, but in fairness I do not think this was so much for refusing the order, which would have compromised the ambush anyway, but for not fitting in with the Rhodesian system. In 1978, the Rhodesian SAS had swelled from being just 'C' Squadron and had become the 1st Rhodesian SAS Regiment. There was therefore a lot of promotion flying about but the political state of Rhodesia at the time and the inescapable fact that the war was getting out of hand made the atmosphere in the SAS (I can't speak for the other Rhodesian infantry regiments) even more embattled than ever.

The number of terrorists operating inside Rhodesia relentlessly continued to rise, from 2,350 in April 1977 when I joined the SAS, to 5,598 by November of that year, to 6,456 by March 1978 and 11,183 by January 1979 (these are the Rhodesian government's own figures).

I can understand that the Rhodesian officers and senior ranks felt cornered, frustrated by political restraints to their military plans, and unwilling to be specially tolerant of outsiders. As you have read, there were plenty of foreigners in the Rhodesian SAS, and a goodly few are buried in the cemetery in Warren Hills by Lake McIlwaine, but we were never fully accepted. Some did better than others, like Steve Kluzniak, but we were always different, and I suppose my own profile has always been higher than most. I know that one of the regiment's later commanding officers, Major Wilson, was a fine soldier with all three of Rhodesia's highest bravery awards, the Grand Cross of Valour, the Silver Cross of Rhodesia and the Bronze Cross of Rhodesia, but he did not approve of me. I suppose my record in the field was spoiled by the various incidents I have outlined above, because he never worked with me on a patrol and he must have formed his opinion by hearsay.

Anyhow, the plain fact was that I was a 37-year-old corporal and did not fit. I tried a transfer from 'A' Squadron to 'C' Squadron in the newly formed regiment but it made no difference, so when an old friend from the British SAS and 3 Para, Sergeant Major Jock Hutton, told me about his work for the Special Branch training blacks as Auxiliaries, I applied for a transfer. Several of us foreigners applied at the same time, for similar reasons, including Steve Cleary, Ron Cook, big Jim McGuire, Frank Tunney who was an Aussie, and Paddy Gibline. We all marched in to see the Rhodesian regimental commanding officer at the time, Lieutenant Colonel Garth Barrett, who lost his temper. "You bastards aren't transferring," he shouted, furious we had taken our own initiative to leave the regiment. "I'm posting you!" With that, he tore up all our transfer applications and threw them in the bin.

What a blow to our pride! This illustrates SAS thinking, because, although we had served as well as any in the field, a 'posting' rather than a 'transfer' meant that we would no longer be entitled to the few perks enjoyed by special forces (which included the Special Branch): this was the $33 a month extra pay called 'SAS Gunner Pay' and 'Special Unit Allowance'. So, the CO struck a terrible blow to our pockets as well! Plus, being posted from the SAS made us available for any general duties and at the mercy of the bureaucrats in Army HQ.

We ended up standing to attention in front of a postings officer in Army HQ in King George VI Barracks in Salisbury. He had clearly been briefed by Garth Barrett. He hardly glanced at our records of service before shouting, "You've arrived at last, eh? Well, you bastards aren't going to Special Branch! If I say you're being posted to Gwelo, you'll go to Gwelo, and if I say you'll be posted to a black battalion, that's where you'll go!"

He was getting into his stride when his sergeant major walked in and I was just thinking maybe we could appeal to him as equals in the time-honoured game between NCOs and officers when Paddy Gibline gave a great sigh of despair. He had battered this sergeant major in a fight in Cally's bar not a week before!

That was a dark day. However, we extracted ourselves from the officer and sergeant major in the postings office and in desperation I contacted Jock Hutton. He arranged an interview with a Special Branch officer called Chief Superintendent Mac McGuinness. Mr Mac, as everyone called him, interviewed us and declared, "I can use people like you!" Without delay, he sorted out the postings office, we were all transferred to his department, and we kept our pay. He was an SB Commander attached to the Selous Scouts, who thought very highly of him. He had the widest connections imaginable, he was never fazed by anything – even in these last chaotic months of Rhodesia's existence – and he always appeared in a cheerful mood, singing and joking. This was indeed a change from the Rhodesian SAS. And so was the work we did.

By the beginning of 1979, the war with ZANLA and ZIPRA was at a stalemate. The Rhodesians had not surrendered a single town or piece of territory to the communist terrorists, but were terribly stretched in all provinces. At a tactical level, the Rhodesians won virtually every engagement and achieved a consistent kill ratio, which never fell below 6:1 and was as high as 2,000:1 at Chimoio and Tembue.

These huge cross-border successes eventually forced the leaders of Zambia and Mozambique to press Robert Mugabe and Joshua Nkomo to participate in peace talks offered by the

British in late 1979. This was a remarkable political result arising from military action, considering the shortage of equipment and manpower always suffered by the Rhodesian Army, but on the world political front, Rhodesia's position was worse than ever. In the words of the Selous Scouts commanding officer, Lieutenant Colonel Reid Daly, "There had been a time when Rhodesia had been held up as an example of fine race relations, but now we were the pariahs. Only black rule, not multi-racial rule, would satisfy our enemies, or even our friends."

World opinion pushed Rhodesia relentlessly into the hands of the hard-liner communists in ZANLA and ZIPRA. Obstinately, Rhodesians tried to shut out reality but, as Reid Daly said, "The stink of political defeat, which always pre-empts a military defeat, had begun to seep like blood poisoning into the veins of the security forces and into the veins of Rhodesia itself." My next year in Rhodesia more than proved he was right.

Mr Mac sent me to Bindura.

There was a definite feel of the Wild West about Bindura. It was a small town about 50 miles north-east of Salisbury in the middle of ZANLA's Mashonaland. A population of about 1,200 whites lived there with some 5,000 blacks but no one knew exactly how many because no one had bothered to count them. The blacks worked mainly in the Trojan nickel mine and in the large citrus plantations all around which gave off the most wonderful scent of orange blossom along the roads in the spring.

The Special Branch base was in a 'fort' made of corrugated iron sheeting, to prevent people seeing what went on inside. There was one wide main street, sleepy and deserted, which featured the usual necessary collection of shops, banks and stores, and behind this on either side the Rhodesians lived in spacious houses in some style. In time, Jane followed me to Bindura and we set up house in a large villa conveniently near the SB fort. There were two parts to the hospital, one for whites and one for blacks, and when she qualified, Jane worked in the black section, where she was the only white nurse. There were a collection of bars – most notably the Bindura Country Club, which was in the middle of town, and the Coach House Hotel at one end – where increasingly frantic parties took place as the white Rhodesians realised they were losing the country they had made theirs.

Far from losing pay, Mr Mac had me promoted to sergeant and I went with some of the others to prepare for the reception and training of black Auxiliaries. On January 30, 1979, Rhodesian whites loyally followed the advice of Prime Minister Ian Smith and 80 per cent voted in favour of a new constitution, which they knew would lead to a black-dominated parliament. Since 1977, Smith had been talking to Bishop Abel Muzorewa, who led the United African National Council (UANC), and the Reverend Ndabaningi Sithole, who controlled the ZANU Patriotic Front. Both these black nationalists had agreed to come back to Rhodesia for elections. Neither had great support and they had no connection with the Chinese-trained ZANLA terrorists of Robert Mugabe and the more conventionally Russian-trained ZIPRA troops of Joshua Nkomo, both of whom refused to participate in the elections.

On April 24, Bishop Abel Muzorewa won fifty-one of the seventy two black seats in the hundred-seat parliament and was elected Rhodesia's first black prime minister. Now, having agreed to renounce terrorism, his UANC and Sithole's ZANU Patriotic Front had to come back to Rhodesia and join the fight against ZANLA and ZIPRA. To further complicate matters, there was another Auxiliary group formed from surrendered enemy personnel (SEPs) from ZIPRA. Frankly, it would be an understatement to say none of these groups got on with each other.

We drove trucks to Salisbury airport in the greatest secrecy to meet the first of Muzorewa's guerrillas. One hundred of UANC flew in from Libya, where they had been training courtesy of Muammar Gaddafi. You can imagine how they felt flying right into the lion's den. Mr Mac had arranged for the civilian aeroplane to unload them in total seclusion at the top end of the airport and they probably thought they were going to be summarily executed.

If that didn't make them sweat, then the heavy overcoats they were wearing against the Libyan winter certainly must have. February is pretty hot in Rhodesia. Their discomfort probably increased by the minute, for the SB men who were detailed to drive the trucks had been drinking bottles of whisky while waiting for the plane, and one drove his truckful of astonished guerrillas straight into the airport gate as we left, flattening it. The final insult to the Rhodesian image was that we had to keep stopping on the dangerous 50-mile drive through bandit country to Bindura, while these drunks relieved themselves up against the wheel. I was to discover that Special Branch employs the strangest people, especially it seems, in the last throes of a military campaign.

The UANC soldiers were put up in an old army barracks in the fort and we started to train them. The object was to use them to fight the ZANLA guerrillas infiltrated among the Shona villagers in our region. This had been tried before, with mixed success, and it was not entirely cynicism which made the Rhodesians nickname these black Auxiliaries as 'tame terrs'. They were not always so tame. One group of black ZANU PF nearly 200-strong had been 'trained' and then ran riot. Out of control, they had seized their white liaison officers and threatened to kill them if their increasing demands were not met. The liaison officers managed to talk their way out but reported the ZANU PF had dug defensive bunkers all round their camp and were in a murderous mood. The Rhodesians had no patience for that sort of behaviour. An SB liaison officer called Pete Donnelly, who was known to this group of ZANU PF, drove out to their camp and issued them all with smart red baseball hats. They were impressed and put them on. Next day, the Fire Force made a full assault, boxed them in with stop groups and swept through the camp while Alouette K-cars floated around above, blasting 20mm cannon at anyone wearing a red baseball hat. In all, 178 stroppy ZANU PF died.

Understandably, there was little mutual trust between the Rhodesians and Auxiliaries, except where groups had personal supervision, and the Rhodesians often gave them AK bullets with only a couple of grains in the charge, in case they went on the rampage.

For me, all this was quite an eye-opener, to find out what had been going on in this war while I had been head down on serious operations fighting with the Rhodesian SAS, on a continuous round of camp attacks, mine laying, ambushing and other cross-border operations. The rest of Rhodesia seemed to have taken an altogether more bizarre attitude to life.

"I've lost the monkey on the operating table," said Steve Hartful to me one evening at the bar of the Bindura Country Club, where several others were sitting around on the upholstered wicker chairs enjoying bottles of Tshumba beer under the slow-turning fans.

"What happened?" I asked.

Steve looked sad. He was an American training to be a doctor in Salisbury and he spent his call-up periods working for Special Branch. The monkey was a tame animal which lurked on the mown lawns round the club and persisted in masturbating on the verandah, ejaculating all over the clean red paint. This gripping sight fascinated the women and Steve had righteously decided such outrageous behaviour was not in the public interest. The solution was, he thought, to put his medical training to the test by castrating it.

He shook his head. "There's a shortage of anaesthetic so I used an elephant tranquilliser and it never came to!"

God help his patients, wherever he is now.

Quite probably the nature of Rhodesian Special Branch work reflected the desperate internal and external security situation which the country faced at the time, and exchange or 'au pair' operations with the South Africans were an example.

Before describing my part in one of these, I should explain a little the serious side of Special Branch or you will think it was all about drunks and animal experiments. In a conventional war, military intelligence collects information about the enemy, sifts it and produces useful intelligence. In a civilian war, when terrorists are mixing with the population inside and outside the country, military intelligence is supplemented by police intelligence, which depends on sources of information among the civilian population and basically comes through two branches, the Criminal Investigation Department (CID) and the Special Branch. However, the police CID is not ideal from a military point of view because its work is too open. Its object is to bring criminals to justice. Everyone can see what the CID does: scene of crime work, arresting, interviewing and giving evidence in court. In counter-terrorist operations, so much of police effort must be covert, for all the same reasons military intelligence is covert, which is where Special Branch fits in. Naturally, when terrorists operate from safe havens over the border in neighbouring countries, SB interests follow them, what they called 'going external'.

There are good historical examples of successful co-operation between Special Branch and the army, notably during the Malayan Emergency (1948–60). There is also an interesting parallel between the British Special Branch working with the SAS and specialised surveillance units in urban Northern Ireland, and the Rhodesian Special Branch working with the SAS and the Selous Scouts reconnaissance patrols who were brilliant in rural Africa. I imagine the principles of operation in both cases are very similar.

In its search of the enemy, Rhodesian SB came into contact with the South African Police (SAP). They had the same enemies. South Africa was supposed to be taking a low profile in its support of the Rhodesian regime, after the United States had put the pressure on, but the communist enemies of Rhodesia were the enemies of South Africa, or allies of their enemies, so the two countries came to a secret arrangement. South African security forces would operate secretly in southern Rhodesia and Rhodesians would do various secret jobs for the South Africans.

The Rhodesian Army was desperately stretched, so the South African Defence Force (their army) had a base just over the border inside South Africa and came over to fight in Rhodesia in some strength, as a complete Fire Force unit. Their targets were ZANLA or ZIPRA terrorist groups located by the Selous Scouts, or groups of the (then) illegal communist African National Congress (ANC) hiding from the South Africans just inside southern Rhodesia.

Of course, the whole thing had to be politically deniable. Naturally, the world would have been shocked to learn the truth of the collusion between Rhodesia and South Africa. So, the SADF Fire Force, its Paras, helicopters and all, was completely equipped and dressed as Rhodesian, and 'sterilised' of anything which might give away their real origin. In return, the Rhodesians carried out secret attacks on targets for the South Africans.

Mr Mac drove from Salisbury to Bindura fort and briefed me on such a task. The South African Police had identified a target in the triangle where the borders of South Africa, Swaziland and Mozambique meet. For years, they had warned Mozambique that if she harboured the ANC, which was carrying out terrorist attacks in South Africa, then South Africa would hit out across the border. Our target was a houseful of important ANC men in Namaacha, just inside Mozambique on the Mozambique-Swazi border, and the whole operation had to be utterly deniable. Mr Mac said the SAP wanted us to come to South Africa for a reconnaissance.

I took Ron Cook with me. He had been in 'A' Squadron in the British SAS. We drove 375 miles south, through Salisbury to Beitbridge on the banks of the Limpopo River which marked the Rhodesian-South African border. We left our vehicle in the Rhodesian police customs post where we met the SAP men. They came across routinely to coordinate customs matters. Hidden in the customs compound, we got into a yellow SAP van so no one could see us and were driven straight over the border into South Africa. No stamps, no passports. Secrecy and deniability were paramount.

They drove us 300 miles to Pretoria and put us in the Hotel

Continental, all expenses paid. I have to say this made an agreeable change after life in Bindura and the rigours of Rhodesia under sanctions. I don't think Ron or I look like your average tourist and I can't believe the hotel staff were fooled for a minute, but they blandly served us very well as we tucked into steaks and a bottle of wine.

The following day, the SAP drove us another 275 miles to Komatipoort, a small town on the South African-Mozambique border some 37 miles north of Swaziland. Here, in the SAP police station, we met a Colonel Slade and received a full briefing on the ANC target. Namaacha was right on the border triangle between South Africa, Swaziland and Mozambique, and only about 60 miles from the important port of Maputo on the coast, where Chinese ships docked with arms and equipment for the terrorist groups. We did not want to risk compromising our intentions by going too close and made a note of the ground as best we could. All round were hot, sunbaked hills, with dusty red tracks and thick, burned-out kneelength grass.

Back in Bindura, Mr Mac heard our report, gave the go-ahead and we made our preparations. I had a team of eight, including myself as commander. We were equipped with non-attributable Soviet weapons and plastic explosives, which had been captured during various camp attacks and could not be traced to Rhodesia or South Africa. Our civilian clothes, jeans and Veldskoen (Chukka boots), and everything else we carried, met the same rigidly deniable specification in case one of us was captured, dead or alive. We would go in completely 'clean'. After a week's training and rehearsals using pictures of the building supplied by the SAP and working on a similar building I found locally, I expressed myself ready. We drove back south in an unmarked civilian Mercedes truck. It was March and the summer rains poured down all the way.

The SAP met us at the border post, covertly ferried us across the border as before and we drove through the rain to Messina, a very hick little town just inside South Africa. We spent the night in the Messina Hotel, which was like the Last Chance Saloon. As if to confirm this, 'Saloon' was engraved in curly lettering in the mirror behind the bar. We took the opportunity to stock up with sweets and chocolate. Rhodesia had had no decent sweets since sanctions began.

Very early next morning, we drove to Pretoria and straight on to Komatipoort to meet four SAP men. They surprised me by asking if we could do the attack at once. Our guide (the SAP 'source'), a black man called Ramon, had turned up to say that the four key ANC men we wanted to hit were in the house and he was afraid they might move the following day.

We were tired after our fourteen long hours of driving in the truck from Messina, but I agreed without hesitation. As the sun went down, we followed them for 30 miles bouncing over rough tracks through the grassy hills to a remote staging point they had already secured in some disused African farm huts about 3 miles from the border. We unpacked our gear, changed into green FRELIMO combat uniforms and totally blacked our faces and hands so we would look like black men. I must admit that I look rather convincing in this disguise, my nose being ideally flattened from too much fighting! We loaded several magazines for our Russian AKs and Beretta 9mm pistols, checked over the explosives, carried out some quick lastminute rehearsals – of cutting the border fence and our actions on the target – and generally talked over the whole operation. Then, at about nine o'clock at night, with Ramon leading, we started walking to the junction of the three borders.

We walked south in the darkness for 3 miles, brushing through the long dry grass, and crossed the border into Swaziland. Then we turned east towards Mozambique. We moved very carefully indeed. There was a guarded Mozambique radar station on one side and a Swazi border post on the other, with a road running parallel to the border inside. Mozambique which FRELIMO used to patrol in vehicles. We opted to cut the fence between the two guarded posts but I held the patrol back in the grass, to watch the fence for a while, to feel 'at home' in the place before we moved. Sure enough, we saw a Land Rover pass along the road. Then I signalled the move forward. Keeping alert to the chance of more border patrols, we reached the fence and worked fast. First, we tied the tensioner wires together to stop them springing apart, wrapped the cutters in hessian and then cut a hole. After we had all come through, we wired it together again so no one would notice.

Inside Mozambique, Ramon walked unerringly through the thick grass towards the lights of Namaacha in darkness a half a mile away. When we reached the slums on the edge, he led us through the back streets to the target. Plainly, Namaacha had once been a pleasant colonial town of attractive flat-roofed two-storey villas surrounded by gardens, but since the Portuguese had left, the place had become a pit. It stank. The gardens were overgrown, we passed empty swimming pools filled with chickens, garbage was strewn everywhere and pools of untreated sewage lay in the dirt roads.

As we picked our way through the narrow streets, we used our Passive Image Intensifier Night Observation Devices (NODs) to see ahead through the darkness. We spotted two guard positions, one at either end of the town, and we avoided a FRELIMO army barracks we knew was on the outskirts. Otherwise, we encountered no one else. If the occasional black saw us in the shadows, he probably took us for FRELIMO troops. Anyway, I doubt any of them wanted to get involved. In a Marxist state, civilians are not encouraged to be inquisitive.

In the centre, Ramon stopped at a street corner and pointed at our target. The house was an old Portuguese colonial villa with terracotta tiles on the roof and it had been built back into the sloping garden so the wall of the house at one end was higher than at the other. After a final check, peering up the darkened streets, I motioned the guys forward. As rehearsed, we split into three groups. The four men with the bunker bombs containing 1½ pounds of plastic explosives moved to their allocated windows round the villa, two men covered the door and I waited with Frank Tunney.

Checking my watch, I blew a whistle. The bunker bombs sailed through the windows and the guys jumped back out of the way. Four seconds later, the bombs erupted, blowing out the walls at each end, and the roof partially collapsed on either side of the door. Deafened by the blast, I dashed forward with Frank to lay a huge 44-pound satchel charge. The first charges were to kill the occupants but this second charge was to rub it in, as a warning to the ANC. However, one of the four ANC was still alive. He started shooting at us with a Tokarev pistol when we ran through the door and I never heard him, still deafened by the bombs. Fortunately, Frank shot him. Then we laid our charge, pressed the percussion igniter and ran for it as hard as we could. I had chosen a thirtysecond delay to give us time to run clear. We legged it through the streets and just before time was up, Steve Cleary shouted, "Down! Take cover!"

We dropped as the villa blew up in a shattering roar, which shook the ground and should have woken all Africa. For minutes, it seemed, pieces of smashed terracotta tiles and brick splattered down all over the town like rain.

We scrambled to our feet and withdrew, at speed. People started popping out of houses to see what had happened and Ramon called out, "Don't worry, everything's fine!" He told them we were a FRELIMO patrol looking for the bombers, which seemed to do the trick. As we left, we dropped pamphlets all over the place telling the locals that the bomb was the work of the Mozambique National Resistance, RENAMO, and exhorting them to join at once! This was the idea of the SAP and seemed to me to stretch credibility rather.

We were all dying of thirst when we reached the border fence. The night was warm and we were literally soaking with sweat after the excitement of the attack and our dash back to the border. We were all dreaming of the water which I had asked the SAP to provide for us on our return.

No one could have missed that bomb, yet the border guards never reacted, nor did the FRELIMO burst out of their barracks to chase us. At the fence, we hid in the long dry grass watching the same Land Rover cruise along the patrol road as if nothing whatsoever had happened to spoil the calm of a hot African night. After it passed, we slipped through our hole in the fence and jogged back over the hills to our start point.

We arrived sweating, dirty, blackened with the stinking camouflage cream and desperately thirsty. And all the SAP could offer was beer. They had heard the explosion and were delighted. I suddenly realised they genuinely thought, *Surely real men want to drink beer after such success?* This, as I found later, was typical of the macho South African psyche.

We drank the beer, because we were gasping. And, naturally, because our bodies were deficient of fluid, we got drunk. Very quickly. The SAP approved. So, there we were, miles from Rhodesia, standing about blackened, hot and stinking on a remote South African hillside a couple of hours before dawn, talking to a group of South African policemen who were ecstatic that we had, as the Special Branch called it, 'deployed' four people into tiny pieces. And everyone was pissed.

Finally, the police saw we really did need water to clean off the black stuff before going back to the streets of Komatipoort. While we stripped off our uniforms to change back into jeans and shirts, they found two battered old galvanised tin baths from somewhere, fetched water from a nearby stream and we did our best to wash off the black. But cam cream is impossible stuff. Glazed with beer and not having any soap, no amount of scrubbing helped. We cleaned the easy-to-get-at bits, but we still looked like a troupe of the Black and White Minstrels or eight black and white stand-ins for Rudolph Valentino, with dark mascara rings round our eyes, up our nostrils and stuck in the crevices of our ears.

This seemed to satisfy our SAP friends, but I cannot believe that the long-suffering staff at the Hotel Continental in Pretoria were duped when we rolled up later that day.

The hotel staff watched impassively as eight of us stamped through the foyer, still reeking of sweat, our ears and eyelashes streaked with black, as rough a crowd as any, with bent noses, cropped hair, bulging muscles and big Jim McGuire's arm covered with a huge peacock tattoo. "All right, mate? How's it going?" we called out cheerfully and breezed into the lift.

"These are tourists," announced our policeman friend.

The hotel staff smiled blandly, as before, and checked us in. Such is South Africa. I expect the SAP used the hotel a lot, so they must have known we were up to something. Anyway, they never blinked when we ordered a large number of bottles of real Scotch whisky, which, like the sweets (which we ordered too) could not be had for any money in Rhodesia. We showered, which was glorious, and ate ourselves stupid on a huge spread of steak, lobster, crayfish and wine, which would have fed the population of Swaziland for a week. The SAP picked up the tab, which for all eight was enormous, we passed through the customs into Rhodesia, all deniable and secret again (and so did our drink), and finally reported back to Mr Mac. He was more cheerful than ever, especially when I gave him a bottle of real Scotch. Rhodesians never liked to admit it, but the proud Rhodesian 'sanctions imitation' of whisky was, frankly, disgusting.

"Call the chaps into my office," said Mac McGuinness, grinning broadly. When we were all gathered in front of him, he said, "You've done well. Mission accomplished and no one any the wiser. The South Africans are very pleased and wanted me to give you this."

With that, he reached into a briefcase on the floor and pulled out wads and wads of money. The highest denomination in Rhodesia was $10 and he counted out $1,000 for each of us. He said, "Look on it as a bonus. But if you'd rather have it in South African rand, then just say."

At once, eight packets of money dropped back on his desk. He grinned widely. At that time in Rhodesia, no one needed telling that Rhodesian money was worthless abroad, even had we been allowed to take it out of the country. He reached into his briefcase again and gave us each 1,200 rand instead. I opened a building society account in the South Africa Bank, and went to see Jane for a couple of days' leave.

These bonuses were a common feature of Special Branch life. I doubt it was widely known but there was an incentive scheme, call it a production bonus if you like, for killing terrorists. Quite simply, if you were out on patrol or ambush and killed up to five guerrillas, you got a $50 bonus, whereas if you killed six or more you got $100. I imagine most people will be rather shocked by this, except possibly my old blue-rinse lady in Salisbury who wanted me to 'get one for her' every time I went on patrol.

Now, the human mind is both devious and never satisfied. The incentive scheme, like most others, had loopholes. A liaison officer called Bob Perch saw a way to cash in. On ambush, he took no prisoners at all and killed as many as possible. Let us, for example, say two. On his return to Bindura, he only declared one, for which he was given $50. The other one he stuffed in a deep freeze, which he acquired specially and kept in his garage. The next day, he claimed another successful ambush, and defrosted and handed in one dead terrorist, at $50, thereby gleaning $100 instead of $50.

I can hear your outraged voice, "This is insane!" You would be right. Bob Perch went berserk one day and threw a hand grenade at the police.

Communist terrorists were also criminally commercial. As the war worsened, so did the poaching of animals for profit. The poachers were black terrorists 'on leave', making some extra pocket money on the side, or 'genuine' black poachers who joined ZIPRA or ZANLA to be on the safe side. This, incidentally, is no different to the IRA protection rackets in Northern Ireland, for example, taking a levy from the black taxis or the builders working with Belfast City Council grant schemes, which makes them no different from Al Capone.

Special Branch produced information about a group of poachers in the south-west. The SB source was a very thin, 60-year-old black man who said the poacher-cum-terrorist lived just inside Botswana on the edge of what was known as the Tuli Circle. The Tuli Circle was a bulge in the border line which arose from the historical arrangements made between the first white settlers and Chief Lobengula, who was the last Matabele king. Lobengula agreed to let the settlers build Fort Tuli and take land all round in a circle with a 25mile radius. Later, when borders were fixed between Bechuanaland (as Botswana was then known) and Southern Rhodesia, the circle remained and provided the poacher with a salient into Rhodesia, where the circle came back to a straight line near the Shashi River.

The moon was up and full when we set off through the trees and scrub towards the border with our elderly black 'source', but soon the sky darkened with clouds and it started to rain. I found a place to cross the Shashi River but the water was up and we had to force our way over, chest deep, holding our weapons high above our heads. Inside Botswana, the rain poured down as we sloshed on through the soaking trees towards our target. We were still some way off the poacher's hide in a cluster of village huts, making slow progress in the pouring rain and darkness, when the old black man in front of me disappeared down a big hole. We dragged him out but he was howling with pain. Then he sat down and flatly refused to go on, forward or back. No amount of cursing helped. I had to abort the patrol as he was the only person who knew where the ZIPRA poacher was living and we had to carry him back to Rhodesia.

The rain never let up and we were pretty tired by the time we reached the Shashi River. By this time, the water was a black raging torrent swollen by the downpour of rain. I cast up and down the bank for the least frightening place to cross and we plunged into the water. This was the last straw for the old man, who began flinging his skinny arms and legs about in terror. Like a good many Africans, he could not swim. We were soon exhausted, washing about in the darkness, fighting to hang on to him and trying not to drown ourselves, till we noticed the old man's fearful thrashing kept him afloat to such a degree that we just gave up struggling, flopped all over him and used him as a flotation raft!

Amazingly, the old fella was persuaded to go back and they got their man. Steve Cleary took the patrol, as I was training Auxiliaries elsewhere at the time. I only mention it because they were rather pleased to be carrying new Romanian AMDs for the first time. These were Russian AKs cut right down, and they certainly looked the part. The wire shoulder butt folded round, the barrel was reduced in length almost to the gas port and a huge muzzle brake – like something off a tank – was fitted on the end. However, when the shooting started in the dark, Steve opened up with his smart new AMD and was immediately blinded by the muzzle brake, which blazed like a flame-thrower and sounded like a huge 12.7mm heavy machine gun. Not good kit for covert work!

Back in Bindura, I soon had cause to involve Jane more closely in what I was doing. We had married, on January 26, 1979, and were living in a villa near the fort. Like all civilians, Jane saw the security situation worsening, but as a nurse she saw more than her share. There were an increasing number of victims of road ambushes brought to the hospital, on one occasion a bus full of blacks had been shot up, and she had lost friends on the civilian Air Rhodesia Viscount which had been shot down with a SAM-7 by ZIPRA terrorists. Her friend, a nurse who had trained with her, was one of those who survived the crash only to be murdered by other ZIPRA who found them injured on the ground. It is perhaps not surprising that she smoked a lot and then got bad-tempered about once a month trying to stop!

I set about training a group of one hundred UANC men on the outskirts of Bindura in some farm buildings at a place called Retreat Farm. The house and a barn were used as barrack rooms and another barn was turned into a cookhouse. The training was basic infantry work with a counter-insurgency lean to it – covering subjects such as individual weapon handling and battlecraft, section battle drills, fire and movement, radio procedure, immediate action drills and contact drills. We fired on ranges nearby and practised every technique we needed to operate against the ZANLA terrorists who lurked in or near most villages by then. I found the Auxiliaries keen and enthusiastic, and their leadership intelligent, though communism had left a commissar-like attitude in their senior ranks. Lieutenant Sam was shadowed by a political advisor.

When the training was finished, I was told to take my band of Auxiliaries to guard Keep Three, which was one of the new protected villages about three-quarters of an hour's drive from Bindura, near Mzura. Rhodesian Keeps were an attempt to copy a plan which had been very successful in the Malayan Emergency, when over 600,000 Chinese squatters were given freehold land of their own and new villages to live in. This encouraged them to support the government and provided the security forces with the added advantage of being able to separate the civilians from the terrorists. However, in Malaya, the British were able to staff, fund and man the Keeps so that the concept worked. In Rhodesia, the Internal Affairs Office had responsibility for Keeps and the plan suffered from a lack of all these crucial elements.

I was not at all keen to go to Keep Three, or any other, but SB officer Brian Pym explained to me that, because Special Branch had overall authority for 'tame terrs', we might be able to win back the 'hearts and minds' of the blacks in the protected village and develop some sources of intelligence among the terrorists in the bush around. If I was reluctant to start with, when I drove down the dirt track through savannah woods and grass to the Keep, I was totally appalled. I found 1,800 black men, women and children squatting about in utter squalor in a ring-fenced village of mud huts with rotting thatched roofs. The Africans just sat about in the dirt, with not a sign of a man between the ages of sixteen and sixty. They seemed beyond the stage of being able or willing to help themselves, and totally at the mercy of ZANLA terrorists roaming the bush around.

Keep Three was in a worse condition than San Antonio di Zaire in Angola had been, and some of the awful scenes of starvation seen on television are reminiscent of it. These people certainly gave up without leadership but, given the political decay in Rhodesia in 1979, they were well down the list of the regime's priorities and I wonder if some whites would be much different elsewhere under similar pressures.

The Internal Affairs Office supervisor never turned up, so I set about organising the place myself. I could see it had been well set up originally, so there was hope. First, I deployed my Auxiliaries on patrols into the surrounding bush to drive off the ZANLA guerrillas and prove to the people inside the Keep that we had the upper hand. Then I turned to the Keep itself.

In order of importance, I discovered six youngsters with Kwashiorkor, the terrible condition affecting small children suffering from severe nutritional protein deficiency, which gives them dry skin, de-pigmented hair and the awful swollen bellies which have become the hallmark of African famine. I drove them straight to see Jane in the Bindura hospital for blacks and she presented them to the doctor immediately. Two were beyond help and died, but the others were saved.

Next, I realised that my own medical skills were quite inadequate to deal with a problem of this magnitude (remember the goat?). So, I appealed to Jane again. She agreed to come at once and, having qualified as a nurse with an 'A' grade in Community Health, she was a great help with her advice.

Plainly we needed a medical centre and supplies. I went to the army, posed the problem and asked for medicines. The quartermaster in charge of the stores said, "It's no good, Sergeant McAleese. You're only allowed one trummel of kit. Standard Platoon Medic Kit. One box only. Those are the rules."

"Fine," I replied, saluting smartly, and went straight round to his stores, where I looted as many boxes as would fit in my truck. I took the best part of twenty boxes I found lying about there. One Platoon Medic Kit was sufficient for only forty people and I had 1,800 desperate people in the Keep.

Jane emphasised that the blacks needed protein. She rang up

a local company called pro-Nutro which made a milk protein supplement for children, and asked for all their rejects, such as the punctured bags which could not be sold commercially. They offered them at cost. I persuaded the Special Branch to cough up $600 from their seemingly bottomless account and we drove off with a truckload.

At the same time, I set up a crèche for the children in the Keep. This probably sounds rather charming, but there was a good reason. In Africa, the women feed the men first, before the children, so when, as here, there was little food, the kids went hungry and got sick. So, I fed the kids at the crèche, telling the older children – who were not more than 15 years old – to make sure they fed the small ones. It was really very satisfying to see the way they perked up at once and began behaving like children again.

I got more enthusiastic than ever as things improved, and looked round for further ideas. Fortunately, Bindura was an important citrus-growing area, with acres of orange trees, and I had noticed a lot of ground-fall oranges in a place called Mozoe. For only $200 this time, we picked up enough to last a full week.

Next, I called a meeting of the headmen on the verandah of the wood house where I lived with the other SB men, and we discussed the feeding arrangements. The Keep was a protected village and everyone had to eat in one fenced-off area, about the size of a football pitch. The idea was to control the food so that they could not give it to the terrorists or, if ZANLA threatened them, they had an excuse for not being able to supply food. However, the place was a serious health hazard. All 1,800 people had their meals there, among a jumble of filthy, stinking grass huts and shelters. There was no running water, no sanitation and no cleansing. I persuaded a doctor called Sandy Kirk, who had been my best man at our wedding, to condemn the place and – quite by chance – one night the feeding area caught fire and was burned to the ground.

"What the hell happened to the feeding area?" demanded a young white man who came rushing down from the Internal

Affairs Office as soon as they heard it had burned down. I had seen neither hide nor hair of any of them till then.

I watched him strutting up and down with his arms crossed over his chest in my little office in the wood house, and my blood began to boil. I asked him, " D'you want to know the truth?"

He nodded arrogantly.

I lost my temper and snapped, "I fucking burned it down! And I'll tell you something else. There are people here suffering hardship because pricks like you won't do your fucking job and run the Keep!"

Seeing the expression on my face, he backed off rapidly through the door and vanished across the verandah, shouting, "I'm going to report you to the District Commissioner, see if I don't!"

I did not see him, or anyone else from the Internal Affairs Office, again.

We rebuilt the eating area. By various means, I obtained sand and cement and we made concrete washing places, with gradients and drains. We piped water in and kept the place clean.

Now that the basics had been fixed, I turned to other ideas. I obtained some white paint and had my Auxiliaries paint the houses in the Keep. This spruced the place up, but again there was another reason. When our patrols were out at night, they could see who was moving about in the Keep much more easily against the white walls of the houses than against the dark wood.

I organised daily football matches to give the Auxiliaries and young lads in the Keep something to do. I got blue and red football strips from Bindura and they played every afternoon.

In the mornings, I made the villagers sweep the ground round the Keep with a patrol of Auxiliaries. This kept them busy, kept the place clean, gave them a sense of pride in their village, and my black Auxs could report if they saw any tracks of people creeping in and out of the Keep at night. These UANC Auxs had been doing their stuff among the people, persuading them to give up support of ZANLA and that we were on their side. This had the desired effect. The headmen approached me, their confidence buoyed by all that was going on, and said they had heard that some ZANLA hiding in the bush wanted to talk. This was exactly what the SB had wanted and all the work seemed to be paying off. I knew that the local ZANLA leader, Teddy Natsango, was away in Mozambique for resupplies, so I gave the headmen a quantity of cigarettes to pass on to the guerrillas and waited for developments. Maybe some of Natsango's less committed terrorists would come in while he was away.

Some days later, a black headman came to my office in the wood house and announced that there were four ZANLA men at the back gate waiting to talk to me.

"Great!" shouted two SB officers, who by sheer bad luck happened to be with me at the time. Ignoring my pleas not to get involved because they were not known by the local people, they ran over the verandah and charged through the Keep to the back gate. Word travels faster than the average SB officer, so the four ZANLA were alerted in time and they legged it back into the bush before I could intervene.

There were no more contacts by ZANLA and I wondered if the people trusted me after that. However, we carried on. All the import ant things had been done in the Keep and now the villagers wanted a church. We made a wood frame structure, complete with a spire, and covered it with long lengths of hessian (courtesy of the quartermaster's stores again) which I made them dip in a mixture of sand and cement, giving the building a wattle-and-daub finish which dried off hard. We painted it white. Our church was no bigger than a large room, but everyone, especially me, was very pleased with it. I can also say it was a Catholic church. Roll on Father Brett.

I knew we had won over the villagers when the men came to me and said that ZANLA had laid a mine on the dirt track leading to the Keep, just for me. I went to have a look and skirted the place every time I drove in and out until a suitable period had elapsed for me to have the engineers lift it without ZANLA blaming the villagers for telling me about it.

The terrorists did not give up easily. They tried again and this time their approach was more devious. Again, the villagers let me know. One morning, they called me over to one of the gates of the Keep. ZANLA had done a beautiful job! They had taken a Chinese POMZ-2 stick grenade and buried it in the ground just under the gate, leaving a fragment of the top visible. This was the fuse, which they had attached with a string to the bottom of the gate. Open the gate, pull the fuse, and bang. It all looked very straightforward, but it was a little bit obvious and I was instantly suspicious.

Sending the others back to a safe distance, I circled it and had a good look. Then I approached the bomb, knelt down and began brushing away the earth around it with my fingers. I worked very carefully indeed, because these Chinese fuses only need a 2-pound weight to set them off. Soon I revealed some of the pineapple body of the grenade and saw that the fuse was a MUV-2 with a three-second delay. Still suspicious, I worked deeper and saw the real danger. Underneath the first grenade was a second. This had an instant pull-type MUV fuse tied to the first grenade. If you fell for their booby trap and lifted the first grenade thinking that was all there was to it, you set off the second beneath it and died. I disconnected the second grenade, took them both away and blew them.

Shortly afterwards, just before Christmas 1979, I was called away from Keep Three to train a new bunch of Auxiliaries at Retreat Farm again. I arrived, formed them up on parade in ranks and I can honestly say I have never seen such a motley collection in my life. Every second man had a pair of mirror sunglasses and the British sergeant major who talks about 'mixed orders of dress' has never seen anything like this! One man wore a black tank suit held in place with bright yellow socks up to his knees, another had a purple shirt and a yellow cravat, and there were jeans, T-shirts and hats of every colour and shape. Their weapons were as varied. I found out the SB had rounded every unemployed freeloader and 'tame terr' in every barracks and camp in the region and sent them to Retreat Farm for me.

"This," I said to myself, "is not on!"

I raided the stores in Bindura again for shorts and khaki T-shirts, paraded this motley crew again and stripped the lot naked. I issued them with two pairs of shorts and two T-shirts each. A Greek called Maggie Christo ran a shop like Aladdin's cave from which he supplied the Special Branch with all manner of odd things. He supplied me with boots. I persuaded Mr Mac to use more SB cash to pay for some very smart camouflage uniforms of my own design to be made up by a local tailor, and soon the men began to look the part.

At my suggestion, Mr Mac obtained some very impressive helmets I had noticed being worn by the police who guarded the nickel mines near Bindura. They looked similar to helmets worn by German troops in the Second World War. I gave them to the largest of my recruits, some of whom were huge men, and they looked splendid in these helmets and new uniforms. I made them my military police section to mete out discipline to the rest as necessary.

The next two months were spent training and they shaped up well. By the time we had finished, I had fifty keen black soldiers ready to start operations. We were to be used to fight ZANLA in the area, to take the pressure off the regular battalions of RLI and RAR in the Fire Force.

I found that discipline round the country was beginning to collapse. Of course, regiments like the SAS, the RLI and the RAR remained solid until the very end, but elsewhere the command and control of operations had begun to fall apart and I certainly found this the case in the places I was deployed with my new Auxiliaries. The terrorists were everywhere and total anarchy was galloping over the hill.

I tried to make sure we were deployed on operations based on good intelligence from professional Special Branch officers, like Mac McGuinness, but increasingly this was not possible. One SB man called Steve sent us to ambush a crossroads in the Matepatepa area, which he was certain was used by terrorists. We approached the crossroads on foot through the scrub in darkness and pouring rain, and I just didn't feel right about it. I changed course and stopped at a local police station. I asked the station sergeant on duty, "Anything going on in the area tonight?"

"Yeah," he replied, waving his hand at his operations map. "Guard Force are ambushing the Matepatepa crossroads tonight."

I was furious. If we had carried on as planned, we would have walked straight into the Guard Force ambush. I turned my black Auxs round and we walked back to base, where I left them to clean up. I set off at once to find Steve. He was, naturally, in his billet stretched out in his bed. I burst in, dragged him off the mattress and demanded an explanation. He apologised. He promised to 'get it right the next time'.

He did not. On a Thursday evening, I was sent with a reaction force of twenty-five of my Auxs to observe an area near Keep Three over the weekend, with a view to attacking ZANLA if we saw them, which was highly likely at that time. Just after lunch on Friday, the comms went dead on my radio. Lying hidden in bushes with no radio, no support and no casevac, I felt this was no way to treat the black Auxs. I left the radio with the others and took one man with me. I washed up in a stream to remove the black cam cream, or as much of it as possible, and we hitched back to Bindura. I should imagine that the driver was terrified when he saw us emerge from the bush. In Bindura, I marched into the ops room only to be told that Steve had pissed off for the weekend. And it was a long weekend.

This was no way to treat men in the field. Without hesitation, I took two trucks and brought the Auxs back in. Typically, when Steve got back after his weekend, he just shrugged.

I should add that by this time, Keep Three had once again degenerated into total chaos. I regretted it, but I suppose I always realised this was inevitable. Rhodesia was a country in its death throes.

I teamed up with an ex-RLI officer called Mike Webb, who was on contract to work with the Auxiliaries. He suggested we use my Auxiliaries in another area, where they came from, because they might know where to look for terrorists and to show a good example. So, we moved across Mashonaland to a small town called Mudzi, north of Mutoko, and there things were really bad.

Mudzi was smaller than Bindura, one main street and the bare necessities, and a great deal more rural. Around, the savannah was characterised by dongas, which were small hills standing out of the landscape, and the area had become virtually a liberated area in Mao Tse-tung terminology. ZANLA did not actually control the region, but whenever a truck left base, the young black boys called Mugibas, who kept watch, passed the word and within minutes a flare would shoot into the sky alerting every terrorist for miles around.

Understandably, morale was rock bottom in Mudzi base. The local SB were on their own, without the discipline of the regimental system, unlike the RLI or SAS, and matters were beyond recall. One ground coverage officer called Trevor, dressed in the usual SB fawn shirt and shorts with Veldskoen shoes, used to sit in his office and play 'darts' by firing his 9mm pistol at the dartboard on the wall opposite.

We did a lot of patrols in this area, characterised by long night approach marches on foot, to avoid being seen by the local Mugibas. We went deep into enemy country, where we manned observation positions to spot the enemy. Our first operation went well. At night, I took ten Awes and we marched across country into a range of hills. By dawn, we were hidden in scrub overlooking a group of huts. As the sun warmed us in our bushes, we watched a black woman go about her daily chores going down to the stream to wash and mix meal. One of my black Auxiliaries nudged me and whispered, "Boss, there are terrorists in the huts."

"How d'you know?"

"Either she's got an awful big family or she's feeding terrs."

"Why?" I had noticed nothing unusual. To me, we had spent the day watching a typical African village scene, complete with chickens and dogs lying in the dust round the huts.

"She's been four times to the stream. That's not usual. She must be making extra food for the enemy."

This was a glimpse of how the Selous Scouts worked, knowing the ground and the people so well they could read every action.

I considered the options. It would take too long to call for troops and the enemy might hear the planes or helicopters coming and run off. I decided to attack. I whispered quick orders, we simply broke cover and swept down onto the village.

We captured two ZANLA terrorists and returned to Mudzi, where they were interrogated by Trevor. His technique was not clever.

He pointed at a place on the wall map and said, "The terrorists with you came from here, didn't they?"

"Yes," said the black man, eyeing a length of rubber hose Trevor was hefting in his other hand. Trevor hit him with it.

He said, "And were they carrying rifles with straight magazines or curved magazines?"

"Curved ones."

Trevor hit him again. "So, they were carrying AKs?"

"Yes," replied the desperate black man, knowing that whatever he said, even agreeing with everything suggested by this mad white man, he would be hit with the rubber hose.

Trevor nodded and hit him again.

At this, Mike Webb stepped in and took over. Mike's technique was more subtle and the black ZANLA terrorist was so relieved he told us the location of a camp which was occupied by ZANLA. One of my Awes confirmed that he had heard the place was used by ZANLA.

The camp was 30 kilometres away and the only possible approach was on foot. It was a long, sweaty night march in full patrol gear and a light pack. We skirted villages and huts but we had to cross a bridge, which I found manned by a small patrol of black Guard Force soldiers. We walked up silently and caught them unawares. They were lying loose-limbed on the earth floor of a small hut by the bridge, smoking dagga and totally spaced out of their minds. "You're smoking dope?" I asked incredulously.

"Er, yeah," replied the black NCO in charge, the whites of his eyes rolling vacantly in the light of my torch. The germ of an excuse occurred to him. "We're, er, testing it. To see if it can be used as evidence." He grinned with embarrassment but made no attempt to hide the great pile of weed on the floor at his feet.

I shrugged. That was the way of things.

We crossed the bridge and continued our long march through the night. Shortly before dawn, Mike Webb disappeared into the shadows with a group of six Auxs to act as my stop groups behind the camp and, as light brightened the sky, I started the sweep with the rest. We advanced methodically but cautiously through the dry scrub, grass and trees towards the camp. It was deserted. We searched the make shift huts and found little hides of clothes and personal belongings, but no ammunition. I guessed the place had been recently occupied and we all agreed that we had missed them by less than two days.

We continued to search the area around and came across some huts nearby, where we flushed out several young men who were ZANLA lookouts or Mugibas. The first young black walked out of a dusty yellow hut and looked up in surprise at the line of soldiers advancing on him through the dried grass and bushes. He was no more than 40 or 50 metres from us, we had all seen him, and yet to our great amazement he casually pulled a Chinese stick grenade from his trouser pocket, unscrewed it and tossed it at us, thoughtlessly, as if he were experimenting with it. It was the last thing he did. The grenade had hardly burst before the whole sweep line opened up on this young idiot and literally shot him apart. Mike Webb captured two others with more sense, trying to escape through the stop groups, and we returned to base.

These ops were not much different to the endless patrolling on minimal intelligence which the British SAS did in Borneo or Northern Ireland in the early days. By 1980, the extent of guerrilla infiltration into Rhodesia was such that the structure of the counter insurgency operation was falling apart at the seams.

There were some excellent professional Special Branch men but there were also men without the necessary training and police background who could not cope under such pressure. Without adequate supervision, their wildest schemes took shape and added to the increasing atmosphere of anarchy and unreality.

Mudzi suffered particularly and Trevor was not the only lunatic. An Internal Affairs officer, who was a tiffy, a mechanic, spoke Mashona rather well and decided on his own initiative to infiltrate the local people disguised as a black man. His plan was to pick up gossip and information about the terrorists. His problem was that his features did not lend themselves at all to this plan. However, ignoring all criticism, he blacked his face and arms and legs very thoroughly and set out on his own to join the villagers. He took a bus and sat down. The blacks looked at him very strangely but no one said anything. The bus rumbled on. A black policeman on duty in uniform got on at a stop, glanced around and walked up to the Internal Affairs officer.

"Hey, you're white! What the hell you doing on this bus?" he demanded in English. "This bus is for blacks!"

"I'm a black man," insisted the Internal Affairs officer in Mashona.

"No, you ain't," said the policeman, pointing at the man's head. "You got red hair."

"I'm telling you, I'm black," said the Internal Affairs man, sure of his disguise.

Exasperated, the black policeman replied, "Then, if you're a black, I'll treat you as a black." And he gave the Internal Affairs officer a sound beating, threw him off the bus and sent him back to Mudzi on foot.

I might mention how this lunatic ended up, as it illustrates just how seriously the system of law and order and democratic values was threatened at that time. He had a friend nicknamed 'Steptoe' and these two Internal Affairs officers went to Salisbury for a day's leave. They got completely drunk, emerged from the bar, climbed into their truck – which had two GPMGs mounted on the rear – and drove back to Mudzi machine-gunning blacks on the road as they passed. They were given ten and twelve years. On reflection, I suppose it is possible that their imagined experiences with Special Branch undercover work at Mudzi made them believe they could do what they liked, and made them think they were above the law.

Rhodesia's political situation worsened. The world refused to recognise the new black majority under Bishop Muzorewa. Robert Mugabe's ZANLA and Joshua Nkomo's ZIPRA were too powerful and could not be stopped militarily. At Mrs Thatcher's first Commonwealth Conference in August 1979, she agreed with Australia, Jamaica, Nigeria, Tanzania and Zambia that Nkomo and Mugabe had to be a part of any settlement. She agreed to set up talks leading to 'genuine' black majority rule and new elections. For their part, the terrorist leaders Nkomo and Mugabe were persuaded by other black African leaders (like Kenneth Kaunda of Zambia) to fall in line with these proposals.

The black Front Line states were not keen on the successful Rhodesian cross-border operations, which were becoming increasingly ambitious and seriously threatened their countries' vital infrastructure, such as bridges, railways and oil installations. In March 1979, the SAS had attacked the Beira oil depot on the coast with considerable success. New Rhodesian elections were eventually set for March 1980.

During this period, I was moved about with my Auxiliaries from place to place, and everywhere the security situation was grim. Without coordinated intelligence, we tended to mount operations on the scantiest of information and ended up going out hoping to stumble on some terrorists by chance. They were certainly there, in droves, but the tribal villages were increasingly under the control of ZANLA terrorists and we constantly ran up against the Mugiba lookout system. This was particularly well developed round Mutoko, east of Bindura, where the villagers sent out their children to clear tracks of our ambushes by walking parallel to the tracks, in the bushes. Of course, they walked straight into us as we lay back in cover watching the track, and at once set up a terrible caterwauling, "The soldiers are here! The soldiers are here!" Within minutes, we could hear the message being picked up and shouted across the valley. Nothing for it, we packed up and went back to base.

There was madness in Mutoko too. One day, a couple of Special Branch men there called me to one side and said quietly, "We've got some gooks here. They're real bad bastards. Murderers. Take 'em out somewhere quiet, Peter, and deploy them, will you?"

Here we go again, I thought. *Why is it people always think I am a pathological murderer? I like fighting, I like soldiering, but I am not the sort who just knocks people off for the fun of it.* Bluntly, I told them, "If you want to kill them, do it yourself."

They acted insulted but tried me again. "Listen, Peter, these bastards killed a couple of men in the RLI!"

"Then get the RLI to do it," I said, and walked out. Rhodesia was falling apart.

Bindura was the same. I took my good friend Dr Sandy Kirk out to the range one day to show him how explosives worked. He used to help me with medical training and supplies, and in return I satisfied his curiosity about demolitions. The place for dems practise was at the back of the gallery range butts, in a secluded wooded area where the explosions would cause no harm. One day, I set out several charges for him on the dry ground, linked them in a ring main – a circle of detonating cord so they all go off at the same time – and set them off with a length of safety fuse. Sandy and I retired to one side of the butts and, after the explosion, came back to see what had happened. To our horror, a hand was sticking out of the ground where I had set one charge. Needless to say, the hand was black. The blast had blown away the earth, and beneath was the rest of the body in a shallow grave; one among several casually buried behind the butts.

No one was too surprised when I reported this to the Special Branch office at the police barracks. "Oh, sounds like Phil Young has been up to his tricks again." I would very much like to give his real name, but he was never charged. Young was obsessive about slaughtering captured black terrorists. He had simply taken them off to the butts and shot them. Nothing was done about it.

Young's enthusiasm for killing the enemy led to him making a Q car. In theory, this was a reasonable response to the constant threat of ambush on the roads outside town, but the elements of overkill and absurdity were typical of the times.

The Rhodesians developed a variety of such vehicles, with armoured sloping sides to deflect the blast of a mine and soft tyres to give minimum low ground pressure (remember the Russian TM mines which set off with only 26 pounds of pressure). They filled the tyres with water to absorb the energy of the explosion. Young's idea was not merely to survive but to give the black terrorists a blasting as well and, typically, he overdid it. He took a police Land Rover, kept its canvas back so it looked perfectly normal from the outside, and fitted bulletproof armour inside. There, he concealed twin machine guns mounted to fire on automatic, several grenade-lobbing devices and twin cannons under the bonnet to fire out the front. The whole device worked electrically, at the flick of a switch near the driver's seat. Young used this Land Rover as his private vehicle and one day parked it in the Bindura Country Club car park in the centre of town and slipped inside for a beer.

In the back of the Land Rover was a can of petrol he had managed to acquire (petrol was scarce) but he had foolishly left it sitting on top of the batteries which powered his Q car arsenal. After a while in the heat, the metal can short-circuited across the terminals. Suddenly, the unmanned Q car exploded into action, shattering the peace of Bindura. The top flew open, the machine guns rose majestically through the roof firing on automatic, the twin cannons roared at the front and grenades sailed out in all directions, exploding all over the place. The car park was turned into a shambles, and cars were riddled with shrapnel and shot to bits but amazingly, no one was hurt. Everyone thought it was a big joke and forgot all about it.

Africa casts its spell on white and black alike. Or, maybe in the death struggle of cultures and politics there is always space between the great battles to hide a multitude of terrible sins.

The annual Bindura ball suffered too. All the whites who were anything in Bindura went. There were 1,200 whites, so numbers dictated the ball had to be in the Bindura school hall. Bindura etiquette demanded black tie and long dresses, while the security situation obliged the men to arrive tooled up with Kalashnikovs and FNs. Ben van Schalkwyk, the hotelier who ran the Bindura Coach House Hotel, arrived in a natty tuxedo and dickie bow tie, toting an FN rifle. The festivities started at a pace, huge volumes of alcohol were consumed – temporarily driving away Rhodesia's grim reality – and the dance floor heaved to the sound of tunes like Abba's "Money, Money, Money" and Ivor Novello and Lena Guilbert Ford's "Keep the Home Fires Burning (Till the Boys Come Home)". Even the poor woman nicknamed the 'Kiss of Death' inveigled some new men onto the floor. The name arose because she had been married to three men, all of whom had been killed in the war, and word was that she was looking for a fourth. Sober men avoided her like the plague, but at the ball that year nothing seemed to matter any longer.

Of course, the opportunity was too good to miss and out in the darkness on the edge of town, a couple of ZANLA fired off their magazines in the general direction of the school. At once, pandemonium broke out at the ball. The men left their women standing on the dance floor, grabbed their weapons and poured out of the brightly lit school hall into the darkness.

"They're over there!" shouted one.

"No, they're fucking not. I saw tracer that way!"

Within seconds, all these drunks in dinner jackets were pouring out a fantastic weight of fire all over the place and, because most carried Soviet weapons like the AK and RPD, a tangle of green tracer patterned the darkness over Bindura.

Elsewhere in the town, the Guard Force heard the racket and tumbled out of their barracks. They were not top-line troops, but one look at the green tracer flashing over Bindura convinced them the town was under heavy attack on every side. Their commanders were grimly determined Bindura should not be the first town to fall to the hated ZANLA, and immediately despatched fighting patrols to defend the place at all costs. Cautiously, the patrols advanced through the dark streets towards the unseen but obviously very powerful enemy, using the American technique of reconnaissance by fire, shooting at everything which made them jump.

At the ball, the increased noise of firing convinced the revellers that matters were almost out of hand. Bow ties were loosened in earnest and someone went for more ammunition. In the garden, a terrified dog skulked under a bush while inside the school hall, an old lady was slumped drunk in an armchair singing loudly at the top of her voice, "Land of hope and glory, Mother of the free …"

The battle went on for hours.

In the morning, when the partygoers were sleeping off their hangovers, a police patrol found the little pile of AK empty cases which had started it all off. ZANLA, if they had a sense of humour, must have laughed till they cried. Bindura looked as though it had been attacked by stormtroopers, with bullet strike on walls, smashed windows, shattered doors and riddled cars, but once again, with the luck of the devil, no one was hurt.

The Bindura school achieved prominence a second time, during the March elections. The British had rushed through their new plans for Rhodesia and soon Bindura was visited by British Army officers for the first time in years, and a British bobby, complete with peaked helmet. These were men of the 1,300-strong Commonwealth Monitoring Force who were charged with supervising the elections to ensure fair play, British style. I must say they looked very weedy, white and British compared to the bronzed, muscled bush fighters of Bindura.

Robert Mugabe had no intention of allowing this golden opportunity to slip away. Here was his chance to be democratically elected in the eyes of the world. There had been a sort of ceasefire since December 28 and in the new year his ZANLA (now ex-terrorists) had been forming up in camps supervised by the Monitoring Force. Mugabe instructed them to make absolutely certain that all the blacks voted for him in the areas they had infiltrated, which by then was all northern, eastern and southern Rhodesia, no matter what techniques were necessary. ZIPRA did their best to do the same in the western areas.

The most appalling stories of intimidation emerged from the bush. Lord Soames, who was the interim governor, was faced with cancelling the elections. He let them go on. Later, he said, "I believed that Robert Mugabe was going to win anyway, and you must remember this is Africa … They behave differently. They think nothing of sticking tent poles up each other's whatnot and doing filthy beastly things to each other. It's a very wild thing, an election."

Amazing as it seems in retrospect, I am told that many Rhodesians did not believe that the communists, Mugabe or Nkomo, could win. Somehow, they believed a Seventh Cavalry would come charging over the horizon to save them and maintain Bishop Muzorewa in power. ComOps and Army HQ certainly maintained this bubble till the bitter end, to the chagrin of the security forces (particularly the SAS, who were actually in position in Salisbury waiting for the code word to assassinate Mugabe).

Of course, the Rhodesians did their best to fix matters too. Don't go thinking the blacks were the only ones busy rigging the election. In Bindura, the Special Branch did their bit. On March 3, the voting took place in the school and a Special Branch group took over a house. Acting in the greatest secrecy, they barricaded themselves in, covered all the windows with thick mattresses – on the pretext that they wanted to prevent anyone using powerful directional microphones on the glass panes to hear what was being said inside – and methodically set about fiddling the result.

A German observer of the elections said, "If this election is fair and free, then I am a Chinaman." He was right. While the famous British bobbies stoically sat out in the heat watching the blacks and whites sliding little pieces of paper in the slot at the top of the voting boxes, the Bindura Special Branch were hard at work in the gym unscrewing the bottom of these boxes to get at the papers. No one seemed to have noticed that all the boxes were wood and the bottoms held in place with only four screws.

The calm of voting day in Bindura was disrupted only once when a ZANLA-inspired black began pointing at the house windows and shouting that the whites were rigging the votes. He was quickly suppressed by the Rhodesian police and dragged away. No one else was allowed near the house and the SB sweated on through the night, but the result was beyond their control.

At 9 AM on Tuesday, March 4, the Marxist revolutionary Robert Gabriel Mugabe was declared the winner, with fifty-seven out of the one hundred seats taken by his ZANU and ZANLA. Within a year, he had declared his intention of setting up a one-party Marxist state.

The white Rhodesians were shocked to the core.

The witch-hunts and purges started very quickly. Mugabe's ZANU were now the majority party and they started trying to find out who had been giving them a hard time in the bush. The SAS, Selous Scouts and of course the Special Branch were prime targets. Black men seem to be more unpleasant to other black men, and it was no surprise that our Auxiliaries were under pressure at once.

I was in the office in the fort in Bindura and the phone rang.

"I want to quit," said one of my sergeants, called Sam, his voice oddly quiet. He was phoning from Retreat Farm and close questioning revealed that my band of black Auxiliaries were in a state of near anarchy. By then, they consisted of all factions in the war: SEPs (Surrendered Enemy Personnel) from ZANLA and ZIPRA, Sithole's ZANU PF and Muzorewa's UANC. Using their newfound power, Mugabe's ZANU/ZANLA were using the well-tried communist technique of denunciation to terrorise the people and consolidate their position. The existing rank structure and discipline was seriously threatened by three junior ranks who were bullying the rest. Worse, many of the Auxs feared that they would be hounded all the way to their tribal villages even if they left the unit.

The poison of this made me mad with fury when I thought of how all these men had so enthusiastically committed themselves to our training and fought well on some very tough and dangerous operations. They were good men and they had worked damned hard in difficult terrain during extraordinarily difficult political times. Tough measures were called for.

I rang the farm and told Alec Lennox to find some excuse to gather the men on the square, without their weapons. Billy Chambers and I jumped into a Land Rover and drove out to confront the three ringleaders, who were led by an enormous 6 foot 4 black man called Israel. We arrived and Billy, who was himself a big man, stepped briskly over to Israel, who towered over us all with a smug expression on his face. Without warning, Billy shattered a knobkerrie stick on his skull and floored him. I had all three arrested and jailed in the cells.

Here I must confess, like a (not so) good Catholic, I was so angry at the insidious and underhand way that Israel and his two henchmen had bullied and terrified the others that I decided to kill him myself. He had, after all, betrayed the others. At gunpoint, I ordered him into the back of a police Land Rover with a lock-up in the back and drove him out to a secluded place on the Shashi Pass between Bindura and Mount Darwin. I left the road at the top of the steep climb, pulled into some trees and dragged him out of the back of the Land Rover. Crying, pleading and begging me not to kill him, he crawled on his hands and knees towards me, his face streaming with tears. Disgusted, I levelled my Beretta pistol at him and prepared to squeeze the trigger.

A flood of memories came back. Of Old Miles's anger killing Germans in the First World War, of the maniacs killing each other in Angola, and of all the murders I had come across in Rhodesia. In retrospect, perhaps I had been sucked into the unreality of Rhodesia in those days, because I knew there would have been no comeback for killing him. By the grace of God, I hesitated at the last moment. The act of killing the man would have been so easy. Pulling the trigger would have been a simple thing to do of itself, something I had done numerous times on operations, but this was murder, without question, and I did not want murder on my conscience. I lowered the Beretta and ordered the big blubbering bully back in the truck. We drove back to Retreat Farm.

The lesson was not lost on the Auxiliaries for a while, but circumstances were beyond small incidents like that and soon the ZANU comrades fingered another two Auxiliaries called Saul and Eddison. I arranged for them to move out to a small place called Sinoa, but ZANU tracked them down and they had to come back to Bindura. These two were ZANLA SEPs and the triumphant ZANLA wanted to settle old scores with people they reckoned had been traitors to the cause. Later, Saul disappeared but Eddison was captured by ZANLA and tortured to death. Life is all a question of perspectives but the victor calls the tune.

Another black man, called Lazarus, who worked for the Special Branch, was a Matabele and had been trained in Russia for eighteen months before changing sides and working for the Rhodesians. He was caught by ZANLA in their post-election purges and taken outside Bindura to the mines. They tied him down, ripped open his stomach with a knife and filled his intestines with burning coals.

It was the end. I spoke to Mac McGuinness, who agreed to disband the whole group. He gave them three months' pay each and they dispersed very rapidly to hide in their tribal areas.

I lost track of them, but I do know that Mugabe used the new Zimbabwean Army's 5th Brigade, incidentally still with a good many white officers, to eliminate all opposition to his plans for a one-party Marxist state. Primarily, the 5th Brigade worked in Nkomo's areas of support in the Matabele lands, where they destroyed ZAPU and ZIPRA. Estimates of the numbers killed vary between 100,000 and 150,000. There were very few reports of this in the Western press, which was still intellectually blinded by its support of Mugabe's glorious fight against the Rhodesian whites-only regime. One lone voice was the editor of the London *Observer*, Donald Trelford, who did write about the massacres taking place after a visit to Zimbabwe in 1984. On April 15, he published an article entitled "The Agony of a Lost People," but the call was not taken up by other journalists.

Did no one care what happened in Zimbabwe any longer? The war had cost some 30,000 lives, mostly black, and yet, true to African form, more died in the settling of scores after the war was over. The pity is that the blaze of world media attention during the war died away when the Africans got down to the serious and murderous business of staying in power.

The media had switched their attention to South Africa.

I recognised that my time in the country that was now Zimbabwe was finished, but the final straw was hearing my name being broadcast by Radio Maputo. Several young men I had trained had been captured on cross-border operations in Mozambique and obliged to tell all.

"The Forces of the Glorious People's Republic must find Mr Peter!" crowed the Maputo radio station 600 miles away on the coast of Mozambique. "He has been training the enemies of the revolution!" For four days, they plugged their line, including an identifying reference to my Dobermann dog.

"The search goes on for Mr Peter!" they shouted over the air.

The incident focused my thoughts. The political situation was fraught with dangers for people like me, who had not merely fought against the new black government when they were guerrillas in the bush, but had served in special forces. Guerrilla propaganda always paints special units as 'criminal' because these units work on specific intelligence and are more successful than regular infantry units. The Rhodesian special forces group had been spectacularly successful, man for man, though fighting what was ultimately a losing battle, and we could not expect any favourable treatment from Mr Mugabe's new regime, many of whom had been on the receiving end of our operations. What focused my concern was Jane, who was pregnant with our first child. I saw no future for us in the New Rhodesia, now called Zimbabwe. I handed in three months' notice and we departed for South Africa.

CHAPTER 9
HIGH STANDARDS ARE THE ONLY ONES ACCEPTABLE

Joining the South African Army was a no-nonsense affair. I decided that at 37 years of age I was more use to 44 Parachute Brigade than the South African Reconnaissance Commando, their special forces unit, and, in June 1980, I went straight to Hallmark Buildings in Pretoria to sign on. The board examined my British and Rhodesian service record and interviewed me carefully. There was no covert surveillance of my private life, as in Rhodesia, nor the defensive arrogance typical of the British SAS, as in, "You don't know nothing but you'll learn it all with us, lad!" The South Africans are accused of being humourless and intense, but they are committed and straightforward too, which is not a bad start for an army.

Without further ado, they signed me on as Number 80021355 in the South African Defence Force (SADF), and gave me the rank of colour sergeant. I was posted to 44 Parachute Brigade at Murray Hill, 40 kilometres outside Pretoria, and within a month, Colonel Breytenbach promoted me to sergeant major. He wanted me to work as chief instructor for Captain Botes, and our job was to raise and train the first regular Pathfinders and form No. 1 Pathfinder Company. This, I felt, was a job in which I could really use all my knowledge and experience, and I can honestly say I started the most satisfying period of my soldiering career.

Jane and I were given an apartment in the Hotel Pretoria, a grand name for a SADF families hostel, where we had our own flat but ate in a communal dining room. Community is a powerful concept in South Africa.

A truck came by at 7:30 AM every morning to take me to the camp. During the last few months of chaos in Rhodesia, my fitness had slipped and I decided to run along the road to camp ahead of the truck. I made a race of it. Every morning, I tried to increase the distance I ran before the truck caught up with me. The fitter I became, the quicker I pounded along, and my best distance was eventually 17 kilometres.

Murray Hill at the time was nothing more than a tented camp on sloping grassland beside some old farm buildings, with a stream running through the camp from a small reservoir. Colonel Breytenbach, a brother of the South African poet Breyten Breytenbach, was the commanding officer of 44 Parachute Brigade and he energetically set about building up his command to prepare it for the fighting on the Namibian-Angolan front. His technique of cutting across red tape and bending the rules attracted a good deal of criticism elsewhere in the South African Army, but he gave us the backing we needed.

Captain Pete Botes and I were told to start picking men for the first selection course for the Pathfinders. He wanted it based on an SAS selection and said, "Make it hard!"

We rustled up twenty-three people to begin with, from all over the world. There were Italians, Belgians, Germans, Rhodesians – who, like me, had come south after the loss of their own country – Canadians, a British public schoolboy called Roy Kaulback, an American banker who was bored of banking, Hungarians, and a Russian who made the South African Special Branch totally paranoid.

Colonel Breytenbach told us to find an area to train in and we ended up at a place called Mabalique, which was tucked right up in the north-east corner of the country, on the border with Zimbabwe. The camp was remote, reached on a dusty dirt road through tall, thorny trees and tough grass, which grew up to the gate through the fence bordering the Kruger National Park. This was a double wire fence separated by a thick planting of spiky, cactus-like sisal plants, impossible to walk through, and it was said the sinews would even foul up tank tracks.

From there, the vegetation softened down gentle rolling grassy slopes to the winding flat banks of the River Limpopo, "great, grey, green, greasy" and "all set about with fever trees", as Kipling described it. Our camp was on a pimple of high ground about 3 kilometres from the park gate overlooking the river. The nearest town was a small place called Pafuri, in the corner of the Zimbabwean, South African and Mozambique borders, and its only claim to fame was that it had been used on location to film *The Wild Geese*. Further away was a spa town called Tshipise.

Needless to say, the living conditions in our camp were primitive, though the place had been used as a camp before. There were two concrete buildings, the kitchen was made of wood and the rest was tented, so we were constantly improving the place as time went on.

The Pathfinders are the vanguard of any parachute unit, and I really worked hard to make our people the best. Colonel Breytenbach and Captain Botes made sure we had the necessary support and I organised the fullest training I could devise.

Our routine was tough but intensely satisfying. We started the day at 4:30 AM with a cup of tea followed by a 7-kilometre run from the camp down to the Kruger Park gate in the sisal fence and back. I was by now so fit that even at my advanced years, only three others could beat me. I have always worked on the principle that I should never ask anyone to do anything I could not, or would not, do myself. After a shower and breakfast, we went to the range, where we trained hard for the rest of the day on the whole spectrum of subjects: shooting practises, section battle drills, contact drills, ambushing, live camp attacks, medical skills, demolitions and so on.

We fired every weapon we were likely to encounter and threw literally hundreds of grenades in realistic bunker and trench-clearing practises. We acquired a box of new Spanish Star automatic pistols and I ran a CQB (close quarters battle) course for extra interest. These pistols were terribly unreliable. Sometimes several fell apart in a day, but we were so involved in our training that when they broke up – usually the slide would shatter – the men just dropped them in the bin, picked out a new one from the box and carried on firing.

We did an awful lot of night training, such as reconnaissance patrols, movement by night and map reading, ambushes and contact drills. We also had to man a picket every night on a hill outside the camp in case of guerrilla incursion. The security situation in South Africa was worsening and now, with Robert Mugabe in Zimbabwe, the African National Congress had another sympathetic communist border to hide behind.

We worked very hard for five months and became thorough in every aspect of our employment. In all my soldiering, I have never seen such a level of training and competence as there was among even the average member of the Pathfinders in 44 Parachute Brigade, but my enthusiasm began to wear the men down!

One morning, I woke up at Reveille, at 4:30 AM as usual, had my cup of tea and started to feel rather groggy. I shook myself and marched out on muster parade with the men lined up in three ranks. When I came to attention to present the parade to Captain Botes, my legs buckled and I nearly fell over.

"Have you been taking drugs, Sergeant Major?" demanded Captain Botes in a shocked voice.

"No, sir!" I replied, trying to stand up straight.

"Have you been drinking?"

He watched, amazed, as I wobbled about in front of him. "No, sir!" I said, furiously trying to get a grip of myself.

"Well, what the hell is wrong with you?"

"I don't know, sir! Permission to lie down, sir!"

"A very good idea, Sergeant Major!"

I wobbled off the parade and spent the rest of the day on the camp bed in my tent, trying to understand what had happened to me. The men took a most gratifying interest in my well-being and kept me supplied with cups of coffee and Coca-Cola. For five days, I suffered, wobbling in and out of my tent as I fought to stay with the training programme.

Hazily, I began to piece together various remarks I was overhearing among the men, like "He seems to be perking up a bit!" and "Time to hit him with another dose!", and I realised the whole thing was a scam. The bastards had grovelled about feeding me with all these drinks just to keep me heavily tranquillised with Valium!

I was not the only one to suffer the combined humour of men from so many countries. Captain Pete Botes was very popular and a good soldier, but he was seriously Afrikaans and had a particular obsession with the fauna and flora of his country. One day, someone backed a truck into a small tree outside his tent and knocked it over. He flew into a rage, muttering about rare plants and damage to the environment, and the men were rather taken aback. Next day, he came out of his office to find the tree had been fitted with a medical splint and bandaged exactly as instructed during the medical lectures. Bemused, he stared at the tree and went off scratching his head.

After five months, we returned to Murray Hill and I made a recce of the Drakensberg Mountains for a suitable place to hold our selection process for non-Paras. You might wonder why we did not select the men before training them so thoroughly, but we were building up the Pathfinders from nothing and we had to wait that long before we had enough men to work with. Anyway, Colonel Breytenbach had no intention of losing the failures, who he planned to post into non-combative roles in the brigade, as drivers or clerks in HQ.

Jane had given birth to a boy, Billy, on November 1. I was delighted. Life was really looking up. Also, we had a place of our own, in a South African Army quarter. Faced with an influx from Zimbabwe, the South Africans had put up mobile homes, which sounds grim, but in fact they were very nice indeed, spacious and extremely well appointed. We had nothing ourselves, but we moved straight into the quarter and lacked nothing. The South Africans treated all the exRhodesian servicemen very well, and we lived comfortably for a charge of only 15 rand a month.

Jane got a job in the military hospital and we settled down to our own lives much as we had in Bindura. She was always very supportive: tolerant of my single-mindedness and the endless stream of friends who came to visit. One day, when I was running selection, the cook in Murray Hill failed to get up in time to give us an early breakfast, which we wanted before driving into the Drakensberg, so I took all the eggs, bacon and sausages and we trucked off to my house where Jane cooked breakfast instead, for two dozen.

I ran the first selection in the Cathedral Peak area and we decided to make it short and sharp. We had already seen these men over all the previous months and formed a good assessment of their capabilities, so a final five days would set the seal. All ex-SAS and Selous Scouts were in theory exempt from this selection, but I took part in it because I was their sergeant major. I wanted to show them we had all been through the same mill.

We started with a march in full patrol gear, with rucksacks and an 82-pound ammunition box which I added as the sickener or 'embuggerance' factor. We carried that box up and down the hills for about 40 kilometres and it was a real ball-breaker. It was a terrible weight, an impossible shape to carry and rubbed your shoulders bare. We slept out at night, as I had done for the Rhodesian SAS selection. This is a fine hardening process, and I have often wondered why the British SAS truck back to camp at nights during their selection.

We left our ammunition box behind after that first day, but the marches continued relentlessly. The Drakensberg are an impressive range of grass-covered mountains rising to 15,000 feet. One day we marched up hill and down dale for 32 kilometres with 50-pound loads. Next morning, the guys groaned aloud when I told them we were going to march all the way back again, and very scenic it was too! The South Africans are obsessed with runs, so we finished off with a short 2.4-kilometre run uphill in full gear and rucksack. We lost a few during this, as you may imagine, but the majority were determined, fit young men and we finished with seventeen good soldiers.

One of the men came in from a march between Champagne Castle and Cathedral Peak, which had taken him sixteen hours, and I said, "What took you so long?"

Understandably annoyed, he said, "I bet you couldn't do it."

"Give me your rucksack," I snapped, and did it in twelve hours, a time which remained unbeaten for a good while. I was very fit then.

We divided our new men into two sections, one commanded by Sergeant Major Dennis Croucamp, an ex-Selous Scout who was excellent on recce work, and returned north to Mabalique. Meanwhile, Colonel Breytenbach was pushing the army 'system' to employ us on the Namibian border with Angola, so I set about perfecting our live camp attack training. I wanted to get the men used to the reality of live fire and the blast of shrapnel overhead. For example, as we crawled up to the start line for a camp attack, which I had put right on the edge of the safety distance for mortar bombs landing on our target 'enemy' camp, I would be counting the number of high explosive mortar bombs landing in front of us. At a prearranged number, say eighteen, the mortar team would switch without a break to firing smoke bombs, and as I counted the burst of the last HE round, I shouted, "Stand up!" and we began to advance, using 'fire and movement' into the camp while the smoke bombs continued to drop ahead of us. It was good realistic training and it was to come in very useful indeed.

In January 1981, after some Christmas leave, Dennis Croucamp took twenty men up to the Angolan border and I followed with another eight in February. We were attached to 'C' Company of the 1st Para Battalion for an operation on a small town 30 miles over the Angolan border called Quamato. This was at a time when South Africa was vigorously denying it had anyone inside Angola!

My new enemy was SWAPO, the South West Africa People's Organisation, which was the liberation and nationalist movement for Namibia. Like other African guerrilla movements, SWAPO was communist, trained by Russians and East Germans, and supplied by the Soviets. SWAPO claimed to represent all Namibia but, like the other African liberation movements I had encountered, SWAPO was recruited tribally. The majority of SWAPO recruits came from the Ovambo tribe which occupied the northern half of Namibia and made up 46 per cent of the population. Now I had come full circle in Africa.

SWAPO guerrilla bases in southern Angola were supported by the Marxist MPLA, which I had first come across during the chaotic days of January 1976 when I was in Holden Roberto's FNLA. Then, in 1976, SWAPO was an emergent and weak force of about 2,000 men, but once the Portuguese left Angola, it grew rapidly in the fertile communist ground of Marxist Angola to over 10,000. More importantly, SWAPO was able to transfer its bases from Zambia, which were well out of the area of their interest far down the Caprivi Strip, to communist Angola and infiltrate Namibia directly over a long frontier. By the time I joined the South African Army, the security situation in Namibia was beyond the control of the South African Police and was firmly in the hands of the South African Army.

All this may sound similar to the problems which had faced Smith's Rhodesian regime, but South Africa had three great advantages: it was not completely surrounded by land frontiers, it had a much larger white population (4 million among some 20 million blacks in 1976) and it had economic strength. The South Africans were horrified by the prospect of a double threat from terrorism, from the ANC in the east, through communist Mozambique and Zimbabwe, and from SWAPO in the west, through communist Angola. The government took special measures. It doubled the length of its national service to twenty-four months and developed its own arms industry.

The result was to produce a standing army of some 70,000

men with a total mobilisation of 400,000 available by calling up the Reserves, the Citizen Force and local 'home-guard' Kommandos. Of these, about 30,000 to 40,000 were employed in Namibia at any one time.

And South Africa took the offensive. They had considerable business interests in Namibia and, rather than let SWAPO keep the initiative, with the usual round of terrorist attacks on police stations, bridges and small towns, the SADF struck out at specific SWAPO bases inside Angola and kept the problem at bay. This was a similar strategy to the Claret operations in Borneo and the Rhodesian cross border camp attacks, like Chimoio and Tembue. Because of its size, South Africa was not so severely affected by the international pressure which had hampered Rhodesia's strategic decisions. There seems to be no doubt that cross-border attacks to punish terrorists in their resting places and training camps is an essential ingredient to winning a counter-insurgency war. When I came back to Britain, I wondered at the situation in Northern Ireland. Of course, this brand of military action is nothing without the political will to see it through.

Cross-border attack does not mean invasion. In 1975, South Africa tried conventional military invasion and found it did not work. The SADF was obliged to withdraw from southern Angola, after a spectacularly successful attack, for very similar reasons which had made the Israelis quit Lebanon. They realised they did not want the ground they had captured and that just pushing back the border did not prevent terrorism. Instead, the SADF employed an effective combination of supporting Jonas Savimbi's UNITA inside Angola; encouraging anti-communist resistance among tribes like the Ovimbundu and Chokwe which were hostile to the Ovambos and the Angolan government; and sudden SADF strikes against SWAPO camps in Angola. Our attack on Quamato was one such strike.

Quamato was a town about 30 miles inside Angola, a miserable cluster of single-storey buildings dotted round a dusty main street, painted in typical Portuguese pastel pinks and greens, lying heavy with neglect in the flat sandy scrublands of southern Angola. Intelligence said that SWAPO and FAPLA (the armed wing of the MPLA) were stationed there.

We flew to the forming-up point near the town in Puma helicopters and advanced into the town only to find the place was deserted. I noticed two things. There was a strong, sweet smell of human occupation, and washing was hanging out to dry, some of it still wet. That meant the enemy had been there that morning, and fled, probably hearing the beat of the heavy Puma rotor blades·, or maybe SWAPO had lookouts nearer the border. As if to confirm this, our pilots informed us that they had seen plenty of uniformed enemy moving out north. 'C' Company was ordered to advance and flush them out.

We found them, all right. About 7 miles north of the town, all hell suddenly let loose somewhere at the front. The ground was flattish and sandy but obscured by thorny trees and scrub bushes, and out of sight ahead the forward elements of the Paras had stumbled across a heavily defended SWAPO and FAPLA camp. The firing was intense, tearing through the branches above us, and stopped the leading platoon dead in its tracks. I happened to be near the front, crouched down with the platoon officer, and heard the company commander's voice on the radio. He was further back, nearer the town, and said we needed to keep up the momentum of our advance to get a foothold in the camp.

The young platoon officer, called Lieutenant du Plessis, turned to me and said, "Do you think you can break in there?"

Dear Lord! I said to myself, staring at the bushes for a moment, where a horrendous volume of fire was ripping the leaves off the trees just over our heads. *Why me?* I thought. The front sections had taken casualties and were pinned down on one side of a wide-open clearing about 150 yards across. Both sides were busy shooting shit out of each other, trying to win the firefight. Plainly, on the other side was a full-blooded SWAPO camp with all the trimmings – trenches, underground bunkers, heavy machine-gun positions, anti-aircraft guns – and beyond I could hear the hollow thwack of their 82mm mortars coming into action.

Crouching, I moved forward in cover of bushes to look at that short piece of open ground between us and them. It seemed like a mile across. I considered zigzagging across or doing it in bounds of fire and movement, and then decided that the best way was just to storm across in a straight line. I said to the young National Service officer, "Can you give me as much cover as you can, lay down a heavy base of fire support? We'll cross as fast as we can."

Du Plessis agreed and I called up my Pathfinders for some quick orders. "Now we earn our pay!" I said urgently, crouching down. In my excitement, my quick attack orders tumbled out in short staccato sentences. I watched their faces and felt proud. I knew how hard they had trained. I could feel their confidence and the momentum of the attack building up. My hand stabbed this way and that. I pointed out the ground, the place where I wanted to force a way into the camp, the location of supporting fire from the other Paras, and I finished with, "We've done this a thousand times before in training, now it's for real! I'm asking you to put yourselves in my hands. Follow me!"

And we charged straight over that open ground. God knows, I've never felt bullets cracking round me like that, but we lost no one and killed six enemy as we burst into their first line of trenches. My grandfather, Old Miles, told me about being on the receiving end of heavy fire in the First World War, and SWAPO certainly poured out a heavy volume of fire. I could feel it in the air around us but, thank God, it was not effective.

We held on in these trenches and supplied covering fire for the follow-up. A 'Valk' of twelve Paras (half a Dakota-load) charged over the open ground behind us into the SWAPO trenches on our right. Immediately, hand-to-hand fighting ensued in which two South Africans were killed, but they held on.

We took another bunker. I threw a white phosphorus grenade inside, stepped aside as it exploded and suddenly a black SWAPO soldier burst out covered in flames. He was understandably agitated and very aggressive. I aimed to shoot him with my South African Galil but the bloody thing jammed! Without hesitation, I swung the Galil and started to beat him up with the butt, to stop him shooting me, till thankfully someone behind me leaned over my shoulder and shot the man dead. These South African Galils, called SAR-4s, commonly jammed with double-feeds (when not one but two rounds are picked up from the magazine) because they used different metals to the Israeli Galil and overheated with heavy use. We certainly gave them heavy use that day.

More Paras poured in behind us and we tried to fight our way through the camp. It was enormous and well defended, and we got bogged down as darkness fell. Our helicopter K-car gunships could no longer support us with their 20mm cannons. To my fury, we were ordered to pull out. All that effort for nothing! The young platoon officer had no alternative. Orders were orders, but as we withdrew, the enemy realised what we were doing and went completely crazy. They opened up with everything they had, pasting us with mortars, and we could hear the thudding racket of large-calibre ZU 23mm anti-aircraft machine guns winched down for use in the ground role. We lost another man killed and two wounded on the way out, and Lieutenant du Plessis was hit in the back with shrapnel near me. I picked him up and carried him out of the firefight on my back.

We retreated, or as the military prefer to call it, withdrew, several miles back from the SWAPO camp to Quamato and dug trenches on the edge of the town. It had been a long and dangerous day, and all for nothing. I must admit I was angry at the waste. We had lost three dead and five wounded in 'C' Company for no gain. I was so fired up inside the camp that I had wanted to dig in right where we were and fight on the next morning, but I said nothing.

The problem was that SWAPO were a tougher enemy than ZANLA and ZIPRA had been in Rhodesia, and the South Africans could not afford the political and social criticism of losing too many men in one day. They had to pursue their aims with a very wary eye on casualties, however committed the white population may have been to fighting terrorism.

Later, another officer, Lieutenant Taylor, told me that he had written me up for a bravery citation. Nothing ever came of it, and later Taylor was himself killed. I mention this only because I have not hidden anything in this account and by now you will have made up your own mind about whether I invent things or not. I would have been very proud to have received an award, especially for Jane and my little son Billy, but the fact that I was not accepted made no difference to my feelings towards the South Africans. They treated us foreigners well and fairly, probably because they were a bigger, more established state than Rhodesia had been and possibly because the Paras were more straightforward and honest about other men's performances – being regular troops – than the SAS. Of course, they always made fun of us, calling me, "Fucking Sotpeel", "Sotty" or "Roineck" as a term of affection, meaning that us foreigners had only one foot in South Africa because the other was still in England and our balls in the ocean, but we always got on well.

Perhaps this was because I felt that I blended into the regular army system here better than I had elsewhere. I had a rank which I could handle, I felt I was functioning properly and had something to offer. Also, the South African Army, with a standing strength of 70,000 men, was much bigger than the Rhodesian Army had been, where the regular combat units were only the RLI, the SAS and the Selous Scouts, totalling some 3,000 men, excluding the two battalions of Rhodesian African Rifles, the Rhodesia Regiment and all the parttime call-up men.

At six o'clock the following morning, the SWAPO enemy camp was given a thorough softening up. We waited in our trenches watching Mirages screaming overhead to neutralise the mortars and 23mm anti-aircraft ZU-23-2s with bombs and rockets. Once the antiaircraft guns were out of action, sixteen Alouette gunship K-cars swooped in with their 20mm cannon and they really pounded the place. The racket carried on all morning, reaching us quite clearly over the trees beyond the edge of town. At midday, we flew in Pumas to attack again.

My Puma touched down in a clearing among the low scrubby trees and we had no sooner jumped out than we were under fire. We caught sight of four FAPLA men hiding in bushes trying to load an RPG-7 to fire at the Puma and blasted them. One died immediately and the rest stuck their hands straight in the air. By contrast to my experiences in Rhodesia, these men were made POWs and taken back for interrogation, more use to Intelligence alive than dead.

We advanced in a sweep line through the dense trees towards the camp and spent the rest of the day clearing one line of trenches after another. The K-cars had really shot the guts out of SWAPO but we still went through our drills, attacking with covering fire, firing, doubling forward, firing again, grenading the trenches and underground bunkers, firing into them to make sure the enemy were dead, consolidating and moving on.

In the darkness of one deep bunker, I was nearly caught out by the enemy's old trick of pretending to be dead. As I moved cautiously inside, shining my torch about, out of the corner of my eye I spotted the glistening skin of a black SWAPO guerrilla behind me. There was no time to turn my body. In an instant, I swung my pistol up and over my left shoulder and shot him three times.

That evening, the Pumas lifted us out altogether and flew us back to a South African army base in Ondangwa, a large town 63 kilometres south of the Angolan border.

Quamato had been a good start for the Pathfinders, but Colonel Breytenbach ordered me back to 'The Republic' to train another batch of Pathfinder recruits. The turnover was too large and he needed more men. This was partly the South Africans' own fault, as they signed men on for only one year. By the time their training was up, they had precious little time for operations, and most men, given the chance to leave the army, behaved like soldiers anywhere and quit. After all, the pay – about 300 rand a month for a trooper – was no incentive to stay. Even Captain Pete Botes left, because there was no career planning for his future in the army.

Some of the men may have found the South African Army rather more formal than the Rhodesian Army, where most of us came from. The regular cadre of the SADF were the senior ranks, staff sergeants and above in the Sergeants' Mess, and captains and above in the Officers' Mess. Many of them did not like to see troopers and privates in the Pathfinders with five or ten years' service who did not jump to attention for them. They were used to dealing with young National Servicemen and Reserves. For example, a South African sergeant major never queued at the camp barber's for a haircut; he just walked in and whoever was sitting in the chair with his hair being cut had to leap out of the way at once and wait till the barber had finished with the sergeant major. In all National Service armies there is a confusion between true discipline and just fucking the men about.

The South African sense of humour was rather strait-laced, with a strong religious bent, so there was uproar when they saw a mannequin's hand stuck up outside one of the Pathfinder's tents with condensed milk dripping off it and a sign which read, "Hand jobs available within". The mannequins, by the way, were part of a deception plan. When patrols went out in a truck to be dropped off near the border, they hopped out in cover of some bushes and the driver's mate, riding shotgun, stayed behind to set up a row of these mannequins, wearing natty camo hats set at a jaunty angle, to make the locals think the truck was still occupied.

On another occasion, the Pathfinders encouraged a young officer to blow up the reservoir at the top of Murray Hill camp. Lieutenant Brown was a very self-obsessed young Permanent Force engineer officer who became fascinated with mosquito larvae breeding on the water in the warm weather. He decided the only solution was to blow a breach in the earth bank of the dam and release the bad surface water. He also wanted to impress the Pathfinders. They thought this was a wonderful plan and egged him on no end. They sat about in the sun on the grassy bank of the reservoir while Lieutenant Brown sweated buckets digging holes in the earth bank, rammed in pounds of plastic explosives, tamped and blew it, sending up showers of earth and newly planted trees. All afternoon, explosions shattered the peace of Murray Hill and officers all over the camp wondered what sort of training was going on. Guy Gibson – as the Pathfinders had by now christened Lieutenant Brown, after the Dam Busters' pilot – slaved on and his breach grew bigger and bigger. Finally, a little water trickled over the top. Exhausted, filthy with earth, but triumphant, he turned to his audience and said, " How's that then?"

"That's great, sir. Great!" said the Pathfinders, thoroughly entertained. "But why didn't you just let out the water with the valve gates over there, at the side of the dam?"

They wandered off, leaving him shell-shocked on the earth-spattered grass bank among a wreckage of tree stumps with water pouring through the cratered, muddy breach. The commandant was not amused.

The new selection course started and an ex-British SAS officer called Major Alistair McKenzie joined the SADF to reorganise and formalise the Pathfinders' training. He preferred to run selection along modern British SAS lines, reducing the weights carried on the hill marches and studying the timings carefully. He also cleared out the nooks and crannies in the brigade where the inevitable freeloaders were hiding. He wanted professional standards and did not care how unpopular he made himself. I admired his attitude and I think we got on well.

Of course, the first lot of Pathfinders complained that the new course was not tough enough, echoing that typical refrain beloved of all specialised units, "Selection isn't as hard as it was in my day. There's no 82-pound ammo box to carry for a start!"

However, the men continued as committed as before and Alistair McKenzie went up to Angola to see what the Pathfinders were doing on operations. While he was there, he blew a road culvert. Shortly afterwards, he left the SADF.

Colonel Breytenbach was posted to a command on the NamibianAngolan border and we had a new commanding officer called Colonel Frank Bestbier. Within days, he decided the

Pathfinder Company should be transferred lock, stock and barrel to join the Reconnaissance Platoon of 32 Battalion. This was the best of the South African infantry battalions, as it was permanently posted on operations and filled with experienced men. It was composed of retrained anti-communist black soldiers, who had been in UNITA and the FNLA, and was used extensively to conduct 'search and destroy' sweeps in enemy areas and to conduct attacks on the Angolan economic infrastructure.

Colonel Bestbier called me to his office and asked me to train groups of Citizen Force Territorials on their National Service, at Murray Hill, and then take them on operations in Angola. I readily agreed. This gave me the best of both worlds, training at home when I could see Jane and my son, and taking the men I had trained on operations. He let me select my training team and I chose four men from the Pathfinders: Derek Andrews, Terry Tagney, Chris Rogers and Jock Philips.

The South Africans called their Territorials up in companies of about 200-strong, and my first group went on operations for two terribly uneventful months during which 2 Para skirmished with SWAPO and FAPLA inside Angola, through Namacunde to Ondjiva. The only thing I can remember about this trip was the convoy out, back to Namibia. I was in the cab of a big 'Kwévoël' six-wheel truck and we were towing a trailer stacked with Commandant Montagu Brett's company office equipment, plus items Brett had liberated to make life comfortable during our stay in Angola.

Only fortune dictates that a whole line of vehicles passes down a road before one finally sets off a mine. We drove over a mine on the very culvert which Alistair McKenzie had blown before. Fortunately, the cab passed over the mine, which exploded under the wheels of the back axle, blasting the trailer up and forward in a huge arc till it crashed down on top of the truck above us in a shower of paper files, chair legs and typewriter keys which fluttered down like sycamore seeds.

The Territorials were so keen to get back after their stint that the column carried on.

"You stay here, Sergeant Major," said the officer in charge of the convoy. "You've got twenty-three men to guard the broken truck and I'll get them to send out a recovery vehicle as soon as I reach the border."

I was not impressed. Recovery might take hours, and I had heard on the radio net that HQ reckoned SWAPO were converging on the area.

I inspected the damage. The two back wheels of our 'Kwévoël' were smashed, but the other two axles, the engine and the winch at the back seemed all right. So, I took the wire from the winch, pulled it up and over the whole truck, and across the underneath of the trailer, attached it to the front of the trailer and started the winch. The wire strained alarmingly but it gradually winched the trailer upwards in a big circle off the top of the truck till it toppled over and bounced back onto the road. We straightened it out and limped south to Santa Clara on the border that same night.

Operation Daisy was busier. Special forces Reece Commando units had reported heavy concentrations of between 1,000 and 1,500 SWAPO in the Techematet area inside Angola, near Cassinga, and a big combined airborne-armoured operation was mounted. The East Germans trained SWAPO as conventional troops rather than terrorists living among the civil population, and they were heavily armed. The South African politicians allowed the SADF to strike at SWAPO camps with all the force they could muster, and concealed the SADF's real impact by constantly denying cross-border operations. Operation Daisy was the deepest penetration of Angola since 1975.

The SADF armoured cars, called Rattels, looked rather top-heavy with their 'Eland' gun turrets, but were born out of necessity and suited the environment. They were wheeled, for the sandy flat country, fired 76mm or 90mm shells, and had a distinct advantage over others because they carried not forty-six rounds, which is normal in a European armoured car, but 200 rounds. They were to drive to the target while the Paras did a parachute drop to pin down the enemy till they arrived.

We were training a Para battalion at the time, in 44 Para Brigade, and the battalion filled six Hercules C-130 aircraft. We were dropped at night, about 2 a.m., over the sandy, scrubby dropping zone (DZ) at an operational jump height of 500 feet, and as usual I was wildly fired up with excitement as I jumped. When I stripped off my parachute on the ground, it made a stirring sight to see and hear the other big aircraft trundle overhead and watch hundreds of parachutes floating down. Our DZ was just off target so we could reorganise without having to fight the enemy at the same time, and, as sergeant major in charge of DZ rallying, I spent a very busy two hours running around in the darkness gathering the men together in their fighting formations, including one plane-load which was dropped off line. I was pleased with the time, and the whole battalion was ready for action, with the leading companies moving off the DZ as dawn lightened the sky.

Each company had a different area, and was to find the enemy by advancing to contact. Each one had a group of my Pathfinders, to lead the advance and navigate. I was with 'C' Company. Mid-morning, we stood among scrub bushes and trees and looked across a wide, dried-up swamp, called a flay. These flays become very soggy in the wet season and tall grasses grow thickly, but then it was dry, open and caked hard. The company moved across in bounds, one platoon covering another onto the slight hill on the other side of the flay. In the lead, I breasted the top of this hill and spotted movement in more scrub and trees 70 metres beyond. I ordered Corporal Sean Wyatt to keep his eye on the place while a platoon moved into a sweep line behind us.

Suddenly, someone fired their rifle by mistake. In answer, the enemy opened up on us from their cover in the trees. Everyone fired back and Sean loosed off his 40mm grenade launcher at the leaves we had seen twitching. We assaulted across the open ground. Moments later, as we fought into the trees, we discovered the grenade had hit a man right in the chest. Of course, everyone else had fired too, so they all claimed a hit. And what a hit it was! The dead man turned out to be SWAPO's second-in-command of logistics. Of course, he was rather beyond interrogation, but there was a goldmine of information in papers we found in a leather pouch on his belt. Strange how the communist system lets itself down in these wars with its obsession for keeping details on its members and their activities.

The second day, the sweeps continued with sporadic contacts all over the area of operations under a baking hot sun. It would be no exaggeration to say we walked and skirmished over 65 miles or so in these two days. As sergeant major, I had a lot of chasing about to do and I must admit I began to feel my age! We were promised five litres of water per man per day, but it did not always reach us. I made every man parachute with extra water canisters, but that supply was long gone. I also ordered them to carry a trauma pack each. The drips carried were usually 0.9 per cent normal saline. Giving sets were also carried. The drip bag and giving set were packed ready for immediate use, with the needle fitted to the tubing and three strips of sticky tape ready to attach the set to the arm. Casualties could be given the contents of their drip without delay and it was also handy because you could drink the fluid if you went down with heat exhaustion.

That day, I suddenly passed out on the march during the afternoon and woke up to find the doctor sticking a needle in my arm to fit me up with my drip. It was strange to feel burning hot one minute, with the sweat pouring off me, and suddenly cooling down as the drip slipped into my veins the next.

The SADF realised they had a serious heat exhaustion problem. Their answer was typical of the universal sense of emergency in South Africa at that time. They stripped a Coca-Cola factory of its 2litre plastic bottles to fill with extra water and choppered them in crates to the troops.

Heat, rationing and water supply were always important features of any operation in this dry, sandy and barren area, as was the disposal of rubbish and soil. On Operation Daisy, I recall being able to smell the pungent sweet stench of the latrine pits hundreds of yards away as we walked through trees into the base camp. Not enough quicklime (even cement can be used) was being sprinkled on the excrement every day. On another operation, we dug a tin pit where we threw our rubbish and scraps, and there was so little food anywhere in Angola that I reckon every bloody dog in the country was in this tin pit at night, howling and fighting in rival packs. We could not get a wink of sleep.

Finally, Derek Andrews lost his patience. He set up a Claymore anti-personnel mine over the pit and soaked the place in petrol. That night, the howling began as usual until suddenly there was an enormous explosion, an eyeball-searing ball of flame which lit the dark trees around, and we lay on the ground listening to the sound of shrieking dogs fading at speed into the night. For days afterwards, we saw nothing but bald, hairless dogs with charred eye lashes wandering about in a daze. For a while at least, we slept fine.

Fresh rations are always vital to troops in the field living off tins. When we were resupplied with bags of vegetables, mainly potatoes and cabbages, the South African National Servicemen simply did not know what to do with them. I took charge and instructed the cooks how to make bubble and squeak. The troopies thought it was delicious and we lived on it for days.

Another operation, Carnation this time, pushed FAPLA and SWAPO back from Ondjiva altogether. Now we could fly straight from the Republic and air-land inside Angola at Ondjiva, which became a staging base for our attacks further inside Angola. The South Africans gave Ondjiva to UNITA to control, treating the area as liberated and turning Mao Tse-tung's own communist principle of a People's War on its head. This allowed the SADF to penetrate deeper into Angola and attack SWAPO bases further away without worrying about the areas near the border. Also, regular SADF troops were not tied down protecting captured areas, as had happened in 1975.

Ondjiva at that time even used South African currency. This was a pity because my Pathfinders and I were digging trenches on the edge of Ondjiva airfield and we discovered a large box which contained no less than 60 million Angolan Kwanzas! Presumably this treasure was being used to finance SWAPO. In theory, the official rate was, as far as I can remember, 32 Kwanzas to a pound sterling, but in practice no one wanted Kwanzas and we treated it like Monopoly money. I divided up the loot among the men, who went round like children with great wads of paper money, like Kelly's Heroes, played cards with it and argued about who was richer than who.

As always, when operations were uneventful, boredom set in and standards began to slip. Only a couple of months of operations is not enough to turn National Servicemen and Territorials – or even regulars – into experienced combat troops, and maybe the intrinsic anarchy of operations loosened discipline too. Whatever the cause, men started to sunbathe by their trenches in Ondjiva, lounging about bollock naked on the sand. I did not agree with this at all. I went round picking men up for being improperly dressed. One of their officers, called Lieutenant Peter de Klerk, objected. He was a student at Cape Town University doing his National Service in 44 Para Brigade. He said to me, "I think you're being too harsh on the men, Sergeant Major."

I looked at him standing there in just boots and yellow underpants and replied, "I don't think so, sir. We're on operations."

"But this is South Africa, land of sun."

It was the land of shorts as well, like Rhodesia had been. At this point, our conversation was interrupted by shouts of alarm. A Soviet BRDM armoured car was tearing across the airfield behind. It was actually a UNITA vehicle, but Lieutenant de Klerk thought it was attacking the position. He grabbed an RPG-7 anti-tank rocket and the last thing I saw was this absurd-looking young officer sprinting after it dressed only in boots and yellow underwear!

There was nothing wrong with the spirit of these young officers, even if their military professionalism was at times a little relaxed. In fact, Lieutenant de Klerk joined us later on the training team.

Ondjiva was a wreck by now, with few civilians left who could stand the constant battering and looting by FAPLA and SWAPO. Our battalion commander, Commandant Monty Brett, said, "I can't have my men sitting on the ground!" All the shops were empty, so he ordered us to liberate vital items from the town, for the proper comfort of the men in the field. Among other things, the Town Hall yielded some rather fine old colonial Portuguese furniture: a set of high-backed chairs and the long table from the council chamber.

Major Jet van Zyl, the company commander, gathered us under the tent shelter he used as a briefing room and gave us orders for our next operation. We were to attack Evale, a town about 40 miles north of Ondjiva and 65 miles inside Angola. After two days preparing our rations and ammunition, we drove north in armoured Buffel troop carriers and four-wheel Samil-29 trucks to form a HAG. This was Afrikaans for a helicopter administration location, like the Rhodesian Forward Air Field (FAF), and contained everything, including bulging bladders of AVGAS, firefighting bowsers, boxes of ammunition and a medical first-aid post.

The plan was for the Alouettes and Pumas to lift us from this HAG directly into the attack. Van Zyl's briefing gave us to expect about eighty SWAPO, commonly called Garden Boys by the South Africans, so we all expected a good day's jousting, as firefights with the enemy were termed. It did not work out quite like that.

I was left in charge of the HAG, much to my annoyance. Even back there, I knew things had gone wrong. When the first lift went in, they found not eighty Garden Boys but 300, with some rather large tools. Like 14.5mm and 23mm HMGs, 82mm mortars and a call on Soviet T-54 tanks at Mupa. In addition, they were well trained and stiffened by large numbers of East Germans. Our choppers ferried the troops in and when they returned to my HAG, we picked up the pilots' shocked radio messages, relayed back via a rebroadcasting station in a Bosbok light fixed-wing aircraft orbiting high above us.

The pilots reported very heavy resistance, and seeing white faces and East German uniforms in the defensive trenches and among the houses of the town. This was confirmed when our ground troops joined battle on the edge of Evale. The ground comms radio buzzed with reports of heavy machine-gun fire and being accurately pasted by the enemy 82mm mortars, which were almost certainly being organised by the East Germans. Within half an hour, we had suffered ten wounded. The Alouette helicopters flew in under fire to lift them out and landed back in the HAG peppered with holes. I could tell things were grim. I detected a feeling of panic and felt very frustrated being left out of it.

Major Jet van Zyl came down in his command helicopter to refuel in the HAG. Over the noise of other choppers landing, I shouted at him, "Sir! The guys are having trouble. They're National Servicemen up there and not many with experience. Can I join them, to see if I can help?" I was fired up and at my most persuasive. I hated seeing them coming back all shot up and not being able to contribute.

"Yes," said van Zyl without hesitation. From his position in the air, he had seen the battle was getting out of hand. A big Puma took me forward, hugging the tops of the trees covering the flat savannah landscape. Ahead, I could see the low rooftops of Evale, where the sky was filled in all directions with streams of green and red tracer and white vapour trails of SAM missiles being fired at the Alouettes, and the noise of battle was fearsome.

Nearer the edge of the town, we saw an Alouette K-car lurching and dropping through the air towards us. We learned later it had been badly shot up when the pilot landed to pick up the crew of another K-car which had been destroyed on the ground by enemy machine guns as it tried to rescue some wounded Paras. The pilot was later awarded the Honoris Crux Gold, the highest South African bravery award. His Alouette was heavily overloaded with Paras and the crew from the other K-car and I gestured at him to land wherever he could. Our Puma set down beside him and we took the men, with the wounded and one dead Para, off the Alouette so it could limp back to the HAG.

Back in the HAG, I shoved all the casualties – by then one dead and seventeen wounded from the company of eighty – on our one Puma. The pilot only just made take-off.

After only an hour, the Paras claimed to have killed forty-eight, but their own casualties were politically unacceptable. We had failed to enter the town and our commanders decided to withdraw until the armour closed up. I must admit I found it strange they never had the armour there in the first place, because Paras on their own never have enough firepower to overcome the sort of well-organised and well-defended positions we found at Evale.

A withdrawal is always difficult, especially after hard fighting, which introduces its own strain of chaos. A sharp explosion went off in the HAG, which I believe was someone accidentally firing a 40mm grenade launcher, and a young National Serviceman in a truck started screaming, "Mortars! Mortars! Incomers!"

Men began to panic. The young National Serviceman in the truck went on screaming hysterically and I did what anyone does to someone with hysteria, regardless of rank. Without delay, I hopped up to the cab and slapped him a couple of times round the face. He quietened, his face white, but things were still on the brink, with men driving about and pushing and shoving. I found Major Jet van Zyl in the melee and said, "Sir! With respect, the attack belongs to the officers, but the withdrawal is ours! Permission for the NCOs to organise this withdrawal, sir?"

He looked at me a moment, at the open space in the trees around us, at the choppers coming in and out, at the piles of stores scattered round and his men running about in all directions and said, "It's all yours, Sergeant Major."

He and the other officers left me to it and I restored order. As the Para sections were choppered back from the fighting round Evale, I calmly lined them up on a track. When the whole company was accounted for, we embussed in trucks which HQ had sent forward for us and trundled back to our camp.

At the same time, the Russian T-54 tanks from Mupa were driving into Evale on the other side of town. We had nothing to stop them. South African radio intelligence confirmed a heavy East German presence, so we left them to it. South Africa's fight was with SWAPO, not with the East Germans. Also, the East Germans could fight.

As the end of my contract with the South African Army grew closer, I decided I must do one more operation and one more parachute jump. I have never agreed with those soldiers who think there is a time when they can hang up their boots and coast along in an easy job to their retirement. The sentiment "I'm running down to demob" is bad enough among cooks, drivers and medics, but in specialist parachute units it is nothing short of disgusting.

My last operation with the South African Army took place on Mupa, some 100 kilometres north of Evale and 200 kilometres inside Angola. The SADF attacks into Angola penetrated deeper and deeper, continuously forcing SWAPO to increase the safe distance between them and the border they had to cross into Namibia. We were allocated to a composite group of 32 Battalion and Paras from 44 Brigade. We were flown in by Puma helicopters to 32 Battalion's area of operations to attack the Mupa FAPLA/SWAPO administration base. To preserve secrecy, we were dropped off some distance from the enemy camp and began the approach march.

I found this excruciatingly painful. I had twisted a knee ligament on a parachute jump not long before and it started to play me up. The South African Army were very good with their medical support, knowing full well how vital it was to let soldiers know they would receive instant care if they were wounded, and they employed doctors down to platoon level on operations. Our company had four platoons, so we had four doctors with us. One of them looked at my knee and said straightaway, "You should be casevaced for this."

I shook my head. "Just give me some painkillers, Doc. I can't call in a chopper just for me. I'll wait till we attack the camp." There were bound to be other casualties then.

We marched 20 kilometres that day along narrow winding tracks through the flat countryside. The grey dusty earth was trodden hard by countless people, local tribesmen and enemy soldiers. On each side, soft knee-length grass grew among the savannah scrub and low trees which stretched for miles in all directions. We stopped and lay on the ground in a defensive position for the night, and by morning my knee was worse, stiff and painful. The doctor gave me a painkiller called Deloxene and I carried on, determined to see the operation through. It was not to be.

At dawn, we continued to follow these narrow tracks through the flat bush and trees. I was at the back of the company and almost 100 men had passed along the track when I heard a soft roaring explosion a couple of men in front of me. One of the soldiers had stepped on an anti-personnel mine. Some of the men near him started to run and help but I roared at them, "Stop! Stand still!"

I had seen this situation many times. The enemy laid these AP mines in the line of march on the track itself, which was only about 18 inches wide. There might have been others not set off yet. One hundred men had walked past this mine and quite possibly there were others they had not trodden on as well. I also knew the enemy very rarely doubled the booby trap with other mines laid out to the flank of the track, to catch people bypassing the scene of the first explosion. I moved several paces directly away from the track through the grass, turned at right angles and walked carefully along parallel to the track, very carefully looking through the grass ahead of me to check for more mines. Then, when I was opposite the wounded man, I turned at right angles again and came in towards him.

His foot had gone. There was just a torn stump of bone and sinew. No blood. The flesh was cauterised by the fire of the explosion. I felt terrible physical sympathy looking at him. He was so young, no more than nineteen. His life had been utterly, irrevocably changed. He stared at me, his eyes wide in shock, and cried out, "I'm an athlete, Sergeant Major. I'm a Springbok!"

I knelt down beside him and held him in my arms. There was nothing else I could do.

I shouted for everyone to search the track immediately around them in case there were more mines.

The young man in my arms began to cry. He was thinking of his career as an athlete. Gone like his foot. In an instant. Happily, he could feel no pain … yet. While the others cleared the track, I could do no more than hold him tight in my arms, like a son, while the tears rolled down his sweat-stained cheeks, and he gazed at the ground, bemused, shattered by what the mine had done to his life.

The doctor came up as soon as he could, moving on the cleared ground, and then someone at the back of the column shouted the alarm. An enemy vehicle was coming. Our track had just led us across a road and the rear party could see a large truck trundling up the road towards us about a mile away.

I lost my temper. The pain of sympathy for that young Springbok turned to fury. I ran back down the track to the road and rapidly deployed the guys at the back of our column in immediate ambush positions. They ran and hid in the scrub at the side of the dirt road. Someone had dug a trench at the junction of the track and the road. There was a large tree beside it. I hid behind the tree and waited, seething with rage.

It's no good ambushing vehicles from the side of the road, because the target is too fleeting. When the ZIL truck came up close, I stepped out in full view in the road and blasted the cab head on with my Galil, smashing the windscreen and raking the three enemy MPLA inside. As it passed, swerving and slowing up, I shot another enemy soldier sitting in the back.

The guys in their immediate ambush positions finished the truck off, pumping a fearsome weight of fire into it as it rolled past them to a stop. One enemy soldier jumped off with an RPG rocket, which he fired off into the bushes. The rocket hit a water bottle on the belt of one of the South African soldiers. He was lucky. I suppose the water absorbed most of the energy of the explosion, but a chunk was blown off his arse.

Now we had two badly wounded men and we needed a chopper casevac as soon as possible. When the Alouette arrived, I went back too. I hated to have to admit it, but my knee was excruciating. They flew us back to a HAG at Nehone, still 100 kilometres inside Angola, where I experienced the South African painkiller called Sosogen for the first time. This stuff is one step off morphine and has to be taken with Stematol to stop being violently sick. I was lying in the medical tent waiting for the Stematol when I felt the analgesic flooding inside my body and I was violently sick. The only container available was a mess tin and I can tell you it was inadequate to the task!

My last parachute jump in my remaining few weeks of service was equally full of foreboding. I joined a basic free-fall course being run for another group of trainee Pathfinders and had a malfunction. I dived from the Dakota at 9,000 feet and after thirty-five seconds' free-fall, I set up in the frog position to deploy my parachute, slightly head up to ease the shock of opening, and pulled the handle. Nothing happened. I beat the parachute pack with my elbows, hit it and kicked it with my feet. Vital seconds ticked past. I hurtled towards the ground. *Nothing for it*, I thought automatically. *Go for the reserve!*

Suddenly, before I had time to move, the main parachute popped open, deployed perfectly and I floated to the ground like a feather.

You must always get back on a horse after a fall. Parachute malfunctions do occur from time to time, and I went straight back to the jump centre for another parachute, strapped it on my back and went up in the next lift for another jump. I did not want to leave the army with a bad feeling like that. This time, it went fine and I felt back on form, but maybe in retrospect I should have taken the warning there and then.

Maybe, but then I'm not very good at giving up like that.

CHAPTER 10
A STATE OF EMERGENCY

My life seems to have been a series of violent swings from one extreme to another. Since things had been going well, perhaps I should have guessed I was due for a fall. However, I left the South African Army on a high, a warrant officer with a good report. I had managed to save some money for once, a remarkable feat for me, and Jane and I were really at peace with each other, happy with little Billy and our prospects. I hoped to get a new job with reasonable pay. A tall, rather serious South African friend called Mark Adams – who had known me from the Rhodesia days when he was an officer in the RLI (and on the Chimoio raid) and later served a shorter time than me in the SADF – recommended I join the company he was working for.

This was COIN Security Group (Pty) Ltd, a big South African security company based in Jan Niemand Park in Pretoria. The name was deliberately chosen to be the acronym for COunterINsurgency, and at first glance it was all there: uniforms, ranks, armoured vehicles, weapons, badges, even medals. COIN supplied manpower, static and mobile guards and guard dogs to protect installations like factories, shops, stores and banks. In the increasingly uncertain atmosphere of South Africa, this was profitable work.

The managing director, John Bishop, interviewed me. He was a dark-haired man, always meticulously groomed with a penchant for good clothes, and his greatest skill was in choosing the right men for the job. He had started COIN only a few years before with a dozen guards; when I joined, there were about fifty, and two years later there were nearly 1,300. With such rapid expansion, he needed men to handle his 'troops', the guards, and realised from his own National Service that the army could provide them. He had been a National Service corporal in the South African Army and been promoted year by year in the Territorials after each annual camp. By the time I met him, he liked to be called 'Major'. I could see that he deliberately cultivated a military image, but that seemed rather appropriate for a security company and he was enthusiastic about having me join the team. So was Yvonne Lottering, an attractive blonde woman in her 30s who was his marketing director and partner.

They offered me a job, and I accepted. A security company seemed to be the closest thing to the army, and at the relatively young age of 41, I didn't want all my past experience and the whole motivation of my life just to be wiped out when I walked through the camp gates for the last time into civvy street.

I set out to serve this new entity with the same enthusiasm I had displayed in my soldiering. I found the company very gung-ho, which I liked. John Bishop wanted to hear all the opinions of new men like me. We had some really constructive meetings about the constant expansion problems, and I felt we all had a common interest in getting on with the job. COIN was expanding rapidly and there was no time for unnecessary paperwork, office minutes, memos and reports.

Jane and I went to live in a rented company house in Jan Niemand Park. The house was a spacious bungalow with big rooms in a quiet residential road called, tellingly, Lamerwangerstraat. A lamerwanger in Afrikaans is a griffin, the symbol of 44 Parachute Brigade. In fact, several old army friends lived close by, from 44 Para Brigade and Rhodesia days, and life was looking up. Jane found a nursing job in the Huis Luzetta old people's home, also in Jan Niemand Park, close to home, and we settled into the local community well. I even began going to church with Jane every week. There is still a lurking vestige in me of the Catholicism of my early childhood, deeply implanted by the iron discipline of Sister Loyola and Father Brett, but I enjoyed the church for its sense of belonging to the community. The South Africans are very strong on religious background and social conformity. Belong to the club and you can go to heaven!

Every day, I went to work in my Daihatsu Charade company car, which I was paying for by instalments, and I worked long hours. We were on the go all the time, running around meeting clients, managing the guards and checking installations. South African businesses needed no persuading of the importance of security. Nelson Mandela was still in prison and the still-illegal African National Congress was increasingly active, with tremendous international support. Car bombs, shootings and riots were common. In this atmosphere and having just left the army, I saw nothing strange about the way we always addressed each other by rank and the other deliberate parallels with army life.

My expertise has always been motivating the people working for me, and the 'Major' gave me a job which entailed behaving as a sort of sergeant major over the black guards in an area. I discovered that the lower the rank in civvy street, the grander the title, so I started as a 'Senior Operations Commander' which we called SOC, copying the army's obsession with acronyms. In practice, as SOC, I was a general dogsbody for an operations manager. Between us, we kept the company's static guard operations in Pretoria running smoothly – including personnel, pay, discipline and guard rosters, timings, routes, locations on factory sites, and so on.

Most of our guards were from the black homelands, like Venda, Bushbok Ridge and Bophuthatswana, and they worked bloody long hours. In theory, they were supposed to work twelve-hour shifts, six days a week, and take one week off in seven. Most people who have worked shifts of twelve on, twelve off, would agree this was hard enough, but in practice they worked harder. All nattily turned out in their smart uniforms, these blacks did nearer fourteen hours on duty, adding the extra time for driving them to work and back, and signing on and off. Also, the company was so busy that the guards were commonly kept on the job seven days a week and their week off was cancelled. They were in no position to argue, apartheid or not. They needed the work to send money back to their families.

Under the DomPas (Domicile Pass) system, homeland blacks were not allowed by law to bring their families with them to the city, so the company provided living accommodation which was very basic. In Pretoria, they lived in a rectangular building made of concrete blocks, and they slept on bunk beds, built into the walls like racking, three high to the ceiling. In Johannesburg, they were racked four high to the ceiling. Sounds rough, but there was no point providing beds or movable furniture because the blacks always stole it and ran away. Also, the company had to pay a tax on each bed-space, so our 400 blacks used the 'hot bed' system, to conserve space like sailors in a submarine. Two men shared one bunk bed: 200 were on duty while the other 200 slept. It's not surprising they were exhausted, and I was constantly sweeping up the problems caused by men literally falling asleep on guard or going absent in desperation.

"Your guard was not on duty!" one client shouted at me. "I came to the premises in the middle of the night and he wasn't there!"

I quickly worked out a system for dealing with this sort of complaint. I called it the 'toilet trick'. With a deference I had not employed since talking to Sister Loyola as a boy, I inquired politely, "When did you visit the premises, sir?"

"At 02:34 hours," shouted the man. "I noted the time exactly!" They always did that.

"And how long did you stay, sir?" I knew civilians like to check up on people, but they do not care to hang about in the middle of the night when a warm bed is calling. There doesn't seem to be any point. It was a crucial part of my ploy.

" Oh, I don't know. About ten minutes, I suppose."

"Did you look in the toilet while you were there?" Crux question.

"Er, no."

"Well, I'm sure the guard was on duty, sir, and don't you think it's possible that he was in the toilet?"

"Maybe."

Then the ace. "I mean, sir, even black men have to use the toilet on a twelve-hour shift."

Grudging acceptance.

This excuse worked particularly well if the client was British. Whereas the South Africans had no interest whatsoever in a 'kaffir's' urinary habits, British clients retained some vestige of fair play in this vital biological regard!

Mind you, the black guards had little interest or pride in their work, even though it provided them with a wage they could not earn anywhere else. Guards certainly got very tired, but they took every excuse to skive off. If we needed more than one on an installation, they decided among themselves who would stay awake while the others slept, got drunk, smoked dagga, or simply went absent.

The 'Major' was tireless in trying to control these lapses of discipline. He introduced fines. Every time a guard was found asleep, he had to pay a fine of 5 rand. I can still recall one black who was caught so many times that after six weeks' pay had been taken into account against his accumulated fines, he owed money to the company!

Guards going absent was a real problem. Clients were always furious, so the 'Major' also introduced a system of reserve guards who could be held back and used to fill the gaps of those who went absent on duty. The 'Major's' enthusiastic plan was that we could merely replace the missing man with another when we were called out, as our great selling line was that we were the only security company which drove the black guards to work ourselves, to make sure they got there. That was fine in theory, but in practice, when the Reserves were stood down waiting for a call out, they got drunk or went absent too.

I became very good at crisis management. I pacified endless

angry clients after some drama caused by the guards not doing their jobs, and the number of clients continually increased in parallel with the worsening internal security situation in South Africa. In September, President Botha announced a new constitution giving limited political power to Asians but nothing to blacks. At once, the ANC stirred up violent rioting in five townships outside Johannesburg in which twenty-nine died, including the Mayor of Sharpeville, who was hacked to death, doused in petrol and burned. This sort of thing is bad news for everyone except security companies, and COIN expanded. We were always under pressure and I enjoyed that. I found the work absorbing in that first year. It was new to me; I liked the man-management aspects and the hours were long. At least, I worked long hours. To start with anyway, we ex-army men were appreciated and I was quickly promoted to operations manager, responsible for all static guard work in the Pretoria area.

I was so busy with the job that I had little time for my family. Jane and I hardly saw each other. Somehow, this kind of commitment works in the army, where operations and even training exercises demand twenty-four hours a day. Service families seem to accept the whole-hearted commitment required of their men, but I had not yet realised that the civilian work ethic operates with a subtle difference. I did not get the chance to appreciate this before fate intervened to stop my life getting too comfortable.

In October 1984, the 'Major' suggested I do a demonstration parachute drop on the COIN annual open day. I agreed at once and found some 44 Para Brigade free-fallers to make up a team. I was delighted to get back to something I really enjoyed. I was to jump in with a COIN pennant flying from my ankle and enthusiastically organised some practise jumps.

On Sunday, October 7, six of us assembled at Wonderboom airfield near Pretoria and everything went wrong from the start. We were jumping with civilian equipment and, as I had no parachute of my own, I borrowed a square from an ex-Pathfinder. As I walked out to the aircraft, the pin popped out and one of the others stuffed it back in with the 'chute on my back while the Pilatus Porter took us up to 9,000 feet.

Nothing matters as you leap out of a plane into the emptiness of the slipstream high over the earth. The moment of commitment is totally absorbing. No matter how many jumps you have done, every jump is a thrill. I was first man out to act as base-man in free-fall, so the other three could close in on me to form a four-man link-up, or star. On exit, my goggles flew off. On the way down, my altimeter, which was supposed to stay strapped down at my chest, worked loose in the slipstream and flapped its way up where I could not see it.

Two others came in and grabbed me to form our star. I tried to see their altimeters. Normally, this is easy, just a question of looking across at them as you hang on to each other, but my eyes were watering so much without goggles, I could see nothing. Conscious that every five seconds we hurtled a thousand feet closer to the ground, I decided to pull at the same time as them. But civilian freefallers pull lower than army free-fallers, because civvies pay for their jumps and want their money's worth. Every second counts, quite literally.

The others started breaking off to find some free air space to pop their parachutes. Once they were away, I took a last, routine, safety check look around as my hand went to pull the handle, and saw a man above me! I could not pull. I tracked away a little. Suddenly, my altimeter popped up in front of my face. 1,200 feet! Too low! I yanked the handle out at once.

The parachute deployed instantly and I can remember to this day what happened next. Instead of the usual jerking halt and gentle safe swing under the canopy, I heard a series of terrifying snapping noises as several rigging lines broke under the opening pressure, like huge rubber bands being cut. One side of the canopy collapsed immediately and the whole parachute started a sickening rotation with me swinging on the end, as if I were on some mad circus roundabout. There is no time to think of consequences in parachuting. You just act. Working instinctively, I fought with the risers and brake lines to get beneath what remained of the canopy and fly it. After two giant rotations, I managed to control the swing and get under the parachute, but I was dropping too fast.

I looked up. Only four of the seven long cells were inflated. Less than half the parachute holding me.

I looked at my altimeter. Six hundred feet. Too low. No time to cut away this shambles and go for my reserve. I was dropping too fast. No time for anything! I had to ride it in. God, it made a noise! The loose material, rigging lines flying, rattled in the wind like a motorbike with no exhaust as I plummeted towards the ground.

Helpless, I looked down briefly. There was no way out of this. I was totally on my own, and maybe it was the Catholic in me, but I said quietly, "Oh God, get me out of this one!"

I smacked into a deep, soft, ploughed field.

I broke up and bounced like a rag doll. It happened so fast but I was conscious of it all, as if in timeless slow motion. I could feel my right leg snap backwards under my body, crushing the bones and ripping the flesh. My left leg shattered at the thigh, and I felt the vertebrae in my back crushed between the impact and the force of my head ramming down onto my chest. There was blood and snot everywhere. Shards of my broken bones had ripped through my muscles and opened an artery, which sprayed the white parachute as it fluttered around me. At once, it seemed to me, I started shaking with shock and broke out in a terrible sweat.

Jane arrived with the ambulance. She and little Billy had watched my parachute come down without any idea I was in trouble. Then, she saw the others running about shouting for the medics and jumped in the ambulance with them. I was conscious throughout as they lifted me very gently off the ploughed earth, and I could hear, and feel, my broken bones grinding as they straightened my right leg. Jane sat in the ambulance and stroked my head over and over to calm me, and I kept saying, "This time

I've had it. I've really had it!" She calmed me and they cut my parachute harness off my body with knives. God knows what my little son Billy thought seeing his father lying in such a mess, or when one child said, "They're cutting off your dad's leg!" Billy was not quite 4 years old. He was silent for weeks afterwards. Maybe he has never forgotten.

After a speed drive to hospital, I was put in an emergency ward in a bed surrounded by curtains, still conscious, and festooned with drips. By now, the pain was really building up. No morphine is allowed before an operation and I was feeling extremely sorry for myself. In the next-door cubicle, I could hear groaning and someone muttering in Afrikaans, and I wished he would shut up! I stared at a large yellow object sticking out over the top of the curtain rails. Suddenly, the nurses busy around him opened the curtains trying to make space for him and I saw he had a huge chunk of a glider's wing stuck inside his chest. I thought I was in a bad way, but I will never forget the sight of this guy being wheeled off to the operating theatre surrounded by nurses trying to stop that great section of yellow wing knocking on the doorway and tearing him apart.

My turn came. The surgeon, van Dyck, was a gentle man with a round face and the kindest expression in his blue eyes of any man I've seen. You notice that sort of thing when you are all broken up and meet your surgeon for the first time. He bent over me, put his hand gently on my head and said, "If you come out of this theatre with both your legs, then look upon it as a bonus." What a bedside manner, but I was past caring!

When I came to in a crisp bed in the intensive care ward, I remembered what Jane had said about plaster casts when we were in Rhodesia. As a nurse there, she had seen a lot of gunshot wounds. She said that if the surgeons left a leg in an open cast, it meant they were trying to save the leg, but that most of the cases she had seen with open casts ended up with the leg being amputated.

I lay surrounded by five drips, glistening plastic tubes, stretched out and unable to sit up. My legs hurt. I wriggled my toes. I could feel them, so I guessed I still had legs. Then I remembered that amputees sometimes imagined these reactions. I had to look. Struggling, I could just twist my head and, to my horror, I saw my right leg lying in an open cast.

Dr van Dyck came to me the next day. He looked as kindly as ever. He examined me and stared at the open cast on my right leg, the one which had been so crushed under me on landing.

Bluntly, I said, "You're thinking of cutting off my leg, aren't you?"

"Yes," he said. That subtle Afrikaans bedside manner again. "The
bone in your lower leg is more or less pulped from the impact."

"Well, cut if off!" I retorted harshly. You see, I have to admit it, the pain was terrible and I hated the feeling of helplessness.

He just looked at me and said, "I don't think you should talk like that. We'll wait to see how you get on." Then he moved on, followed by a starched procession of matron and a cluster of nurses.

We are never as badly off as we think. An American friend of mine, an ex-Pathfinder called Dave Barr, came to see me. He had lost both his legs, on a mine in Xangongo in Angola. He tottered into the ward on his tin ones clutching a signed copy of Douglas Bader's biography, *Reach for the Sky*. This pilot, of course, had become his role model.

"Sergeant Major McAleese!" he boomed in his wild American drawl. "I've come to see you, as I believe you're in deep trouble!"

I nodded. There was no denying that.

He gazed at me with an intense expression in his eyes. "I know you read the Bible," he said, and I remembered how we used to sit about on the sandy ground in the bush during quiet periods on operations, with a brew of coffee, and talk about Catholicism, life and death. He said, "It's written there in the good book, 'If it offends you, cut it off!' I did just that with my legs, and look at me. I'm fine!"

With that, he swung round awkwardly on his tin legs and

staggered out, teetering from side to side, a legless wreck! Was I going to look like that? But he was a good man and hard as nails. He had refused to give up free-fall parachuting. He jumped with his tin ones and learned how to land on his arse. In fact, one day, one of his legs blew off in free-fall. Someone found it in their back garden and returned it by post.

So, next day, I told Dr van Dyck to 'cut it off'.

He was a kind man and serious, for he knew I was vulnerable, lying in bed helpless and broken, and he replied carefully, "I need to go away and think about this. I'll let you know."

We resumed our conversation during his next visit. I noticed the excitement in his voice as he said, "Peter, if you can stand the longterm pain, I think I can do something to keep your leg."

I said, "How long?"

"Two years, maybe two and a half." Doctors are kind, but they are hard too. What a casual little sentence to describe so much misery!

I agreed. What else could I do? But there were times later when both Jane and I seriously wondered if we had done the right thing.

For twelve days, I lay in intensive care being tended every fifteen minutes by a continuous series of wonderful nurses, drifting in and out of consciousness on a heavy dose of painkillers. My legs stayed on. At the end of this period, during a daily visit Dr van Dyck stated, "He's stabilised." And forthwith, I was taken out of this clean white haven and thrown into the general orthopaedic ward.

Here, I have to explain the South African hospital system, because it, as much as my accident, was responsible for changing my life. All those complaining about the British National Health Service pin their ears back and listen closely, and all those cheapskates who refuse to pay full medical insurance cover think again!

In South Africa, the State paid for all the expense of the initial emergency treatment, the operation and the intensive care, but as soon as you were moved into a general ward, you got the same level of nursing care, but you paid 20 per cent of the cost of everything. They noted down every single paracetamol and tissue. As I said, I was broken physically and, soon enough, financially.

This first time in hospital, in the orthopaedic ward, we were all in a bad way. We tried to control ourselves like good manly Afrikaners and not complain about the pain, but, come night time, everyone lay in bed in their own private hell and listened for the Midnight Express. I cannot describe the joy of hearing the squeak-squeak of that night nurse's trolley which brought our painkillers. When the needle jabbed, I passed out at once and never stirred till morning, when a nurse wiped my face with a wet face towel. I slept so deeply, it seemed like one second they were drugging me to go to sleep, the next beating me about the face with wet towels to wake me up. It was like doing interrogation training on a combat survival course.

Jane was a great support to me, both as a wife and as a nurse. I imagine she will be less than impressed if I mention that she helped me with a rather vital suppository matter after leaving intensive care. I felt terribly embarrassed lying helpless on my bed, unable to relieve myself and hating to ask the nurses. With typical gentle determination, she organised the whole thing for me, told the matron she was a nurse, drew the curtains and, after thirteen days, what a relief! I apologise here and now, as those poor nurses carried on working on the ward when we were all desperate for respirators!

I left hospital only twenty days after the accident, because it was costing 100 rand a day for the bed alone. The ambulance men stretchered me into my house, dumped me on my sofa and walked out. I do not know how Jane coped, because I was completely immobile. It was not till much later that I could use a wheelchair, which Mark Adams bought for me. I have no idea what happens to people in South Africa who cannot afford a wheelchair or do not have friends as I did. In the meantime, I got around on a garage trolley, of the sort used to slide underneath cars. I rolled and heaved my plastered legs off our bed, flat onto the trolley, shoved myself around at floor level with my hands, and could just drag myself back onto our low bed.

This was one of the most miserable times of my life. I had to accustom myself to being an invalid and I was not very good at it. I am a man who loves action. I like to take the lead and all of a sudden, I had to learn how to ask people to do things for me. I was bored and intensely frustrated. I took ages to come to terms with my injuries. I never have, really. I still hanker to be the man I was before my accident.

Our marriage was under pressure at once. With typical calm common sense, Jane put Emelda and Billy in a crèche and carried on working, because we needed all our income to pay medical bills, while I was stuck at home. I had never taken the advice of Captain Harrington-Spear, Royal Anglian, all those years before, and found a hobby. I was bored, helpless and almost incapable of movement. Even as I mended, I resented being stuck in my wheelchair and hated my crutches. I frequently lost my temper and threw my wheelchair across the room. Not only did we have to come to terms with my injuries and pay huge medical bills, but Jane and I were simply not used to having me moping about at home.

I went back to work as soon as I could, ten weeks after the accident, on a stretcher in the back of a TUV truck. Of course, the 'Major' was delighted at this show of gutsy determination. This was just the sort of attitude he liked to see among his employees. He was right to say I would be better off with something to occupy myself, as I was terribly depressed stuck at home. His remarks were encouraging. I felt the business needed me. Yvonne agreed, with a wide smile of 'Welcome back', though I have a feeling she was thinking more that if I stayed home, I was being paid sick leave for doing nothing.

Later, I learned to lever myself in and out of my car. I wheeled my chair near the passenger door, opened it and lifted myself and my plastered legs inside with my arms. Then I wriggled over to the driver's seat and finally leaned back to collapse the wheelchair and pull it in.

So, I went round Pretoria in a wheelchair, trouble-shooting for the company. I suppose I was driven by frustration and pain, but I learned how to manage and delegate to such a degree that I was promoted, wheelchair and all, to branch manager. Now I was responsible for running armed bank guards as well as the static-guard business sector. With promotion came increased perception of the facade civilian security companies present when exposed to hard commercial realities. The company was expanding so fast that the armed black guards on cash-in-transit duty, responsible for moving huge sums in armoured vehicles from bank to business clients, wore imitation flak jackets. The jackets looked good but were actually made of nothing more protective than thick canvas. I guess it has changed now, but it was a good thing the ANC never found out. Maybe that kind of thing would not have mattered in the army, where doubtless we would have thought it was a huge joke, but it mattered in civvy street where it seemed to be a con, not so much of the black guards, but of the clients.

I was determined to get fit again, but I was impatient. I started heaving weights in a gym, working on my upper body, but the screws holding the steel pins in my legs loosened and suddenly I had raging septicaemia. My leg swelled enormously and I had to go back to the hospital. I lay on my bed back in the orthopaedic ward gritting my teeth, my eyes watering with pain. I refused to admit how much it hurt, partly because I was stupidly trying to be manly and partly because of the cost. A nurse noticed and insisted I take pethidine. I sank into a trouble-free sleep which turned into drifting consciousness. I became aware of a really beautiful nun in a blue habit sitting beside me.

"I'm in heaven," I said aloud.

"No, you're not," said the nun, smiling. "I'm a health visitor." She had noticed from my record card that I was a Catholic and she chatted quietly to me about charismatic faith and faith healing. My God, it was an appropriate subject. The pethidine wore off, my swollen leg felt as if it was about to burst apart, and suddenly it did just that, sending a geyser of pus from the broken skin.

I cannot remember the sequence of going in and out of hospital. I had ten operations on my legs over nearly two years. I was in and out of my bloody wheelchair, on crutches, back in hospital, and had footlong steel pins fitted, then taken out. Dr van Dyck was tirelessly patient with me, gradually pulling my right foot round straight, grafting bone he took off my hip to repair my pulped lower leg, and fixing it with bone props. Once he showed me an X-ray of how the bones were healing and I got enthusiastically over-confident. I went to answer the front door at home on crutches and, as I stood talking to an ex-British Para called Bob Phillips, my leg suddenly collapsed in an 'S' bend, like soft spaghetti, and one of the pins pressed out through the skin. Back to surgery. My treatment went on and on for two years, just as Dr van Dyck had predicted, so for fear of being boring, I will leave it at that. Suffice it to say, I had, and still have, my share of pain and I have nothing but sympathy for people in hospital.

Jane and I were constantly paying medical bills and my promotion with a little extra pay was very welcome. Our savings were eaten up very quickly, and still the State-run debt-collecting agency called Med Collect kept sending big, burly men to the house demanding we pay our bills.

"You owe the State money for medical aid," said one, towering over me in my wheelchair. He was very official, and handed me a large invoice.

I sat and read every minute detail of my treatment, and my mood changed to raging despair. Helpless, I tore the bill to shreds and flung it at him. The pieces fluttered about at our feet in the doorway.

He was appalled and shouted aggressively, "You could get into a lot of trouble for that!"

I gestured at my legs and bellowed back, "What more can happen? D'you want my bloody legs too?" The visit ruined the day.

Actually, Jane and I did once seriously wonder if it would be easier to cut the bloody leg off. Just to hire the operating theatre cost 1,800 rand but I am forever grateful to Dr van Dyck, who showed us unending kindness. He never charged for his surgery on my legs, but we had to pay for everything else. Faced with 20 per cent of the total, we were always broke.

Another problem was that I could not afford any post-operative care or physiotherapy. Actually, to be honest, I had one session. A very nice lady asked me to raise my legs off the bed to her hand, which she held out above me, and when I had done that six times with each leg, she gave me some good advice and departed. For which the State charged 20 rand. I could not afford it and I think I am still suffering the consequences. Other people who have had similar injuries say their time spent in physio learning how to walk again was as important as the operations.

My therapy was going to the office, where I had to learn a new approach to my life. I had always been the sort who goes to the other person if there's work to be done, and at first, I continued to do this in my wheelchair or on crutches, up and down corridors and so on. By the end of the day, I was exhausted. My arm muscles looked like Arnold Schwarzenegger's. I took a lesson from Mark Adams, who had been made a COIN director by then, and like a typical officer always seemed to have people coming to him. I began to learn new communication skills along with the art of delegation. The 'Major' cannot have thought too badly of me, even crippled. He promoted me again, to area manager, and put me in charge of all the company's operations in Pretoria, Johannesburg and the Transvaal. This new post gave me responsibility for all the company's security operations: static guards, all bank cash-in-transit work, and now the armed national keypoint guards. These last were run on infantry lines, armed with the .223 version of the M14 and shotguns, and they were used by the State to protect important installations, such as electricity stations, railway facilities and warehouses. They supplemented the overstretched South African Army and Police.

On July 20, 1985, President Botha declared a State of Emergency as the security situation in the country worsened. In February that year, eight had died in terrible riots in the Cape

Flats area, in March another seventeen blacks had been shot by the SAP at Langa township on the anniversary of the Sharpeville massacre, and hundreds of black suspects were arrested under the new emergency measures. Five hundred were estimated to have been killed in the previous twelve months.

My problem was that our national keypoint guards were treated much the same as the other guards. Most South Africans saw nothing in this, but I thought they really did work horrendous hours. Therefore, they were ripe targets for subversion by black activists. During a period of rioting in the townships, they decided to go on strike. I was in the company offices, as usual, and heard about it on the company's radio network.

"There's a mutiny in the blacks' hostel!" crackled the radio in panic. This illustrated the white South African attitude, especially that of the ex-army types we had in the company. The black guards were not seen as 'strikers' but 'mutineers'.

I struggled into my car and drove off for another test of my crisis management skills. I was on crutches at this time and stumped into the hostel to find blacks and whites facing up to each other in the concrete sleeping rooms, and everyone looking very nasty indeed. The company had formed an all-white internal 'police unit' to check on black guards. A group of these, in their dark glasses and American style police uniforms, were confronting a crowd of angry blacks. Outside, the night sky was lurid with burning buildings, we could hear shooting and everyone knew that there was chaos out there in the city. Blacks were fighting blacks, killing each other with burning tyre necklaces, and everywhere the SAP were trying to enforce control with riot sticks, guns and arrests. Tempers in our hostel ran high.

The senior white COIN 'police officer' waved his sub-machine gun at the blacks as if he was dying to use it, and shouted that they should be arrested for stealing Revlon makeup. "Red-handed," he bellowed down at me on my crutches. "I found it in their lockers." A quick glance at the packets showed the Revlon makeup was all rejected stock. I told him, "You can't arrest them for taking this stuff out of dustbins." The blacks were always rootling about in the dustbins to see what they could find.

Disappointed, he paused. Then he perked up and shouted in an outraged and triumphant voice, "But they are smoking dagga!"

"They always smoke dagga, don't they?"

Puzzled, he retired, trying to reconcile the truth of this logic with his president's new Emergency laws. Once I had defused the situation and made sure there was no immediate chance of slaughter occurring, I was able to talk to the Volkani Black Guards Union rep, a woman called Grace. I asked her, "Why are you on strike?"

She shrugged carelessly. "I don't know."

Another black striker was only slightly more forthcoming. He said, "'Cause we've been told." This was equally compelling logic.

It was black nationalist logic. The ANC were busy creating unrest in the townships and the last thing they wanted was our national keypoint guards to take the pressure off the overstretched army and police units. So, ANC activists inside the unions tried to take us out of the picture with a strike.

My loyalty was towards the company and, whatever the problems in the black townships, my job was to keep COIN in business. I found various posters in the hostel, a bizarre but typically African mix of religion and communism, of Christ on the cross and Karl Marx. I showed them to the security police, who studied this sinister mix as if their worst fears had been confirmed.

"That," they said in a tone of high moral indignation, "is proof of nothing less than Christian liberation theology!"

They agreed at once to help. Having defused the strike, I posted all the guards to their installations on the divide and rule principle, and then sent the police out to several posts where five particular guards were on duty who were always causing trouble among the rest. The police simply arrived, slapped on the cuffs and marched them off, no questions asked under the Emergency measures, and we were back in business.

These sort of strikes occurred a good deal and I thought the blacks had good cause to complain. The company had one black in the management, a nice man called Michael Kgabo whom Jane and I got to know well. He often joined us at home with other friends who came round to see how I was getting on. He had been educated at a missionary school but since the political system did not allow his wife Agnes to join him and live with him, because he was black, he ended up finding another woman, which broke up his marriage. I do not pretend to know about the origins of apartheid, but it caused a lot of problems.

Having been educated at missionary school, Kbago was dedicated to being a good, upright member of the community, but South African society abused his dedication, using his commitment to work but spoiling his life and that of his wife through stupid regulations. Of course, a good many of the guard force were idle by anyone's yardstick, let alone Kbago's, but the apartheid regulations did not encourage them to give their best. In Rhodesia, my black ex-guerrillas worked hard on operations with me, on equal terms and through equal dangers, and I don't think much of any system which condemns a man before he's started, because of his colour or creed.

I was normally the man the management used to handle strikes on the ground, as I am not afraid of having things out face to face with the men. I learned a lot about man-management, and dealing with black grievances in the particularly trying conditions of South Africa during the State of Emergency at that time was no easy task. It began to wear me down. One aspect I found hard to bear was pretending that we were an army unit, and not accepting that we were quite simply a company in the civilian marketplace with a manpower product to sell.

In between operations on my legs, I spent a lot of frustrating time at home. I do not suppose I made very good company and the strains on our marriage worsened – keeping pace, you might say, with the increasing trouble in South Africa. However, life was not all bad news. We kept open house for army and ex-army friends, and had numerous reunions which we both enjoyed as old friends passed through. Jane never complained at people being in the house so much. She was always busy cooking and looking after them, she was extremely supportive of me during this period and I was delighted when she told me she was expecting our second child.

Having friends round was a mix of emotions, plain good fun, a desperate Celtic-inspired wake for my old army life, and a release from my current troubles. I had been through every grade of painkiller – morphine, Omnopon, pethidine, Sosogen, Deloxene and paracetamol – and finally drink. There were always friends who came to see how I was and I drank with them through sheer frustration.

One Sunday during the State of Emergency, a crowd of mates from 44 Para Brigade came to take me to a reunion in my wheelchair. I was not feeling too crisp, but they brushed aside my reluctance, promised Jane they would look after me, and drove me to a house in Hillbrow, in Johannesburg. The guys parked me in my wheelchair in the sitting room and the party began. Several beers later, the front doorbell rang and Ossie Overall went to answer it.

A South African policeman stood on the doorstep and raising his voice over the racket of the tape player, he said, "The neighbours have complained you're making too much noise."

"Fuck off," said Ossie pleasantly.

The policeman took a long hard look at the beer we were drinking. There is a terrible streak of old-fashioned self-righteousness in South Africa and the Afrikaner does not approve of drinking on a Sunday. He said, "Do you realise there's a State of Emergency on, and the people who come after me may not be so reasonable?"

"We do. Now fuck off."

Actually, there was no law against drinking in your own place, it was simply that the South African Police did not approve. The policeman went away, Ossie shut the door and the party continued.

An hour later, we were sitting there still singing army songs and drinking, when suddenly to my amazement a gas grenade shattered the street window and burst on the carpet near my feet. I spun my wheelchair to escape to the back of the room when another grenade came through the back window. Within seconds, toxic white CS gas filled the room, and police smashed through the front door and charged all over the house shouting, "It's a drugs raid!"

"You've come to the wrong place," burbled Ossie as he was manhandled away. "There's no drugs here. We're all alcoholics."

Amazingly, the police were not wearing respirators against the gas and brought a sniffer dog with them. I can only surmise its mere presence justified their drugs allegation.

While this battle took place, I sat in my wheelchair, choking on the CS gas which hung thick in the air, and was totally ignored. Dismissively, one policeman explained why as he dragged off one of the guys. "We don't arrest spastics."

What misery. I could not stand, let alone walk, and there were stairs front and back which I could not manage in my wheelchair. I was a prisoner. For an hour, I sat coughing in the wreckage in the sitting room.

Then, to my immense relief, all the guys came back, shouting with laughter. Down at the police station, the SAP had routinely offered them Admission of Guilt forms which they all signed at once, and, still drunk, they had all been released to sin again. Such is the South African system.

The party carried on. Inevitably, because this was South Africa during the State of Emergency, the police raided us again. In strength. More CS gas grenades sailed through the windows, and hordes of camouflaged officers poured into the house, beat everyone up and dragged them off to the police station. Once again, I was totally ignored.

I spent the whole night slumped in my wheelchair in the smoking wreckage of the sitting room, feeling terribly sorry for myself. When some of them were finally released and came back to rescue me on Monday morning, I was totally exhausted.

The South Africans must have wasted a lot of energy trying to stop people drinking on Sundays but they seem to have been driven by the Afrikaner mentality. Illegal shebeens sprang up all over the place. On another Sunday, on my crutches this time, I was in one of these with Pete Donnelly, an ex-Pathfinder, when Pete said, "What's that funny noise?"

We could hear a harsh burring noise on the other side of the front door and, as we looked, a chainsaw burst through the wood, carved the door in half, and as it fell apart into the hall, several fanaticallooking policemen in black overalls burst in, all carrying buzzing chainsaws. They paid no attention to us at all, so we carried on drinking out of their way at the bar, while they charged all around the house carving up all the furniture. Minutes later, they ran out, leaving the place looking as if someone had blown it apart with a satchel bomb, and all they said to us was, "Drink up and leave!"

The pseudo-army facade of the company began to get me down about the same time the 'Major' and his partner Yvonne veered away from their enthusiasm for recruiting ex-soldiers. This was probably due to the expansion of COIN and their realisation that army men do not necessarily make good businessmen. Anyway, there were various little signs which told me my future in the company was not as rosy as I had been led to believe.

As the company expanded, there arose an obsession with paperwork, reports and request forms. Of course, all these had to be managed, so people were employed solely for this non-productive work while the likes of me went on seeking new clients and keeping the old ones happy – not an easy job, as you have seen. One new paper merchant of this type stood over my wheelchair one day and lectured me in a superior tone on how experienced he was. He ended by saying, as if I were a wet-behind-the-ears youngster, "Don't you realise? I was a policeman!" He neither knew nor cared about me or, it seemed, anyone else.

Here lies the rub. Security companies are full of ex-soldiers because they think it is the next best thing after service life. They think all the things they enjoyed about the army – like the camaraderie, the safety of the structure, knowing your place, the emphasis on sports and fitness, the training and the action – can be found in civilian security companies. Well, it can't. The civilian world is, well, just different.

In fact, I find it sad that some of the finest things that motivate a soldier in army life, like loyalty to the unit and self-sacrifice, can be (but are not always) a distinct disadvantage to him as a civilian. There will no doubt be an outcry among civilian managers and employers reading this, but let them honestly answer a couple of questions.

Who in civvy street really has a true interest in your promotion and career development? Does the company, or your boss (whose job you may want and he knows it), or is it really a case of looking after Number One?

And second, who is employed body and soul for twenty-four hours a day but only paid for eight? The soldier is, but not the civilian.

The ex-soldier goes on thinking like a soldier when he becomes a civilian. I found the transition painful, and concluded someone was cashing in on my soldier's attitudes and taking advantage of me. Sadly, I don't think my case is unusual.

I was not pleased when Mark Adams told me that several people who worked under me in my area were being paid more than me. Did all those hours I was working overtime count for nothing?

I think the last straw was when Jane went into hospital to have Catherine. It was Saturday, July 20, 1985 and I was out, as usual, driving round Johannesburg after working hours checking key-points and guards at their installations. Quite by chance I pulled up on top of a hill and listened to the company radio network.

"Call for Papa Mike," crackled the radio. "Your wife has been taken into hospital in Pretoria."

At that moment, I suddenly realised I was in the wrong place.

I had been so obsessed with my own work that I had ignored Jane at a very important time for her. I dropped what I was doing and raced in my car to the hospital. I was too late. I missed the birth. I arrived just afterwards and saw Jane with our tiny baby in her arms looking really quite beautiful. I was over the moon but at the same time miserable. Here was this woman who had done everything for me, looked after me when I was ill, helped me through all the highs and lows, cooked and washed for all the lines of friends who called on us, understanding me so well and finally having my baby. And what had I been doing? Mucking about on some routine job for a company which was a pale facade of the armies I had served and fought in and which couldn't care less.

There were other things. There always are when a person decides to leave a country. One was hearing my little son calling a black man "kaffir". He had picked it up somewhere as it was not a term I used. Of course, there are lots of slang words like that, which have never meant anything to me because my feelings for someone have nothing to do with their colour. However, I suddenly realised Billy was going to grow up attributing something fundamental to such slang and I asked myself what sort of society did I want him to grow up in, with what sort of ideas?

Jane went back to England with the three children.

Almost at once, the company showed its true colours. The 'Major's' partner Yvonne served notice she wanted me out of the job and our house. I was prepared. I had learned a few things about civilian life during the previous two years. When I told them I was going, I left my car, which by that time was three-quarters mine, at the company's garage and quit the house, which was immaculate. I had completely repainted all the rooms as new and given all the furniture to a neighbouring Afrikaans woman who had cleaned every nook and cranny. The final pathetic indignity was the company man, who was sent round by the 'Major' to collect the keys of the house. He found me sitting on a packing case in a house empty of everything except memories, and later boasted that he had been obliged to strangle me to take the keys off me. I suppose he thought the lies would do him some good at work.

I joined Mark Adams at the Palms Hotel and was enjoying a drink to celebrate freedom again when the 'Major's' brother tracked me down. He said, "Yvonne wants me to say you've got four days to apologise and we'll forget you ever wanted to leave."

Actually, I liked him, but he had wasted his journey.

Briefly, in drink, I reflected on the previous two years, the start of my civilian life. God, what a thing is retrospection! I've never feared the future, as I've no fear of death, but God spare me the pain of regret.

I drank up and left Africa.

CHAPTER 11
GUARDIANS OF
THE MYTH

I had been away for ten years. I came home with a wife and three children to find the United Kingdom changed. After seven years of Mrs Thatcher's government, people were pleased to be called British again, but bureaucracy still ruled.

We lived with John Carey, a good friend from Parachute Regiment days, and his family near Birmingham, so I put our names down for a council house and Jane tried to get a job as a nurse.

The council said to me, "Where are you living now?"

"With friends. But only till we find somewhere of our own."

"So, you're in a house already?"

"Yes, but it's not our house."

"Ah yes," he said, cunningly triumphant, "but if you're in a house, you don't need one from us, do you?"

What can you say?

I don't know what we would have done without the kindness and support of John and Julie Carey, who looked after us all those first difficult months. You really find out who your friends are when you are broke and have nowhere to live.

Jane went out to find nursing work and ran aground on the monolithic National Health Service. Officials refused to recognise her qualifications from Rhodesia, though I am sure the medical standards of the Rhodesian health service were as good as anything in British teaching hospitals. However, Rhodesia had lost and was now Zimbabwe. The Andrew Fleming Hospital in Salisbury, Rhodesia, had become the Parirenyatwa Hospital in Harare, Zimbabwe, to which she wrote for confirmation of her papers. While she waited, she found a job in an old people's nursing home pending confirmation of her nursing status. Finally, the Zimbabwean health authority confirmed her qualification but the British NHS then decided not to accept it. She lost her job. Desperate, she asked what on earth she could do to convince them she was a qualified nurse and they told her to write to the South African hospital where she had worked. Four months later, the answer came back and the NHS graciously accepted her as a qualified nurse. To cap it all, another organ of British bureaucracy then chipped in to say she couldn't work anyway because she was born in America and did not have a work permit.

During all this time, I found it impossible to find a job. The security firms in England which cater for ex-soldiers like me did not want a man on crutches. John Carey suggested I went on the dole. He drove me up to Small Heath in Birmingham, where he was the publican at the Gunmaker's Arms pub, and left me at the Social Security office. What a humiliating experience. I queued for ages with a roomful of others from the district, all Pakistanis and Indians, and all I could hear around me were incomprehensible conversations in Pushtu, Parsee, Arabic and Hindi punctuated by that single vital word, 'giro'.

English instructions to sign on were at the bottom of a list of foreign hieroglyphics. I filled out a form, signed and felt like a beggar. The dole staff were surly and antagonistic. They assumed that since I had come from South Africa, I must be rich, and they were cross because I wouldn't admit where I had hidden my money. Maybe they felt like that because of the strong anti-South African feeling at that time, influenced by the campaign to free Nelson Mandela and the anti-apartheid movement, or perhaps the British just think that hot, sunny countries are some sort of paradise where everyone is well off.

Finally, I got a sniff of a job with KMS, a security company in London. I had been in touch with the company through a good friend called Joe Lock, but I was given to understand that they did not like 'my mercenary past'. That was an unfair label if ever there was one, and based, I hope, on ignorance rather than knowledge of the facts.

However, my sister Molly rang up from Hereford, where she lived, to say that Norman Duggan was trying to contact me from Uganda. Then David Richards, an ex-SAS officer who had recently left the army and joined KMS, called to arrange an interview with the company in London.

I felt terribly self-conscious. After two years with my wheelchair and crutches, my legs were wobbly, and I was unfit and overweight. In complete contrast, David was tall, elegant, fair-haired and really rather athletic. He met me at the door of the offices and said with enthusiastic bonhomie, "How's it going, Pete?"

"Fine, fine," I said, sucking in my belly and shaking hands like a man.

"Good. Let's go and have a brew," he shouted, disappearing up the stairs three at a time.

"Yes, yes, of course," I growled, hopping desperately after him. At the top, out of breath, I stuck out my chest and tried to slow things up, walking about like a short, fat John Wayne.

"We have a team of five guys on the Uganda job," David explained, lounging in an executive chair. "The contract is with Brussels and they look after official EEC visitors who go to Kampala. God knows why, but the EEC make substantial grants to Uganda and they want to keep track of what happens to the money. One of the guys on the team called Dougie Measham, who you may remember from 'G' Squadron, is sick, so we need someone to take his place for ten weeks and it means leaving tomorrow. Can you do it?"

"I think I can spare the time," I said casually. I needed the work so badly, I would have crawled to the airport on my hands and knees, there and then.

I flew out to Kampala and Norman Duggan met me at the airport. We had not seen each other for over ten years, since those ludicrous days with John Banks. He took one look at me and said bluntly, "Pete, if I had known you were in such bad physical shape, I'd have said you couldn't come."

I swallowed it. I needed the money. Actually, Norman was always a bit of a physical prima donna, but he was right. My legs were still giving me trouble, and still do, but I set about training again. After ten weeks, I could do a whole twenty sit-ups! It may not sound much, but I was delighted. I had never thought I could again, and I kept going.

Once Norman realised why I was in such bad shape, he was a great encouragement. We used to play badminton as best we could every day, and gradually my fitness began to return. In the event, my contract was extended for a five-month tour and I managed to regain my alertness, balance and fitness. I even managed to beat Norman at badminton a few times by the end. Better still, I took 6 inches off my waistline. In some ways, this exercising took the place of the physiotherapy for my legs which I had never had in South Africa.

The job was really rather boring. We grandly called ourselves an armed response team but in simple language we were minders and drivers for the EEC officials visiting Kampala. We kept one man in the office, three with the officials and one was always on leave. Uganda is rich in farming land and the EEC grants were to help rebuild the country, which was in a dreadful state of corruption and decay after a series of selfish and destructive post-colonial leaders. The EEC money was for agricultural buildings, new coffee plantations, dams, digging new wells, water pumps and so on, and the officials were required to supervise how the money was spent. We had to make sure everything ran smoothly for them, and during the whole time I was there we had no trouble.

Don't misunderstand me. I am not complaining about the job. I was very pleased to have it at all and I think I did it well. I worked hard, got fit and did my best, but there is no getting away from the fact that bodyguarding is relentless, boring routine, even for the busy teams who look after a VIP like the US President or the Saudi Arabian oil minister. In Uganda, we were doing nothing more complicated than driving round some nice and rather naive chaps from the EEC, and the biggest threat was of being robbed by a drunken black soldier.

The Ugandans had these EEC chaps weighed up. We drove them out to a farm in the countryside to see how European money was being spent. On arrival, we found a heart-warming scene of large numbers of black men grafting away digging for all they were worth. Our EEC officials were suitably impressed, asked a few learned and technical questions and we left. Whereupon the black men downed tools and forgot all about the hole till the next visit. We saw the place often and nothing happened between EEC visits.

Uganda was in a terrible mess. The British granted independence on October 9, 1962, when the exchange rate had been 25 Ugandan shillings to the pound sterling, leaving a beautiful, rich and stable country. When I was there twenty-four years later, the official rate was about 900 shillings while the black-market rate was 33,000 shillings and the place was in ruins. We used to go out to buy the week's groceries with a quarter of a million shillings in our pockets.

Kampala must have been a beautiful city once, but was now the epitaph of so many decayed post-colonial African towns I had seen. It had become a shambles, with potholed roads, broken vehicles just rusting away where they had finally ground to a halt, decaying buildings, garbage in the streets, poverty, high unemployment and equally high street crime.

Of course, the British were blamed, but we did not cause the misery which afflicted Uganda. Tribal Africa caused that. After independence in 1962, there were endless coups and counter-coups, each one followed by arrests, torture and murders. In February 1966, Milton Obote seized power, followed by Idi Amin in January 1971. Amin's terrible dictatorship is well documented. He placed members of his Kakwa tribe in all the senior government posts, threw out 50,000 Asians in 1972 and is reputed to have eliminated 250,000 Ugandans by June 1974.

One mass grave in Idi Amin's notorious Luera triangle had become something of a tourist site and occasionally we had to take our EEC visitors to see the piles of skulls on display. The Ugandans made a big show of blaming the 'others' for these atrocities, but I have seen enough slaughter in Africa to wonder if the circumstances leading to such terrible things are ever truly in the past in Africa.

In March 1979, Amin's rule was so outrageous that the Tanzanians invaded to overthrow him and make Yusuf Lule the new president. Lule promised a return of the rule of law. He did not last. In 1980, Obote seized power again, was himself deposed by Tito Okello in July 1985, and finally in January 1986 Yoweri Museveni took over with the People's National Revolutionary Army.

Ugandans must have been utterly fed up with the thieving and depredations of one army after another. One wonders if there is a link between the size of an army and the chaos in a country. Under British rule there were 1,800 soldiers, mostly men in administrative jobs, and 6,000 police, and the country worked. Since then, armies have rampaged over the country and Museveni's personal powerbase, the NRA, was 40,000 strong. An army this size in a country as bankrupt as Uganda is a scandal. The NRA was badly paid, poorly equipped and ill-disciplined. However, President Museveni wanted to do something about it. He must have read a book about the Duke of Marlborough, who insisted his army had proper boots for his victorious marches back and forth across Europe during the War of the Spanish Succession. Museveni must have decided what was good enough for the Duke of Marlborough was good enough for him and his troops must have boots too. He rang up the Belgian Bata shoe company in Kampala and demanded, "I want boots for the NRA."

"We don't make army boots," was the reply. "We only make Wellington boots here."

"That'll do fine," said the president, doubtless impressed with the name of another famous British general, and Bata produced

40,000 rubber wellies for the Ugandan Army, all in different colours for different regiments – for example, blue for the signals and red for the military police.

On the first anniversary of the People's Revolution, Museveni ordered a grand parade and full-dress review of his troops. I have never seen anything like it. For several hours, the bemedalled president stood grinning on the saluting dais while his personal band played the famous military march "Scipio", over and over and over again because it was the only tune they knew, and rank after rank of coloured wellies flicked in the air as his soldiers goose-stepped past. I do not know if the good president was trying to emulate the goosestepping shock troops of Soviet Russia or the Nazi SS, but you just don't get the same effect with wellies.

Another time, Ugandan television showed a propaganda news report of the NRA storming ashore from Lake Victoria, still proudly wearing their wellies. We were not shown whether they had to stop once ashore, to empty them of water, but Uganda is on the Equator and we wondered at the smell.

Sometimes we took our EEC officials to the game lodge in the Queen Elizabeth National Park on the banks of Lake Edward in the Western Province, and here was proof of the transient nature of Ugandan politics. The map on the wall of the lodge had been cautiously updated with tacky little name-tags stuck one on top of the other, as the lake had been retitled with the names of all the successive presidents since independence.

Everywhere was evidence of role-playing in the name of modern politics. At the airport, President Museveni made a televised speech to cheering crowds while near us his police were beating up co-opted Ugandans who were not cheering loudly enough. During another television broadcast, the president was making an important speech to his nation when suddenly a head appeared round the side of the picture on the left, wearing a baseball hat and peering into the camera lens. The president ignored the head till it was suddenly pulled away.

Corruption was endemic, and I believe it was encouraged by

the influx of aid. For example, some supervisors of a project all bought status-symbol Pajero land cruisers, one each, which left very little for the project itself. A large open-air market outside Kampala was full of aid clothing (not made in Uganda) which, instead of going to the starving and poor, had been flogged off to the market traders for general sale. I found a tin of herring from East Germany in a shop and asked the man if he had any more. He showed me a crate.

Aid was as much part of the Cold War as arms sales, with the Eastern Bloc competing against the West, each one trying to outdo the other. The Ugandans readily took everything on offer. In addition to East German herring, they received Soviet 'Hip' helicopters, two of which crashed at the airport where they were left to rot for want of repair, and communist China sent 7 million hoes, enough for one between two of the population. This must have puzzled the Ugandans because in the past they exported hoes themselves.

I suppose some readers will call me a racist and accuse me of making fun of Uganda. My answer is that I treat all men equally, friend or foe, but that you cannot escape what goes on in societies as corrupt as Uganda was then. I imagine some black Ugandan politicians are trying to drag their country forward but, having systematically destroyed the place over a quarter of a century, they have a long way to go, and just pouring in aid is not the answer. Uganda is a beautiful country and I hope they succeed.

In the end, I stayed on the team a year until the end of 1987, when the group split up because a non-ex-SAS man was put in charge. There was no reason why a non-SAS man should not have been in charge of such a routine job, particularly when he had seen it going on for many months, but what irritated some people was that this chap pretended he was ex-SAS. He would cunningly say, "I'm from Hereford, so let's just leave it at that." This dishonest implication that he had served in the SAS was making good use of Joseph Goebbels' maxim, "Good propaganda must contain a grain of truth." Which was that he had lived in

Hereford, though never serving in the SAS. He was a civilian mechanic and the sum of his military knowledge was based on having been a National Serviceman in the RAF. He gave himself airs and his appointment as team leader caused argument which filtered back to London, where it made a poor impression. There was no question of excessive behaviour. While I was there for a year, none of us drank much at all. I for one felt better for it and enjoyed the self-denial. However, the London office decided to change the team and maybe they were right, after so long, to make radical changes to a small team like that. What was wrong was putting a weak man in control.

After Norman Duggan left, the team needed good leadership, although there was absolutely nothing secret, exciting and dangerous about that job. In all that time there was only one drama, which happened before I arrived. An unfortunate Ugandan ex-soldier was shot dead trying to burgle an EEC house. I will never forget it because the guys who killed him never stopped talking about it during the long, empty hours we had together. The trouble was that this and other not very interesting military events, like a parachute jump on Hankley Common in Surrey, seemed to be the most important war stories they had to talk about. I think they knew I had kicked around in Africa, but they were not interested, even perhaps a little jealous and embarrassed to ask, and carried on as if I had nothing to offer. I think that was foolish. Boasting is all very well, as long as you don't take yourself too seriously, but there is always someone else around with more experience than you. This is especially true in Africa with all the wars there have been there, and, thinking of all the Rhodesians and South Africans I had served with, I found it hard sometimes when some 'old SAS hands' carried on as though the British Army's experience of Northern Ireland was the height of soldiering and the only thing that mattered.

Actually, on reflection, I think it may be true that there is more blind arrogance in the SAS about other people's experience than in any other unit I've encountered or served in.

Perhaps we were too old for such a job, too set in our ways. The system was never questioned. Perhaps the system was not necessarily wrong, but it was not necessarily right either. For example, we carried 9mm Browning pistols and the Model 12.9mm Beretta SMGs, but we never practised on a range! This ran against the grain for me, and I made suggestions which fell on deaf ears. The arch-pacifist, Hermann Hesse, said in *Demian* in 1919 just after the First World War, "The man who is too lazy to think for himself and to be his own judge accommodates himself to existing laws, such as they are, and has it easy." We certainly had it easy.

I think that was the root of the problem. The job was undemanding and trivia became important. People would sulk if the security lights were turned out a few minutes late in the morning (even though they had been on for the previous twelve hours), one person was always grumpy at breakfast and refused to talk before he had had a cup of tea, behaving but certainly not looking like a Guards officer, and there was always bickering about who was in charge of whom.

I started an incident book in the office. The innovation was accepted and the man on duty used it to record and pass messages which came in by telephone or radio. One day, I logged a radio message and shortly afterwards another of the team called Pat walked in.

"I've just got a message on the radio which you'll need to know about for tomorrow," I said to him and explained the detail.

"You must tell Norman to tell me," he said. At this time, Norman Duggan was the team leader.

I took a deep breath and said, "Pat, I am going to tell Norman, and I am also going to tell Norman to tell you."

"Only Norman is allowed to tell me what to do," he repeated petulantly.

"I am telling you, Pat," I said with heavy irony, "because you are standing here in front of me, and just in case Norman forgets to tell you himself!"

I found this attitude incredible. The man was 45 years old, one of the super fighting men of the Parachute Brigade which the media so love to admire. What had happened to him? How had he changed so much from the time when he was a serving soldier in the British Army, when he certainly had all those qualities I have talked about before – commitment, loyalty, self-sacrifice and the willingness to get on with a job for very little reward?

Quite simply, the answer is money. He and so many like him have become prisoners of the economics of modern special forces. Of course, you'll be saying, all soldiers have to wise up about pounds, shillings and pence when they leave the army. Blame the civvies. But the fault is not all civilian. The rot sets in before they leave the army. Long after I left, the British SAS were given special pay, to reward them for the extra responsibilities of special forces work. That may be proper, but whereas my contemporaries and I joined the SAS (and the army, come to that) for the life – the pay was appalling – there are now, without question, soldiers who join the SAS for money. Most of them own houses now, which they could never have afforded in the past. So, instead of being committed to the life, too many are committed to large mortgages, and when they leave they have to earn similar sums to maintain them. They are prisoners of their financial commitments, just like civilians, but they are flung out of their employment after twenty-two years and told to start again. Ex-SAS soldiers turn to these civilian security jobs, quite understandably, because they pay much better money and are much easier than driving trucks or digging holes or retraining to start a new career completely afresh.

Of course, everyone has to pay the bills, me included, but what I could not stomach was finding so many of them with absolutely no belief in what they were doing. I firmly believe if you agree to work for someone then you must be committed to that job lock, stock and barrel. I have always worked on that principle and always will. However, some of these guys were only interested in counting their money and totting up their expense accounts. They were actually not interested in the job at all. The real criteria now were pay and extras, not personal commitment, let alone risk. And God, what a pity. What a waste of good men. What prostitution! They are masquerading, pretending that they still have a devotion to duty (which in the civilian world means the job), whereas in fact they are saying, "My loyalty, my dedication and my devotion to duty can be bought!" For that, whether civilian or soldier, there is no excuse.

Are these, then, the real mercenaries so beloved of the media? Set up and paid up by the British government, then signed off, written off and launched onto the civilian market?

CHAPTER 12
HIRED BY MEDELLÍN CARTEL'S GACHA

I can never resist a challenge. When Dave Tomkins told me in the Booth Hall pub in Hereford that he had met a Colombian Army officer who wanted a FARC headquarters attacked, I told him I would go and look at it.

Dave was an old friend from Angola days, a civilian without any previous military experience who had been wounded and shown a lot of guts. In June 1988, he rang up out of the blue and we met in the Booth Hall, an old SAS haunt in Hereford which is now boarded up. We had not seen each other since Angola, so God knows what he thought of the way I looked, but he was as ever a picture of studied elegance – tall, lean, tanned, his hair very grey and fashionably long. I was amused to see he had developed a habit of rather ostentatiously fingering various gold ornaments as he spoke: several large rings, a heavy bracelet and a medallion on the end of a thick gold neck chain which he pulled out through his open-necked shirt.

He explained the background. He had made good capital out of his contacts in the Angolan affair and become something of an arms dealer. During the course of business, he met the Colombian Army officer who had asked if he could provide some foreign soldiers to attack a FARC headquarters.

The FARC (Fuerzas Armadas Revolucionarias de Colombia) is a communist organisation which receives political support from Cuba, in the form of Cuban commissars who train and indoctrinate the FARC military groups, called 'frentes'. Over the years,

FARC has carried out some dramatic attacks on the Colombian government, perhaps the most spectacular being the total destruction of the Justice building in the centre of Bogotá, along with all the documentation on FARC and other anti-government groups. In accordance with Mao Tse-tung's principle of liberated areas, their control over the vast scrublands of the Llanos and the jungle further south is almost complete. They are so confident that they run big training camps where they openly fire weapons on jungle ranges, bring in local villagers for indoctrination and film shows, and drive about in Russian vehicles, all in broad daylight under cover of the jungle canopy. The Colombian Army and police enter the area at their peril.

"The army is fed up with the politicians," Dave said. "The Colombian government is trying to pursue talks with FARC, and the army is caught in the middle. Every time the army wants to attack, the government says dialogue is more important."

"That sounds a familiar story."

"So, they want to teach FARC a lesson. They want to attack a place called the Green House, Casa Verde in the local lingo, where they reckon the FARC politicos have meetings. The place is about 80 kilometres south of Bogotá, in the mountains. They want outsiders to do the job so the army can't be blamed."

"Non-attributable."

"Yes. The whole thing is very secret. This officer and three others have been taken away from their units, supposedly on a staff course, to handle the project."

"I'm interested," I told him. "But I want to look at it first."

This story rang true, especially the part about the army being piggy in the middle, but I had to be careful. Every year Colombia produces 185 metric tons of cocaine, which converts to a street value of US $44 billion, most of which is controlled by Colombian cartels. This staggering figure excludes profits these drug traffickers make from marijuana and an increasing trade in the even more profitable cultivation of heroin poppies.

It also excludes the profit made by Colombian cartels from

trafficking cocaine grown south of Colombia in Peru and Bolivia. Legal national exports simply cannot compete. The World Bank values Colombia's legal production of coffee, bananas, cut flowers, clothing, ferro-nickel and coal at a mere US $5.8 billion. Is it any surprise that the corruption of drugs infects every layer of Colombian society, and that most, if not all, important government appointments are affected by it? I wondered how the people backing this job fitted in.

I knew FARC was involved with drugs. An essential for all guerrilla groups is adequate finance. Like communist movements all over the world, FARC extorts local 'taxes' from tribesmen and villagers, and they still kidnap whites from time to time, but their main source of funds is growing cocaine in the jungle and selling the unrefined base to the drug cartels in Medellín. This produces a cash flow which must be the envy of every dissident group in the world.

However, I have been fighting communism in one form or another all my life, so the job sounded good enough to me and we flew out to Bogotá to have a closer look.

We arrived on Friday, July 1, and stayed in the Plaza Hotel. Dave's Colombian officer, nicknamed 'Ricardo', arrived. A large, likeable man, well-muscled but slightly overweight, he was dressed in a snappy grey suit with rather too much silver thread in the weave, his jetblack hair was smoothly brushed and he sported the typical neat South American moustache. After introductions and a short chat, he said, "I must go now. I'll be in touch."

We waited in our room for three solid days. The hotel staff must have thought we were a couple of gender-benders! Determined not to miss a phone call, we stayed in, lying on our beds, reading, using room service for a constant supply of club sandwiches and Cokes. We took it in turns to leave the hotel, just to get some fresh air and have a walk. Bogotá is at 8,600 feet above sea level, usually overcast with grey smog, and sprawls from a soaring concrete and glass highrise centre to filthy, poverty-stricken barrios on the outskirts.

Nearly 5 million people live there, on streets bursting with noise, murderous traffic and unbelievably filthy poor men, women and children who will try and sell you anything from a stolen gold watch to a single cigarette. We convinced ourselves we were being followed, and went into an absurd and elaborate counter-surveillance drill, cutting back and forth through shopping malls and supermarkets, certain there was a spy round every corner. The fact is, South America was new to us, there were no watchers and we soon settled in, suspicious still, but behaving less like a couple of trainee agents in MI6.

Finally, Ricardo appeared again, apologised and said, "The men you need to meet aren't available yet, but I'll be back in twenty minutes."

This was South American *mañana*. We stayed confined to our room for two more days when Ricardo returned with three darkhaired, swarthy friends with the usual neat moustaches and wearing rather tight-fitting double-breasted suits. He introduced them as the others who were involved in the project.

One man said very little, but the other two were ex-army officers and supplied me with a full set of 1:50,000 maps and air photographs of the Suma Paz area. They understood when I asked to make a helicopter reconnaissance of the target. We flew from Bogotá's old airport in a small Cessna, but bad weather typical of the valley forced us to land at an army airstrip in Villavicencio, where we hung about in the canteen and drank excellent Colombian coffee. Conversation was stilted, as we spoke no Spanish, and, as usual in these situations, revolved around the most impressive thing they could remember about their guests, which was the world-famous British obsession for 'fish 'n' chips'!

We took off in the Cessna again and were soon lost in a maze of mountains, flying in and out of cloud which clung to the dark green trees and thickly carpeted the mountain slopes below us. The Colombians excitedly pointed down, saying, "Casa Verde!" But, from the air, the jungle is like an endless field of undulating broccoli and, as there were no tracks or clearings, I found it impossible to orientate myself between the gaps in the clouds.

We tried again the following day with more success and flew slowly up the Suma Paz valley with huge mountains towering above the little Cessna on all sides, rising to the peak of Cerro Nevada at 15,048 feet. Casa Verde is hidden in trees near the top of the valley, on a distinctive plateau rising steeply above the valley floor. So, I was not able to see the house itself, but I formed a good impression of the ground conditions for our approach and withdrawal. On the other side of the pass, the foothills gradually slope away into the flat river plains of the Llanos in Meta Province and the vast Amazon jungle beyond.

Later, in our room in the hotel, Ricardo and the others spread out a very detailed sketch map. At once, I recognised the layout as a complete communist base. He described the principal locations, showing the secretariat and the training camp, and casually mentioned that the main accommodation was built for 400 guerrillas.

"This is the target," he said, his finger on the secretariat.

"Then where's the Casa Verde?" I asked, puzzled.

"Over there," he replied, pointing to a place on the map about 6 miles beyond the FARC communist camp.

"Well, which do you want me to attack, the Casa Verde or the secretariat?"

At this, they exchanged glances among each other and eventually said, "Can you attack the Casa Verde?"

"Yes, certainly," I told them. I wondered why there seemed to be some confusion about the mission, but put it to the back of my mind. I was happy to attack either place, but clearly the secretariat of a huge FARC camp was a different proposition to a few communists in the Casa Verde, and I wanted the mission quite clear before I started work.

They left and, using their photographs and maps, I prepared a full set of orders, and built a model of the Casa Verde based on their sketches. The house was small, a typical Colombian mountain farmhouse of wattle and adobe with terracotta tiles on the roof. When I was ready, I was taken by Ricardo to an office near a large department store in Bogotá. The other three men were there, all in civilian clothes, with an older man who wore a suit but had the unmistakable stamp of a South American general. I spread out my plans, set up my model – just as the Army School of Instruction tells us – and ran them through the whole plan.

I got quite fired up myself as I talked, visualising the attack, and, when I had finished, they exclaimed, "Bueno, bueno!"

The general asked me how many people I needed. The fees for all this were Dave's area, but I chose the most men I thought they could afford. Ten, plus Dave and I. Twelve men gave me flexibility in patrol numbers, two sixes or three fours, and extra firepower. All support, food, accommodation, training areas, helicopter air support and transport would be provided by them. They said they would think about my proposition.

At the time, Dave was out of the country organising non-attributable military equipment to be brought in to do the job. The weapons were being provided by the Colombians. There must be more unsourced weapons in the country than people.

However, we had the basis of a plan and, after two weeks, relations with the Colombians were very good. One day, two of them arrived in great excitement, without Ricardo, and dragged me out of the Plaza Hotel for a long drive over terrible roads into the countryside in their Land Cruiser, which is the status symbol of every successful Colombian. They kept pointing at the tree-covered hillsides and talking about, "Sumapaz!" and "Bandidos!" so I assumed we were on another recce and concentrated on the task. After more than two hours of bone-rattling terror while the driver played chicken with oncoming traffic, we stopped in a large but scrappy riverside town called Girardot on the Río Magdalena, which looked like a Hollywood set for a Wild West movie, with tall facades fronting the single-storey buildings along the main street and dusty alleys between the houses at the side.

They led the way into a café and shouted at the owner, who listened expressionlessly, nodding and wiping his hands on a soiled apron. He came back a moment later with aguardiente, the local firewater made out of distilled sugar cane. We sat drinking and I was ready for anything except what happened next. Suddenly, the café owner reappeared with a flourish and I found myself looking at a large plate of fish and chips.

My Colombian friends grinned delightedly. They had planned the whole, long day solely to please me, for a single plate of fish and chips, convinced I would be suffering withdrawal symptoms after so long away from my national dish.

Back in Bogotá, word came that they had agreed to my plan and the numbers I wanted, and the senior army officers were satisfied with Dave's prices. I was told, "Go and get the men."

On Friday, July 22, I flew out of Bogotá's Eldorado airport and went through Heathrow, London, to Johannesburg, where I took a room in the Carlton Hotel. I had some friends to meet.

I chose men I knew well, whom I had trained and seen on operations in South Africa's 44 Parachute Brigade. I met them all individually to put the proposition to them in private, and not one refused. With Dave Tomkins and I were Terry Tagney, Dave Borland, Dean Shelley and Gordon Brinley, all men who had also served in the Rhodesian Light Infantry; Gerry O'Brien, who had also been a French Foreign Legionnaire; Jock Moore, who had also been in 15 Para Territorials in Scotland and the Rhodesian Engineers; Mark Griffiths, who had also been in I Para in the British Army; and Roy Kaulback, who had been to the famous Scottish public school, Gordonstoun, before seeking adventure in South Africa. The last man was Alec Lennox, for whom I had found a job training Auxiliaries in Rhodesia, which made eleven, as the twelfth never actually joined us.

They wanted to know whether I thought there was a connection between the job and the so-called drug barons, and I told them, "This task seems as straightforward as anything in Colombia. I've been dealing with army officers; we've been to an army barracks and they produced photographs and sketches with army classifications. I'm sure the Colombian secret police, the DAS, are involved too, because my passport entry visa was stamped without any of the usual bureaucratic hassle and the DAS will have to help Dave bring in the military kit through customs. The whole thing is top secret but I really do think we're working for men in bona fide government agencies, and the bottom line is that we're attacking commies, or drug producers, or both." That was good enough for them all.

I arranged a complicated flight plan for each of them to arrive in Colombia separately. The last thing I wanted was a milling crowd, like John Banks's recruits setting out for Angola with journalists asking to be taken on the coach which drove them to the airport. This operation had to be totally secret.

While I was doing this, Dave Tomkins was buying radio equipment. I returned to England and bought other stores we were going to need from an army surplus shop near Hereford – boots, rucksacks, carabiners, gloves, abseil rope, camouflage clothing, binoculars, compasses and so on. As a precaution, I kept receipts for all this equipment and, as a matter of interest, at no stage did any customs official in England or Colombia stop me at customs.

On Thursday, August 18, I flew back to Colombia, arriving at 6:30 in the evening. I took a room at the Hotel Dan, and the others joined me over the next three days, flying in via Germany, the UK, Uruguay, Madrid and even direct from South Africa. We behaved like regular tourists during this time, going out in ones and twos to visit various sights in Bogotá, like the Gold Museum and the famous church at Monserrate, 2,000 feet above the city at the top of a railway funicular. Standing overlooking the whole city, I thought this was not the first time British soldiers had fought for the Colombians. When Simón Bolívar fought to evict the Spanish and created Gran Colombia, including the present-day countries of Colombia, Panama, Ecuador and Venezuela, he had two battalions of British adventurers fighting with him, called the 'Battalion O'Neal' and 'Battalion Ferguson'. One of our Colombian friends told me they were filled with men who had been fighting in the Napoleonic Wars. Anyway, they sounded plenty Celtic enough for me!

Once all the guys were together, Ricardo arranged for them to be bussed down to Puerto Boyocá, and they drove about 200 kilometres on winding roads through the mountains of the Eastern Cordilleras, dropping all the time until the road joins the flat river valley of the great Río Magdalena. This enormous slow-moving river is a main artery of the country, flowing from its source high in the cold Andes 1,200 kilometres to the Caribbean coast. Halfway, at Puerto Boyacá, life on its banks is hot, sweaty and truly tropical.

I was busy in Bogotá, organising maps and so on, and joined them by Land Cruiser with Ricardo the following day. The Colombians had chosen a very discreet spot for our training. Our base was on a small island about 150 metres wide in the middle of an oxbow lake hidden some way from the Río Magdalena by thick trees. To cross to the island, you had to call up on the radio, and they sent over a boat. It was a four-minute ride over the smooth dark waters to the island.

It was rightly called 'Isla di Paradiso'. We were catered for by a delightful couple who lived in a whitewashed villa with a low terracotta-tiled roof which sloped over a verandah shaded all around with colourful bougainvilleas and hibiscus. Wooden steps dropped off the verandah to the ground, and a barbecue and bar stood near the jetty. Our accommodation was in several similar buildings behind the villa.

They fed us very well, on rice and chickens, which until they were needed scratched about in the yard for food. A small black pig which rootled about by the house was destined for the pot too, but I grew very attached to this pig. By holding my hand up like a witch-doctor, I could make it lie down in total submission and, after I mended a nasty cut on its nose caused by a brass ring (I'd learned a trick or two since the goat in Aden), the pig followed me everywhere. Our South American couple thought this was very charming, but when it came to barbecue time, I had to plead for my pet pig's life and felt very British.

"Don't kill this pig," I pleaded. "I'll get you another pig, but don't kill this one!" The little black pig was reprieved but what happened after we left I would rather not think.

We had no weapons yet so I started a fitness regime. Every morning we got up at five o'clock, took the boat over the water, which was smooth as black glass at that time of day, and ran as fast as we could for 7 kilometres along tracks through the trees. Each day, we tried to cut the time down on the day before.

Our Colombian liaison officer at Paradise Island was an called Julio who had been in a Colombian detachment with the United Nations in Sinai, and we got on well together. Like me, he was always busy checking things, making sure equipment was in place and keeping an eye open for trouble. After some days, he took me to Puerto Boyacá in his Land Cruiser. We pulled up at a seedy-looking hotel in the town, he led the way into the bar, and there I met a man called Henry. I have no idea if that was his real name, but that is what we called him. We sat down to tea. Henry wore a smart shirt and slacks, and showed the exaggerated good manners of a celluloid villain. I discovered later that he was in charge of the peasant militia and took a very strong line indeed rooting out communists in his area. In other words, he killed them. We sat down to tea with a couple of Colombians in army uniforms and talked about the weapons I needed for the job. At this point, another person appeared and without ceremony dumped a large canvas tool bag at my feet.

"For you, my friend," said Henry, grinning evilly. "Take your choice!" Puzzled but keeping my cool, I bent down and opened the bag. It was full of pistols. Fantastic, large-calibre pistols. There were Smith & Wesson .357 Magnums, Colt .45s, a 9mm automatic Smith & Wesson Model 59, and a truly massive .44 Magnum Israeli automatic called the 'Desert Eagle'. Every single one was chrome-plated. *This is like the movies*, I said to myself. *What have I got myself into?* I picked the Model 59, thanked Henry politely and Julio drove me back towards the island.

I reflected on the Israeli connection. Julio had spent time in Sinai, where he had had contact with the Israeli Army, there was the Israeli .44 Magnum automatic pistol which I knew had only just been produced by Israeli Military Industries, and there was a pair of slippers I had found in my billet on Paradise Island which were marked "Made in Israel". There was certainly an Israeli link through arms sales, but did it go further?

Some days later, Julio turned up with a Land Cruiser full of German G3 rifles and I could see at once that they were all in extremely bad repair. In my ignorance, I put this down to the South American factor and told myself, and the guys, that at least there were plenty of them for spares. I amended our daily routine and wrote a programme of training similar to the one I had organised for 44 Para Brigade in Mabalique. We were up at five o'clock and after our run, a shower and breakfast, we started training at nine o'clock. To start thinking like soldiers again, we began with basic weapon skills on the G3, dry and live, followed by immediate action drills, fire and movement, battle skills, patrolling and navigation. There was scrub and trees on the island for dry training and small exercises, but we went over to the mainland for shooting practises and longer fourman patrol exercises.

I made them work against each other, one patrol tracking or ambushing the others in the jungle undergrowth, as I had learned to do with the SAS in Borneo, and I made them all do written military appreciations of various situations which we might face – on the way in, on the target, on the way out and so on. We talked through all these, picking holes in each other's plans. Every day I put a different man in charge of the patrols, mixing them around, so everyone had a chance to demonstrate their version of command and influence the decisions, and also face criticism from the others.

There was method in this. I was not dealing with a group of men whose minds and attitudes were disciplined by the structure of a regular army. These men had all served in a regular army but now they were civilians, working solely for money, as true mercenaries. I had seen in Angola, South Africa and Uganda what can happen to regular soldiers left to their own selfish devices in the civilian world. The last thing I wanted on our job was bickering, because, by the law of that great scientist Murphy, it would happen at a crucial time when least wanted.

I wanted to bring all the gripes, complaints and feelings into the open, and iron them out, which we did. Finally, to set the seal on this process, I gave them all paper and pencil, and told them to go away somewhere private and assess all the others on marks out of ten in a variety of qualities which I listed for them, such as leadership, initiative, self-discipline, knowledge of soldiering, personal skills, weapon handling, shooting and so on. Later, we sat round by the barbecue and each one read out their assessments, so everyone knew exactly where they stood with the rest. Out of this came a wellunderstood pecking order, a rank structure or line of command which I felt confident would work when I was not around. I know this sort of democracy is out of the question in regular armies, though the Americans use something like it which they call 'bayoneting', but it worked for us. I hardly need tell you I came out top!

I had asked Henry when the operation was going to begin and one day, about four weeks into our training, we were on the mainland during a shooting practise in the trees when four very smart Land Cruisers drove up the track, gleaming with chrome fittings and one with a GPMG machine gun mounted on the roof. They skidded to a halt, hard-faced Colombians leapt out, all heavily armed with pistols and clean new Armalites, and immediately took up rather professional-looking protective 'bodyguard' positions round about, facing outwards. Two even carried 66mm Light Anti-Tank rockets slung over their shoulders. Two others ran off to one side, where they rapidly set up communications with a radio. Rather impressed with the discipline of all this, I watched Ricardo, our liaison officer from Bogotá, step down from a Land Cruiser with another man, who was casually dressed in shirt and slacks, like the others, but his clothes were expensive. Everything about him spoke of command.

Ricardo introduced him, "Mr Peter, this is one of the sponsors of our operation, Mr Rodríguez Gacha."

At the time, his name meant nothing to me, but two things were perfectly obvious. He was not in the army and he was a big Mafia man. This confirmed the suspicions I had had all along that army officers, no matter how well motivated, could not fund the sort of operation we were doing. There was always a financier and in Colombia that means drugs.

For the righteous who are scandalised by my working for a Mafia and drug baron, it is worth explaining something more of Colombian society, which, it must be said with massive understatement, is rather different to our own quiet and structured life in Britain.

Colombians live on the edge of violence, and always have. The country was born in violence. In the seventeenth century, Spanish, French and English adventurers murdered each other up and down the Caribbean in their quest for the gold of El Dorado, officially encouraged by the kings and queens of Europe. The Spanish finally settled the northern coast of Colombia and set about murdering the Indians who stood in their way, ably assisted by their priests, who burned more Indians in the name of the Spanish Inquisition and for the good of their souls. Plenty more people died in slavery.

This tendency to murder was maintained throughout the eighteenth and nineteenth centuries and into the twentieth century, with ruling settlers fighting each other in endless civil wars. The last ended as recently as 1960, after sixteen terrible years referred to as 'La Violencia', when rival Conservatives and Liberals between them killed some 300,000 people. In a land consumed with machismo, without doubt bitter memories live on.

On top of all that, and in addition to the usual dose of South American corruption, the country nowadays is torn between the FARC communists, with all the typical violent communist tricks of indoctrination and terror, and the brutalities of the huge drug business. There are at least 15,000 murders every year, most drug-related. In Colombia, there are no shades of grey on the fragile front line between life and death, and this violent history seems to have resolved itself into a straightforward conflict between the Right and the Left. The Colombian Army is basically right wing, and allows the local militia – a sort of home guard – to do what it likes as long as it suppresses communism. In other words, in stark Colombian terms, the army condones, at minimum, the support of the drug business as long as the worse menace of a communist takeover is defeated. So it was that our invitation and weapons came from the Colombian Army or militia, while the cost of the operation was borne by Gacha.

Gacha behaved like a visiting general. "How's the training going?" he asked me, speaking broken English. I briefed him, just as I had so many officers over the years past, and then he had a few words with the men, along the lines of "Boots fit? Food all right? Getting the mail?" He looked a fit, tanned man and showed a serious interest in what we were doing. Then he left. His screen of armed guards piled back into their Land Cruisers, never taking their eyes off the area around, and the cavalcade disappeared down the track in a cloud of dust.

Two nights later, we were relaxing on Paradise Island after the day's training, sitting under the stars round the barbecue fire, when Julio called me up on the radio from the mainland. His voice crackled, "Mr Peter! Come over. I take you to visit. Bring your plans."

Ricardo and I crossed the dark lake in our boat and Julio drove us in his Land Cruiser to a secret meeting. The journey there took four hours. At the end of a long road through grazing land, we stopped at a luxurious Spanish villa with a wide verandah all around, ornate wrought-iron screens on the windows and whitewashed walls shining in the bright security lights.

Half a dozen bodyguards were standing on the verandah, one armed with a 66mm anti-tank rocket, and others lurking in shadows in the garden. Inside, we were taken to a spacious dining room with an enormous circular table in the centre. Suddenly, there was a tremendous noise of horses clattering up outside and a crowd of men burst in wearing classic South American gaucho clothes, ponchos, pointed riding boots and hats all spattered with mud. Everyone was shouting and laughing. Gacha was among them, in great form, and we shook hands. Beers were served all around, they lounged about on chairs and equally suddenly, they all fell totally silent.

"Tell us your plans," said Gacha, his black eyes deadpan.

In an atmosphere of great seriousness, I detailed the plan to attack the Casa Verde. No one said a word till I finished, and then they all burst out shouting, "Muerto a los communistas! Bueno! Bueno!" and stood up, their ponchos swirling, punching their hands into the air.

"We will set up a radio station and tell the world that Casa Verde is destroyed!" shouted Ricardo, wildly excited.

Rather detachedly, I observed this frenetic South American enthusiasm and noticed Gacha was not joining in. He had the most penetrating eyes. I wondered how many people he had murdered to get to his position. Suddenly, the room fell silent again. Gacha looked at me straight in the eye and said, "Could you attack the secretariat?"

"How many communists are there?" I asked at once, remembering the same question in Bogotá and realising this was what they really wanted us to do.

"Seventy men."

"I'll do it," I said, feeling their enthusiasm beginning to affect me too. "But I need more men."

"Bueno!" said Gacha. "Go away and make your new plans, and tell me later what you need."

Whereupon, the place erupted again with ponchos, hats and shouts of, "Kill the communists! Kill the secretariat! Bueno! Bueno!"

I had obviously said the right things, as it only took three-quarters of an hour to drive back, and I reached Paradise Island in great excitement. "We are going to war!" I told the guys, and we all felt better with the news that matters were now on the move.

In a subsequent meeting with Ricardo and Julio, I said I wanted another twenty-five men to attack the secretariat, with all necessary ancillary equipment including helicopters, to which they replied, "We will give you a number of our men to train up and you can select what you want from them, but we don't have any helicopters in this area."

I knew I would have to come back to the problem of helicopters later, but training local men was all right. I had trained so many indigenous troops in the past – in Guyana, in Rhodesia with the Auxiliaries and in South Africa – so I agreed. It was obviously a cheaper option for them, but I was relieved too. Finding another two dozen men from Europe and South Africa would have delayed things terribly and might have blown the whole thing to the press.

Sadly, Paradise Island was too small for these developments and we moved. We drove for two hours on bumpy country roads north of Puerto Boyacá and turned away from the Río Magdalena into the hills. I noticed there was a shack on every little road junction selling sweets and lemonade, although there seemed to be no one around to keep them in business. We stopped at one for a drink, and inside I saw the reason. Behind the counter was a smart new radio. All these little shops were part of a network which covered the whole area so that nothing could move without being reported.

Our road climbed towards the foothills, winding between grassy hills cleared of the mixed secondary jungle trees and palms until we came round a corner to find two bunkers, one on each side of the road, both in disrepair and unoccupied. I dismissed them as typical South American Army show-pieces but when we drove into the small valley beyond, to find our new base, I saw bunkers above us on all the high ground around. Furthermore, when Dave Tomkins and I went for a walk to look for suitable training areas later on, we visited these bunkers and I found they had all been very well sited by professionals, and were mutually supporting, with at least two being able to provide interlocking fire with each of the others.

I was impressed. While we were up there, I was looking round and something caught my eye in the treeline near the top of a ridge of hills about 3 kilometres away. It looked like a man-made gap in the trees. Curious to know more, Dave and I walked up there, brushing through the grass and sweating in the hot tropical sun. We found a Bell 214 helicopter hidden under the trees. Enemies could not see it from the air, but we had spotted it from below, where only friends were expected to be. They had cut a tight clearing into the trees, enough to push the helicopter under cover on wood planks, and left it covered with a silvery green tarpaulin. There were no houses nearby and the unusual atmosphere was enhanced because there were no guards. Plainly, Gacha's people felt so secure in this area, the chopper was quite safe on its own. Perhaps there were more hidden in the trees beyond, which we never found. Anyway, so much for there being no helicopters.

Our camp was laid out among the trees in the bottom of the valley by a stream, where the scrub had been cleared and the grass cut. We lived in wooden huts with wriggly tin roofs covered with atap leaves, and Julio turned up to help us settle in.

I faced him and said, "I thought you had no choppers?" watching carefully for his reaction.

For a moment, he looked puzzled, wondering how I knew. "There's one up in the hills over there," I said, pointing.

At that, he just smiled, giving no explanation, and carried on with his work.

The fifty or so Colombians I had been promised turned up dressed in shirts and slacks, and armed with a variety of weapons in reasonable condition, like Armalites, AKs and pump-operated shotguns. I gathered them together on the grass near the huts, briefed them on our training programme and we started work. I must admit that these chaps, who came from all over the country, were keen, but they were very right wing indeed. Every morning, before we set off on our run, they would stand rigidly to attention on the grass and their squad leaders shouted at them from the front, "Colombia patria mia!"

The response came bellowing back, "Colombia patria mia! Colombia patria mia!"

Then they turned smartly to the right and doubled off on the run. Our valley was at the end of the road, under the foothills of the higher sierras beyond, and needless to say there was another shop where the track petered out in trees and scrub jungle. Now who on earth was going to buy anything from here? Of course, it had the usual radio inside and we used to wave cheerfully at the man every morning as we ran past.

When it came to military training for these new men, I had to start with the basics. They looked good enough toting their weapons for the cameras, but the cohesion of trained soldiers which I wanted for this operation was not there. We started with basic infantry training and individual jungle shooting lanes. There was nothing specialised about this. We were not training drug cartel 'hit squads', as various frenetic journalists later accused us of doing. For that, these Colombians did not need our help. With 15,000 murders every year, men like these grew up with guns as kids and it is the most ludicrous arrogance on the part of our press to think they need outsiders to tell them how to kill each other. Colombians have had plenty of grim practise at that, as anyone who looks for five minutes into their history will see.

In the first couple of days, I noticed Colombians coming and going to one particular hut in the valley near our billets and went to have a look. Inside, radios were stacked on tables and it served as base station for all the little shacks we had passed on the drive in, manned twenty-four hours a day. All in all, the area was well defended and I wondered at the reason for all this organisation. I was sure the setup was not just for us, because there was an air of permanence about it, and I guessed at various possibilities; that Gacha or one of his family had a house nearby which needed defending, or, more probably, that there was a cocaine laboratory close by.

I never found out, and our training never really got started, because, after only four days, Ricardo arrived in the evening and said, "We have to move from here at four o'clock in the morning."

I was not best pleased, especially when he could not say why, or where we were going. All these changes in the plan were beginning to affect the guys, and changing tack all the time is no way to run an operation. However, we were all committed and the following morning we rolled out of our narrow wooden cots in the dark, trucked back down the dusty roads we had come on, and arrived at the broad muddy waters of the Río Magdalena in the first grey light of dawn. The river was turbulent, deep and muddy, flowing terrifyingly fast, with great logs and trees bobbing swiftly past in the brown water. They squashed as many of us as they could into long thin wooden boats with outboards, and I was relieved when we reached the opposite bank. Then we were led a short distance through some trees beside the river bank to a grassy airstrip where we sat down to wait.

There must have been some coordination in these arrangements, because we had not been there long before the familiar shape of a Dakota appeared in the clear, early-morning sky, landed and taxied towards us. We all piled aboard, sat about on our rucksacks all over the floor and the plane took off.

All this may sound very adventurous, and it was – being taken round the country with a group of swarthy Colombians armed to the teeth – but we had no real idea what was going on. Nor, when I asked him out of earshot of the others, did Ricardo. However, some of the guys began to grumble and I gripped them very quickly. I told them to stick with it, whether they liked it or not. My earlier efforts to establish the right pecking order paid off, and they shut up. Frankly, they had no option. We were in the middle of Colombia, entirely in the hands of Gacha's militia. Besides, they were being paid, and at least Ricardo was still with us, so we could communicate with the others.

I found it very frustrating, having no map and no plan. Perhaps dogs feel like this when their masters take them in the car for a drive. They just follow on faithfully, or fatefully. All I could do was note the time, and stare down through the window at the craggy mountains passing beneath us. They faded into lowlands and then into flat, rolling cattle grasslands which spread to all points of the compass. After about an hour and a half, we landed on a laterite strip beside the Rio Yuri, in Caqueta Province.

Once again, the organisation showed itself. The Dakota refuelled at once and took off, leaving us on the strip. Here, we were fed for the first time that day, with a mess of unappetising cooked maize, like couscous, covered by a thoroughly suspect vegetable sauce. There was more muttering. We all missed our nice couple at Paradise Island and the food deteriorated every time we moved base. Little did we know how much further there was to go.

Shortly afterwards, three Cessnas landed. Feeling like mobile parcels by now, we were ferried south. Soon the grasslands gave way to jungle and after another hour and a half I saw the glistening silver reflections of a huge river meandering through the green carpet beneath us. The Cessnas turned and flew across another airstrip, hacked out of the forest by the river. Later, I worked out that this was the Río San Miguel, and that we had flown about 800 kilometres that day, deep into the inaccessible jungle of south-eastern Colombia on the edge of the Amazon basin, as remote and inaccessible a part of the world as any.

We held off landing for several minutes, and I could see little figures scurrying about on the ground beneath, pulling aside coils of wire which had been tied across the strip to stop unauthorised planes landing. It was a neat trick, but quite who they were trying to keep from visiting them, I never did discover. There are many competing government departments in Colombia, as well as the militia and Mafia leaders, and the geographical diversity, with all the principal towns separated from each other by great Andean Sierras, only serves to enhance these power divisions, as it always has done.

The man who met us was a very piratical figure indeed. He was large, muscled and grinned hugely under a great wide floppy army camouflage jungle hat, tied up Australian-style on one side. He wore a filthy green T-shirt and army trousers. In the crook of his arm, he carried an SLR, with ammunition pouches filled with extra magazines festooned round his belt. A vast curving cutlass hung at his side, but for me, the pièce de résistance was his green wellies, which reminded me of Uganda. In fact, wellies in the Amazon jungle, which is always subject to flooding, are not so silly as they may sound. We christened him 'Mungo', from the film *Blazing Saddles*.

To my disgust, Mungo cheerfully announced our travels were not over and led the way to the riverside, where he showed us several quite small fibreglass boats crewed by swarthy Indians manning the big outboards at the back.

"How far in these?" I asked without enthusiasm.

"Not far," he replied, shrugging good-humouredly. There was no point pressing him. *Mañana* again. Time and distance means very little in areas as large and remote as these.

As the afternoon faded, they ferried us all down the river in several lifts, and I went first. The Río San Miguel was very wide, perhaps 200 yards across, and the banks were lined with palms and lianas hanging from the great jungle trees for mile after mile till we pulled in on the right-hand side at a muddy slipway above which was a twostorey wooden house on stilts, painted blue. The ground round the blue house was clear, sandy earth beaten flat and dotted with a few silver-barked trees. A sagging basketball net was strung between two trees.

By the time all my team and some of the Colombians had been brought down the river on the boats, it was evening. We were all exhausted having been on the move since three o'clock that morning and we were all very hungry. This was when we met 'Typhoid Mary'. Typhoid Mary was the cook. I never knew her real name, but that's what we called her, because I can honestly say that of all the meals I have eaten over the years, she served the filthiest. She was ugly and unpleasant, a dumpy, black-haired Indian woman of indeterminate age who smiled with a mouthful of rotten yellow teeth. Hygiene was not her strongpoint. Her 'kitchen' was a wooden hut, with a cutting table on one side and gas rings on the other. Dirt and the fatty grime of past meals were caked everywhere, and the place reeked of smoke and the sickly smell of decay.

Her first offering, from a large battered aluminium pot, was truly awful and there was nearly a mutiny. She produced a sort of maize gruel with bits of 'meat' floating in it. Ricardo sat on a log just looking at his metal plate with doleful, shocked eyes, for he was a man who loved his food and I could see he was wondering how he would survive.

Fortunately, our attention was diverted at this moment by a strange sight on the river. Another of the small boats was back with a live cow in tow. The poor beast had been swimming desperately all the way downriver, dragged along by a rope tied to the boat, its head and wet nose just above the surface. The Colombians have a very casual attitude to life, especially towards those who are about to die, for this cow was fresh rations, Colombian style.

The Colombians pulled the animal, slipping and sliding, up the bank and a crowd of men surrounded it on the basketball 'court'. Quite possibly Typhoid Mary's meal had catalysed everyone's interest in this new source of food, as they all began shouting and arguing about the best way to butcher the animal. In the end, Mungo stepped over with a purposeful look on his face, put his SLR to the wretched cow's head and blew out its brains. The cow had scarcely dropped to the ground before its belly and throat were slit and several men were jumping up and down on its stomach. Others swiftly held buckets to catch the gushing blood. We foreigners sat on the verandah of the house, starving, and watched this grisly scene in amazement while Ricardo explained they used the gore to make a sort of black pudding. Before long, the Colombians had literally hacked the cow to pieces and there were blood, guts and limbs all over the sandy ground around. It looked as if someone had opened the cow's mouth, slipped a hand grenade down its throat and blown it apart.

Typhoid Mary was in her element. For two days, her kitchen belched steam and smoke and food emerged vaguely recognisable as cooked meat, but the tropical heat and flies did their work, so by the third day she was trying unsuccessfully to conceal the extreme deterioration of her ingredients with a doughty mincing machine and dirty handfuls of chilli peppers.

Napoleon was right to say an army marches on its stomach. All we had to drink was the cheapest Colombian coffee and I was the only person with teabags, so by the third day, the British were desperate. Dean Shelley came crawling over the verandah on his belly towards me, gasping out, "A teabag! A teabag! Anything for a teabag."

Eventually, all the fifty Colombians and our stores had been brought down the river and I tried to start up training again, with weapon handling and fire and movement drills.

I sat them all down in a circle and described, through Ricardo, how to do a camp attack, and that night we did an approach march in the jungle area behind the river and walked through the whole sequence. Of course, we were tired after that, but a cockerel behind the kitchen kept me awake all the following night. At two o'clock in the morning, I lost my temper. I sneaked over the sandy earth to its cage and battered the wire with a stick. The bird stopped crowing and eyed me suspiciously. I went back to our hut and lay down on my bed again and, just as I passed into deep sleep, I distinctly heard a strangled squawk. The following day, Typhoid Mary gave me a very reproachful look as she passed me on the way to the river with the dead cock in her hands, to throw it in the water. It was her prize-fighting cock! I looked round the guys, who all looked quite innocent, except Shelley. He must have been watching me in the shadows when I thumped the cage, and then nipped over to administer the stretched neck treatment.

We were in one of the remotest parts of the world, stuck on the bank of a jungle river, feeling very cut off and living in a degree of squalor made bearable only by an increasingly strained sense of humour. Then, I suddenly discovered that Mungo had a radio telephone. In broken English, he started talking about our contract in such a way that I was suspicious. Through Ricardo, I pressed him and he showed me the radio-telephone which linked onto the International Direct Dialling network. He had been using it to call our backers in Bogotá and elsewhere. He did not discuss our contracts anymore and a day later he said, "We must move again."

"Why?" I wanted to know, fast losing patience.

"Communistas! They are close by," he said excitedly, grinning.

"Tell me where, and we'll go and attack them!" I retorted in exasperation.

He wouldn't have it. I argued but he insisted that we move again. To my annoyance, we only moved twenty minutes down-river, which hardly seemed sufficient if they were really afraid of a threat from a large group of FARC. I was convinced there was another reason. Worse, Typhoid Mary came too.

Stuck in such a remote place, our original purpose to attack the communist secretariat at Suma Paz was losing its immediacy. However, I was determined to stick to the plan and went for a walk to find a place to train. Mungo led me across a large open grass field, which had been cleared back from the river for cattle, and we entered the jungle on a track. I had never seen such a path design before, but it made sense. Thick tree bark had been laid on the ground, supported by wood stakes driven into the earth, to form a walkway which lots of men could use time and again without churning the track into a mudbath, which is inevitable when rain and flooding make a jungle track impassable. At the end of this track, in the gloom of the jungle about 100 yards from the open grass field, was a camp.

The place was unoccupied, but all the huts were extremely well made, of wood and roofed with palm fronds, and it had not been allowed to get overgrown by the encroaching jungle. "What's this place?" I asked Mungo.

Off-hand, he replied, "Communistas!"

"Yeah," I answered coolly. I pointed at the largest hut in the centre of the camp and said knowledgeably, "That must be where the commissar gripped them all with his political lectures." You couldn't tell me anything about communist camps, not after Borneo and Africa.

Mungo looked puzzled but nodded politely.

On reflection, he must have thought I was very thick. The right wing and the communists do fight each other, but in Colombia this classic power struggle is not for the hearts and minds of the people, but, quite simply, to control these remote jungle areas where Erythroxylum coca novagranatense is grown. So, while Mungo visualised coca leaves being processed and the warm air filled with the sweet smell of kerosene, I imagined a circle of keen sweating faces being harangued about Karl Marx.

I poked about the huts some more and realised my mistake. In one hut I found powerful electric heat lamps, in another there were microwave ovens for the laboratory, in a third mixing machines like large dishwashers – probably for breaking up the leaves – and finally in a fourth was a stack of hundreds of polythene bags. None of this was anything to do with Karl Marx.

Back in our camp by the river, where the lads were settling in to several more wooden huts on stilts, I confronted Ricardo. "What the hell is going on?" I demanded crossly.

"The communists were here," he said blandly. "And they were driven out."

I gave him a long look and left it at that. None of these Colombians ever mentioned drugs – cocaine or any other – at any stage during the whole time we were there. Which in itself was odd. Some of the Colombians with us disappeared off in the boats every day and came back in the evening, or they went off in groups into the jungle, but there was never any explanation. Even on reflection, I have no idea what they were doing (though I can make a suspicious guess, of course), nor can I decide what they thought we were supposed to be doing for them. I concluded there was nothing for it but to give our training one last attempt.

I started lectures again, using one of these large wooden huts as a lecture room, and Ricardo translated. All the Colombians squatted on the floor and listened attentively to my fiery presentation of a camp attack. At intervals, I stopped and they turned to Ricardo for the translation. The impact was lost at once. Ricardo was gently swinging back and forth in a hammock slung across between two windows in the corner of the room. This was South America, but not the conditions for good lectures as recommended by the British Army Methods of Instruction manual.

There was a streak of madness in the place. No one got any sleep inside the huts, because all the Colombians liked hammocks. They slung them beneath our hut, between the stilts, and all night long the whole hut shook back and forth, squeaking with the sibilant hiss of stretched cord.

Another time, a Colombian called Mia Farrow (literally) caused uproar. He had been a communist in the FARC before switching allegiance to the DEA and was typically self-important, full of South American machismo. The FARC had sent him to Russia, where he had trained as a medic, but he preferred to call himself a doctor. I suppose a trip to Russia was quite something to these young Colombians. They were so impressed with Mia Farrow's stories that they agreed to let him circumcise them. For a fee. He said it was good for their love lives! The first we knew was the camp echoing with screams all one morning and for several days young men minced about in agony, swearing they were going to kill him. Of course, my medic had to fix them up with penicillin injections to stop the inevitable infections.

Another night, we were all woken by the sound of firing. Everyone tumbled out of bed and started shooting into the darkness, only to find two of the Colombians had gone off drinking in a little shack further down the river bank and, returning drunk, had decided to fire off the contents of their magazines to see how alert we were. We calmed them down and I came back to find Ricardo standing by our hut and staring at a grenade he was holding uncertainly in his hand. He said to me, "I'm not cut out for this, Mr Peter!"

He was right. We had had enough. The food was terrible, we couldn't sleep, and the training was constantly interrupted. Far from training drug cartel 'hit squads', we weren't getting any training done at all, and the plan to attack Suma Paz seemed more distant every day. I couldn't see the point of being in the jungle. I became convinced that we were being kept out of the way, just to have us handy if they did eventually decide what they really wanted us to do for them.

I needed to talk to the men in charge in Bogotá. Ricardo and I used Mungo's radio-telephone to make the arrangements and we left. The guys were fed up. They thought I was leaving them and I went to some trouble to explain before they understood I had to speak personally to the men in Bogotá. The fact is they had never served in command positions in the army, and couldn't easily see the officer's, or management's, side of the coin.

Of course, Ricardo was overjoyed. He was fast losing weight and was dying to get away from Typhoid Mary. He was a genial man, an army officer who hated living like a 'jungle Indian'. We took a boat back upriver and flew from the airstrip to Bogotá.

In Bogotá, my suspicions were confirmed. Ricardo set up meetings with the army officers I had met before and we thrashed over the whole business.

I was still prepared to attack the communist secretariat if they could get their act together. I wanted to put pressure on them, because I made up my mind that if we had to call a halt to the whole business, it would be their fault for failing to produce the logistics and support, not ours for lacking the guts or determination. I had discovered that there had been an Israeli connection, as I had suspected in Paradise Island. My 'made in Israel' slippers had belonged to one of a team like ours, but, unlike us, the Israelis had refused to go into the jungle.

I said, "We need helicopters." To drop us off short in Suma Paz and make the final approach to the target on foot.

"It's too high. The altitude is too much for choppers," they said.

This was nonsense. During the talk, it became obvious that the whole problem was simply that they could not make up their minds what they wanted us to do. I made up their minds for them and insisted they fly all my guys back to Bogotá.

We stayed in the Hotel Dan and during those few days in Bogotá, Roy Kaulback chatted up a female journalist only to discover she had photographs of the Casa Verde. Unknowingly, she showed him an article in a glossy magazine with pictures of the Green House in Suma Paz. It was nothing more than an attractive little green-painted mountain farmhouse. Roy, who had studied Sociology at Cambridge University, did not require his 'A' grade on a South African Army intelligence course to realise that our plan to attack the Casa Verde had been a bait to hook us in the first place.

I told the Colombians they must make up their minds what they wanted us to do. Finally, they admitted there was no real plan and advised us to go home.

Back in the hotel, there was some muttering about pay. I hate this, but it's an inescapable facet of civilian life. From being soldiers of fortune in the jungle camp, they had instantly turned back into civilians in the luxury of the hotel. Actually, Dave Tomkins was excellent over this, dealing with the Colombians who paid up all that was due very fairly.

At my last meeting with the Colombian Army officers, I said, "Please, next time, let me come out here to set up everything myself, and then, when I'm sure the job is ready, I'll bring over the men. That way there'll be no confusion, which is bad for the team, and you won't be paying men to hang about."

They agreed.

On Friday, November 18, Dave and I flew back to England. When I got home, the first thing I did was to send Murray Davies £500, to repay his loan which got me to Rhodesia all those years ago in 1976, with a bit of interest.

CHAPTER 13
CALI CARTEL'S MISSION TO KILL ESCOBAR

I am pleased to say that there was no security leak by any of the team on their return home. We all felt there was more to do in the future and they kept quiet. A breach of security to the press or any of the secret services, of Britain or the United States, would certainly have blown the operation apart at that stage.

So, on Tuesday, February 21 the following year, 1989, I was sitting in the Booth Hall at lunchtime wondering why British security firms would not employ me. I had more soldiering experience than most and I had successfully managed the whole spectrum of civilian security operations for COIN in South Africa – static guards, cash-in-transit, and keypoint armed militia operations – all of which covered a region of South Africa bigger than England. My brooding was rudely interrupted when the swing doors of the bar crashed open and Dave Tomkins burst in.

"Here we are, fellers!" he shouted, posing theatrically at the door. He looked typically debonair, in fashionable clothes, gold watch, bracelet, neck chain, coiffured grey hair and his expensive Cartier sunglasses, for which I had paid half as a present on our way home three months before. In his right hand he held a wad of dollars.

He sat down and I quietly said, "Dave, what the hell is going on?"

"They want us back," he replied, slipping into a conspiratorial whisper to match my tone, "and there's a real job to do."

Three days later, on Friday, we left England's cold winter and flew to Panama, where we spent two days in sweltering heat sightseeing the Canal, which surprised me by its narrowness, and waiting for our connection to Colombia. We were going to Cali, Colombia's third-largest town after Medellín, in the Río Cauca valley where a pleasant combination of tropical warmth and altitude among the high sierras produces a rich and beautiful agricultural region, principally noted for its sugar cane.

On Sunday, February 26, we landed at Cali airport, where we were met by our gallant Colonel Ricardo. He was delighted to see us and, after many heart-warming handshakes and greetings, he ushered us smoothly through the usual police checks. Later, when we were on our own in an enormous apartment in the smart part of Cali, he announced dramatically, "Mr Peter, they've made up their minds. They want you to kill Pablo Escobar."

This was a task!

"And you'll get all the backing you need." He gestured round the spacious flat and went on, "You will be based here. Treat it as your own. I shall be your liaison officer and interpreter as before, and they want you to start the planning at once."

Of course, there was the usual *mañana* to cater for, with Ricardo going off for several days at a time, but it was soon obvious to me that they were serious. Pablo Escobar, perhaps the most important cocaine drug baron in the world, was our target.

I have no doubt that the idea of employing a team of ex-soldiers to assassinate someone is repugnant to many people. They may be offended by the savagery and illegality, but I imagine they will be reading this in the comfort of their homes in more civilised places than Colombia. Instead of being outraged by people like me, let the moralists think of Escobar's guilt in the cocaine and crack drug trade, and the misery he has caused millions of addicts and their families. He has consistently evaded justice and perhaps men like me offer the only realistic way in which men like Escobar can be dealt with in places like Colombia.

This won't satisfy some. They will call the argument superficial, the first step onto the slippery slope to a lawless society, totalitarianism and corruption. Georges Bernanos was right, they will shout, "Beware 'the end justifies the means'." All I can say is that Escobar was a big enough villain for me, and the twelve of us were prepared to take the risk.

He was closely protected by seventy guards and lived in a heavily defended villa called Hacienda Nápoles on the road between Puerto Triumpho and Medellín. Puerto Triumpho was about 10 kilometres from Puerto Boyacá, where we had been the year before, and Ricardo arranged for me to fly over the place in a twin-engined Caravan airplane on a normal flightpath. Armed with a Canon lens with a fast film setting to avoid camera shake, I took hundreds of photographs and made notes. The house was enormous, ringfenced, guarded by towers and barrack buildings, and I knew from the previous year how all the road junctions and approaches in the whole area around would be covered by a network of lookouts and mobile patrols all linked by radio. This was not going to be easy and the thing fascinated me!

While I set about working up a plan in our apartment, Dave Tomkins flew out to collect the team. They arrived as before on different flights via various countries so that operational security was maintained. There had been one or two changes. Four men were unable to come back: Gerry O'Brien, Roy Kaulback, Mark Griffiths and Gordon Brinley. In their place were two ex-22 SAS men, Don Milton and Stuart McVicar, an ex-21 SAS Territorial called Andy Gibson, an ex-British Para called Ned Owen, and two men from the South African Special Forces Reece Commando called Pete Donnelly (also ex-Scots Guards and Rhodesian Light Infantry) and Billy Potts (also ex-Royal Marine and Rhodesian Light Infantry). All together we were twelve. They joined me in our comfortable apartment in Cali, which was a great change from the filthy conditions of the jungle, and I gave them a briefing. "We shall be moving on to another place soon, where we can start our work-up training," I said, "but while we're here, don't go round in more than twos or draw attention to yourselves. Behave in a touristy manner."

Two of them let me down. I was well ahead with my plans, so I went for a walk in the middle of Cali, to have a look at the shops. The town was less busy than Bogotá, more provincial, but the streets were still noisy with traffic, hectic Spanish pop music and poorly dressed people selling cheap shoes, plastic buckets and little gold trinkets of frogs and Inca gods. There is a strong religious undercurrent in Colombia, the rich gold-leaf of Catholicism beaten over the hard wood of older beliefs, and I nipped inside a couple of big churches to absorb a little good homely Catholic atmosphere.

On my way back through the main square, I glanced through the window of a bar and recognised a distinctive yellow tracksuit one of the guys had been wearing. I went inside to join them. Two of them were slumped, nearly unconscious, at a table on which there stood not just one but two empty Black Label whisky bottles. I thought they were joking, to wind me up. They were not joking, for they were totally blitzed, but they did wind me up. I was furious, especially as one was ex-SAS and the other ex-44 Para Brigade. They were jeopardising the whole operation. I got them back to the apartment, and next morning when they sobered up, I fined the two men $200 each and left everyone in no doubt that drink was out. The two culprits were not bad men. In fact, they were good soldiers, but idiots with drink. It is one thing being good on operations out in the bush where there is no drink, but quite a different self-discipline is required in a civilian environment when there are bars all around. I enjoy my drink, but when I am working, on a job, I do not drink. However, the lesson was learned. From then on, we drank crates of Coca-Cola and I had nothing but 100 per cent commitment from them all.

I prompted Ricardo to move us to our training base immediately.

Actually, if I had hoped for something spartan, in sympathy with our training, I was in for a surprising disappointment. Our flat had been very comfortable, but our new quarters were exceptional. We were housed in a really beautiful white villa roofed by acres of terracotta tiles in the hills behind Cali, surrounded by landscaped gardens, mown grass lawns, ornamental trees, a swimming pool, a tennis court, a football pitch which served as a helipad, a running track round the perimeter, forests of security lights and fencing, and the largest satellite dish I have ever seen. Inside, the rooms were enormous and luxuriously appointed, we were very well catered for by a few discreet resident staff, there was a weight training room downstairs, and the place was empty, all for us. What a contrast to the year before.

I did four flights in the Caravan past Escobar's house and accumulated plenty of information. I briefed the guys on the job, but I gave no names, nor the location of the target at that time. I said I would wait until we had all the kit together. There were weeks of training ahead and the last thing I wanted was someone to produce an excuse to bottle out and find some scoop-hungry journalist with a loose hand on his chequebook.

I had good reason to be cautious. Only two weeks after arriving, Terry Tagney came to me to say he really could not face the sort of attack I had told them we were going to do. This was the kind of weakness which might destroy the team, but, as I could not hold him in Colombia, I covered up for him by telling the others that his wife was threatening to leave him and sent him home sworn to secrecy. I am sorry to say that once he was home, he went on television and described his reasons for leaving in a rather different light, by saying he had had "a sudden attack of common sense" which obliged him to return home.

The rest of the team stayed solid. I gave them enough information for our training to begin even without the weapons which Ricardo was still organising for us. I must say that I tried to put all my experience of years into this training. We started with the usual but necessary – and I may say terribly satisfying – programme of fitness. As a result of spending so much time in a wheelchair and on crutches, and drinking too much during those two frustrating years without being able to exercise, I was still carrying too much weight, but my leg bones had set firm, for better or worse, and, while I was no longer as fast as I had been, I was strong and became very fit. We warmed up running round the track, then worked out on a speed circuit in the weights room and soon pulled into shape. I lost a lot of weight and built up muscle to the point where I was no longer embarrassed to take off my shirt!

We also saved time by working on our military skills, dry, without weapons. We practised our immediate action contact drills on the football pitch, where we also worked on helicopter emplaning and deplaning drills. I set out chairs in the same configuration as the seats inside a helicopter and we practised covering each other getting 'in' and 'out'. This was going to be a vital part of my plan.

When the weapons arrived, I really knew our backers meant business this time. A shiny aluminium box van, like a refrigeration unit, drove up the paved driveway to our villa. Almaro, a thin, wiry paratroop major who was another interpreter and liaison officer, hopped out and, with a flourish, unlocked the bars at the back to reveal an Aladdin's cave of brand spanking new weapons, ammunition and explosives. These were not the shabby German G3s we had been given at Paradise Island, which had been cast-offs from the militia. This was the crème de la crème. There were American M16s, pistols, plenty of magazines, pouches, gleaming ammunition in boxes, grenades, M72 66mm Light Anti-Tank rockets, pounds and pounds of PE4 plastic explosives, time pencils, detonators, safety fuses, primers, detonating cord, switches and so on. Everything was in excellent condition.

My training programme was progressive, building one thing on another, because I wanted no mistakes on the attack. Rather than rush off to fire our new weapons on the range, I made everyone practise weapon handling in the villa first. For a week among other subjects, we concentrated on stripping and assembly, handling and stoppages, and we got very slick.

I was very pleased with the men. The atmosphere was excellent and a complete contrast to the previous trip. Everyone was committed and enthusiastic. They could all see that we were getting the equipment and support I was asking for, and everyone worked hard.

Actually, we had so much good equipment, some of the guys began to complain they wanted more. This is typical of soldiers, and I suppose people in general. As ex-soldiers in the British, Rhodesian or South African armies, all these men had been used to getting the job done with what was available, or as they used to say in the British SAS, "Just a length of paracord and a piece of masking tape." Now, faced with a veritable hamper of goodies, they wanted more.

All special forces units risk falling into this trap and it leads swiftly to operational castration. A unit's success brings attention, which brings finance, which pays for new equipment, which is when the rot sets in. Like children, soldiers ask for more and their expectations rise for more expensive and complicated equipment. Just having the best personal weapon is not good enough. They want all the other things too, which would be nice to have, just in case, like SatComs for everyone. Of course, special forces soldiers will say they need equipment variety to meet all the demands put on them. However, swamped with fancy gear, soldiers lose the skills of basic soldiering and become dependent instead on exotic technological equipment. Finally, they begin to think the job cannot be done without it. And that's the kiss of death to any unit.

My team had ten 30-round magazines each, pistols, grenades, LAW rockets, plastic explosives and machine guns, and I called a halt to their demands. I said, "There are only seventy or eighty enemy guards on our target and you've got enough ammunition between you to kill over 3,000 people!"

We started firing live ammunition in a secluded spot in the scrub hills beyond Cali. My programme was similar to our previous one in the woods at Paradise Island, except now we had a definite target which I had seen from the air and we could tailor our range practises to suit.

We were well supported now but we could never relax. On

the way to the range one day, we stopped at a little café beside the road, the Cafe Sello Rojo, a long, whitewashed concrete shed, with the usual terracotta tiled roof and several shuttered doorways which were rolled up for ventilation. The weather was hot but overcast and humid, and we were always thirsty. We parked our red and white Toyota Land Cruisers and the aluminium weapons truck and stood about drinking Coke when a policeman passed on a Harley Davidson motorbike.

He was straight from Hollywood, dressed in black leathers patched with fancy badges, a holstered revolver, gleaming black sunglasses and a white helmet. My heart sank when I saw him checking us out. Our boots had caught his eye. We were all in T-shirts and jeans but trying to wear in our army boots, and that single uniformity looked decidedly strange in Colombia. Sure enough, the cop turned round and cruised back. He got off his bike, propped it up, straightened his black leather jacket, adjusted his sunglasses and sauntered over with an unpleasant expression on his face. When he found we could speak no Spanish, he was at once suspicious. The atmosphere became tense, he became aggressive, probably because he knew the police station was only 100 yards away in the village, and the guys began to edge towards the aluminium weapons truck. Almaro strode up and waved his major's identity card in the policeman's face. This produced a vicious clash of Colombian machismo. The traffic cop had no intention of being bullied by an army officer and spat out a torrent of Spanish, the gist of which was clearly, "Get fucked!"

Dave Tomkins floated round the two, smoking and smiling genially, trying to pour oil on the troubled waters, saying, "Everything OK then?"

Almaro glared at him. He was furious that in spite of all our backing, he was helpless faced with this obstinate small-time traffic policeman. The story is the same the world over.

Almaro's response was also a world beater. He bribed the cop. He walked out of earshot round the side of the café and made a call on his radio to our base station. Then he ignored the policeman's demands for an explanation with dismissive gestures till, about twenty minutes later, another man on a motorbike roared down the road and pulled in. This was one of ours, the 7th Cavalry, but more subtle. The rider handed a small bag to the major who handed it to the traffic cop, who took it with ill grace as his due. He swaggered back to his Harley and motored off. Such is South American life, but I wish I could have been bribed to stop every bit of trouble I've ever been in.

A cement Virgin Mary and Child watched over this corrupt little scene from a stone plinth. Garishly painted blue, pink and yellow, with her concrete child balanced fantastically on one hand, the Virgin stared blandly over our heads at the fields of sugar cane. The more violent a society, the more it parades its religion. Our helicopters arrived. A glass bubble Hughes 500, which I wanted to use as a command K-car over the target, was flown by Tiger, a really keen police pilot. A troop-carrying Bell 204, the Huey of Vietnam fame, was flown by a very experienced and sober Colombian called Pablo.

Our villa was an excellent base, but I wanted a larger training area where we could fire without restrictions. They had already prepared something for us. We flew in the helicopters over the 10,000-foot peaks of the Western Cordilleras behind Cali and dropped down the other side over the thick flat jungle plain north of Buenaventura, which is a large port on the Pacific coast.

Our training camp was at La Gagua, which was nothing more than a wide bend in the Rio Manguido. The river provided us with an important training ground. On the bend were several hundred yards of flat, firm pebble and sandbanks which we could use for range work, to build models of our target and to land the helicopters. When the helicopters were not being used, we pushed them on their slides along hardwood slats into hides built under cover of the jungle trees. They were covered with tarpaulins and camouflaged with atap leaves to cut down the reflection of the glass and wet tarpaulins. The technique reminded me of the one we had found on the hillside near Boyacá the year before. Our reason for doing this was 'Opsec'. These helicopters were sterile in that, although the Bell was painted in Colombian police green and white colours and the Hughes was painted Colombian Army plain green, they did not come from any government or police department. Although our backers certainly had the clout, taking away an aircraft from an official source might have provoked unwelcome questions. Our choppers had been flown into the country specially for this job. We hid them in case real police or army flying over our jungle camp saw them by sheer bad luck.

We lived in a solidly built wooden cabin painted ochre red with a verandah overlooking the river. There was a good clean kitchen and we were well looked after by several cheerful Colombian Indians. These men had been preparing this place specifically for our operation, laying out the helicopter slats, building up the hut, and the aircraft mechanic was the hardest-working grease monkey I have ever seen.

There was as much room to rehearse here as I needed. We built a scaled-down model of Escobar's house with wood poles cut from the jungle and hessian for walls, and the ex-SAS men were in their element. Their training on 'house-clearance' learned in the SAS anti-terrorist team combined well with my own experience of fighting real enemies through bunkers in Angola. We built up in stages, first on dry practises, then firing live ammunition, from one man entering a single room, through two men, to a group of four working through several rooms one after the other. The weapons handling was good from our earlier work, and the guys' confidence increased as they improved with each element of my plan. Though we were only twelve men against eighty, we all felt increasingly bullish about our chances.

Our training continued through March and into April. We went back and forth between our villa in Cali and our training camp at La Gagua. Day after day, we practised so that the guys became fluent with every detail of the skills they would need, building on room clearance, firing the 66mm anti-tank rockets, throwing grenades and blowing satchel charges. All the time

Colonel Ricardo kept me informed about the likely date of our attack. Obviously, they did not want us to attack until they were 100 per cent sure Escobar was at home, and equally I told Ricardo that I did not want to commit the team until I was 100 per cent sure we were ready. Ricardo, and our sponsors, completely understood and I began to tie in our skills to my plan of attack.

Back at La Gagua, I briefed them on each phase of the whole plan: how we would fly onto the target, how both choppers would suppress two guard towers each with machine-gun fire, the hover-landing to offload the assault team, the order of taking out the guard positions, attacking the guard accommodation, clearing the villa itself and so on. We began to practise each phase of the plan over and over, firing thousands of rounds, blowing pounds of plastic, and throwing dozens of grenades until everyone knew their own tasks so well that I made them change round. I was determined everyone would carry a clear picture of the whole attack plan in their minds so they would understand what was happening around them while they were busy on their own tasks. This also meant everyone could react flexibly if things went wrong. Each man had a personal radio with an earpiece and I would command them from above in the Hughes helicopter K-car where I could see the action, direct them and be in touch with every man on the ground.

The programme was going too smoothly. One day early in May, we were back in the villa in Cali and Dave Tomkins phoned his home in England to chase up some more equipment we needed. His wife, who was helping, said, "James Adams of *The Sunday Times* has been trying to contact you."

This was the last thing we wanted to hear. If details of our plans hit the newspapers, the operation would be cancelled. Simple as that.

Dave's wife gave us Adams's home telephone number. He had guessed that the story was extremely sensitive, if he knew little else, and preferred to keep any conversations away from *The Sunday Times*' offices. Dave called him at home.

"I would like to talk to you about your recent trips to Colombia," said Adams, coming straight to the point. Dave had been going back and forth to buy equipment such as radios and webbing, and had probably clocked up six trips by then.

"What are you talking about?" Dave replied guardedly.

Bluntly, Adams then detailed Dave's flight numbers and dates, one by one. He said, "I know you're in Colombia now and that Alec Lennox, Peter McAleese and Dean Shelley are there." He finished by saying, "I'm about to print the story tomorrow morning. D'you want to make a comment?"

With hindsight, I believe we should have called his bluff and said, "Go ahead and print." Adams did not know enough of the facts to have damaged the operation. I think he told us all he knew, to create the maximum impact, but we had no control over his speculations. We had no idea then how he had come by the information or who his source was and we felt vulnerable. Perhaps we should have had a PR front, someone briefed by us, in England, to handle people like Adams, to keep them at arm's length and issue statements as necessary, but, as soldiers, we were head down in our training and were taken off-balance. We were at a disadvantage and he knew it.

With difficulty, Dave persuaded Adams not to publish but only on condition we met him in Panama.

On Thursday, May 11, Dave and I flew from Cali to Panama City, the day after a riot in which General Noriega's paramilitaries brutally hospitalised opposition leader Guillermo Endara with iron bars and beat up his supporters, who were demonstrating against vote-rigging. Panama was very unstable in the build-up to the American invasion. We went to our rendezvous with James Adams in the Hotel Continental in the evening, as the sun was setting. We found the patio bar by the swimming pool and sat down at one of the tables with a beer each to enjoy the last of the sun with other guests. We were in casual shirts and trousers, Dave his usual suave self with sunglasses and gold chain. Adams was late. He may even have been watching us from somewhere in the shadows, but he appeared on the patio with the studied elegance of a seasoned world traveller. Tipping a rolled-up copy of *The Times* at us in recognition, he walked over at a stylish pace between the other tables and sat down. He was in his mid-40s but deliberately cultivated a youthful image, with smoothly coiffured blond hair, à la Jason Donovan, and a trim figure. He was wearing a carefully crumpled fawn cotton suit with the cuffs turned back on his wrists, *Miami Vice*-style, which he probably thought appropriate in Panama City at that time, a pale yellow open-neck shirt and comfortable slip-on shoes with no socks. After introductions, he called over a waiter with a flick of his hand, allowed us to order another couple of beers, and then asked for a long drink I had never heard of, which turned out to be filled with fruit, topped with a small paper umbrella, and strongly alcoholic.

There ensued a few minutes of verbal strutting while we metaphorically circled each other like rival dogs. Dave Tomkins had learned his tricks at a rather different school to James Adams but they both put in a lot of mirror time and I was treated to the sight of two professionals at battle. Dave toyed with the gold Kruger, and talked about the important arms deals he had done, while Adams swanked about Frederick Forsyth using his material for his books. Adams's trump was his threat to print a story about us and he set about softening us up.

"What you're doing is an open secret in Whitehall," he said in a hard, knowledgeable way, hoping to frighten us.

This did not have the impact he wanted. I felt quite sure people in Whitehall in the secret services and narcotics departments would be only too delighted if Pablo Escobar was off the scene. Besides, the fact is that the CIA and the DEA have more clout in Colombia than Whitehall, and we had already been in touch with a. contact in the CIA, to test their reaction to what we were doing. Unofficially, the CIA response was, "Do the job as soon as possible and get out."

Actually, I very much doubt that anyone in Whitehall had any idea what we were doing in Colombia at that time.

However, Dave and I were on a damage-limitation mission. We were not sure how much he knew. The last thing we wanted was Adams to print a story which would blow the security of our operation, so we felt somewhat at his mercy. He took full advantage. He suggested we all had dinner together and we went to a seafood restaurant where he continued a barrage of questions for more than an hour. His technique was relentless and, in retrospect, not very subtle. He taxed us about Dave's arms deals and my soldiering life, typically concentrating on the short mercenary period in Angola rather than the years of my regular service in Rhodesia and South Africa, and every few moments suddenly switched back to our 'job' in Colombia. He could tell we were on to something really big, but we gave nothing away. It was clear he felt very much in control and, though he finally agreed to hold back his story on condition that we gave him an exclusive on our operation, his cool and rather arrogant bonhomie got me down. I decided to go on the attack myself.

"James," I said, leaning forward over the table and looking him straight in the eyes, "why don't you come with us?"

"What d'you mean?" he asked, startled.

"Come with us. On the operation. With me. You'll see the whole business first hand."

Non-combatants get in the way, but here was a chance to take a journalist with me into the fighting, really involve him in what we were doing and be correctly reported for once in my life. I felt quite enthusiastic. "I'll fit you into the team," I told him encouragingly. "Then you'll be able to write a real story!"

He declined.

His attitude changed abruptly after that. Altogether more cautious, our conversation continued on more equal terms. He had lost his opportunity to be part of the action, and relegated himself to reporting it second-hand.

We met him the following day but nothing further emerged. We flew back to Cali on May 14 reasonably satisfied that he would not blow the story until we had finished the job.

However, our trip to Panama made me realise how delicate our security might be. We had been so involved in our training in our villa and in the jungle at La Gagua, where security was excellent, that I had lost sight of what was going on elsewhere and plainly more people knew about us than needed to.

Outside Colombia, we were confident Adams would not publish for the time being, but how long would he hold off, for fear someone else might steal his scoop? Would his source who had betrayed Dave go to someone else? On our return, Dave made numerous calls back to England and finally pinned down this Opsec leak to a man who had supplied radios to us. This charmer had bugged and taped his telephone calls with Dave and bubbled us for a handful of filthy lucre, carelessly putting our lives at risk. Knowing we were in Colombia, it had been easy for Adams to find out Dave's flight details after that. The circle of people in the know would have spread wider if Adams had told anyone else himself, because people love sounding important by talking about stories like this. The circle would have spread still wider if our contact in the CIA had told anyone. I guessed he had made a report to cover his own back (and enjoy the kudos) and who knows what circulation of readers such a report might have?

There was reason to think that the CIA knew more about our operation than they let on. The concept of 'decapitating' drug cartels, which means removing the leaders, was gaining favour in Washington over the more resource-draining effort of stopping the thousands of little men in the lower echelons of the drug trade. That year, the US Attorney General Dick Thornburgh said, "You take an Ochoa, an Escobar, a Gacha or a top money launderer out of the operation, it disrupts them."

There was reason to worry about security inside Colombia too. Our contact in the CIA had been quite blunt. He warned us not to speak to anyone in the DEA (the US Drug Enforcement Administration). The DEA are very active in Colombia but they use a good many local undercover agents and the CIA man told us they would be as liable to corruption by drug money as anyone else in Colombia. Actually, I do not think the DEA knew, but I had no control of our backers or who they might tell.

For certain, the more people who knew, the greater the chance of a catastrophic leak to Pablo Escobar himself, in which case we might fly in to an enemy who was ready and waiting. The bottom line was that our lives were forfeit and the risk increased day by day.

I told Ricardo we wanted to attack as soon as possible and set about the final dress rehearsals with full equipment and live ammunition. We flew down to La Gagua and gathered in the cabin for the full orders briefing I had prepared for the attack, with our pilots, Ricardo and all the team dressed up, tooled up and ready to go. I set out my models of the target house, aerial photographs – which I had fitted together in a panorama – sketches, diagrams and variously scaled maps, and painstakingly took them through every detail. "This is Operation Phoenix," I told them, calling it after the South African 44 Parachute Brigade arising from its own ashes. "And our mission is to kill Pablo Escobar." Several photographs of him stared down at us from the wall.

After my orders, I questioned them on details, to make sure they understood precisely what each and every man was to do, and we went out to the helicopters for a full-scale dress rehearsal. We wore camouflage uniform, boots and combat waistcoats which contained ammunition pouches stitched across our chests, and other pouches for extra equipment, grenades and ammunition round the back. Every man wore a thin black balaclava helmet with a large phosphorescent yellow cross sewn on the top so I could identify our men from my K-car in the air. This was a trick I had learned in Rhodesia. Anyone else on the ground would be fair game. They also carried extra bags of grenades, 66mm anti-tank rockets and explosives. I climbed into the Hughes helicopter. Dave Tomkins, Almaro the army major and Ramon, another interpreter, sat in the back. Our pilot Tiger gunned the engines and we lifted off the river bed, up and over the jungle canopy. With the beat of the blades, I could feel the excitement building.

We had trained so hard, perfecting every detail, for weeks, and I was supremely confident. Below, the other pilot, Pablo, pulled the larger Bell 204 off the sandbanks and we began our simulated flight to the target. This was not merely a question of wasting time. From this moment on, we were totally engaged.

Operation Phoenix Two only had four phases. I have always believed a plan of attack should be simple, direct and aggressive. There was Phase One, the flight from La Gagua in the jungle to our refuel point at San Diego, codenamed 'Kiko'; Phase Two, the assault; Phase Three, the reorganisation; and Phase Four, the withdrawal.

On this rehearsal, we practised everything from Phase Two on.

As we flew over the jungle canopy, Dave Tomkins behind me was busy sorting out the big satchel charges we were going to use to bomb the guard houses. He needed them ready to hand. He checked the safety fuse igniters on each short length of three-second-delay safety fuse and pinched the splayed tines of the safety pin ready for pulling. By trial and error, we had found that when we dropped the charges, the safety fuse trailed behind and acted as a stabiliser, which kept the charge upright. Then we found that if we packed the charge with a bag of air at the bottom, the airbag burst on landing, cushioning the charge and stopping it from breaking up.

Flying above us and to one side, the men in the Bell were going through in their minds the important features we expected to see on our fly-in towards the target. I had detailed simulated points to maximise the realism of our attack.

They started with the terracotta roofs of the small village of Dora Dal, which told them there was only six minutes' flying time left. Tiger took our Hughes frighteningly low above the trees on the final run-in. We saw the Akatamia airstrip, the San Diego refuelling point codenamed Kiko, Escobar's own airstrip which he could not use because it had been cratered by the Colombian Army, then the target villa, Hacienda Nápoles, and then we were over the wire perimeter fences round the football pitch and over the villa complex itself.

I opened up with my pintle-mounted GPMG at two wooden guard towers on the perimeter fence, one after the other, to neutralise them, while the men in the Bell fired on the other two towers.

Then Tiger swung the Hughes round to bomb the buildings. As we flew over the guards' quarters and along the length of the target villa, Dave leaned out, pulled the pins from his charges, yanked out the extractor handle to ignite the three-second fuses and dropped three 6-pound explosive charges on the side of the villa furthest from the Bell, which was swinging round over the football pitch firing and checking there were no obstacles to landing.

The Hughes rocked as the charges went off one after the other beneath us and we flew on to drop a further two big 10-pound charges on the other guard billet at the main entrance. The helicopter shook as the Tiger swung round out of the blast. We pulled into a higher orbit over the house and gave covering fire while Pablo took the Bell down to hover-land over the football pitch and all the guys tumbled out with their equipment.

As soon as Pablo's Bell banked away, the assault team split into the support group and the house clearance group. Stuart McVicar's support fire team opened up at once on the guards' quarters with a 7.62mm G3 on automatic and with 66mm anti-tank rockets. The building disappeared in flames.

The house clearance team began to move towards the villa, each of the two call-signs leap-frogging each other with fire and movement. Covered by the men on the ground, Tiger dropped in to hoverland too. Almaro, Dave and Ramon leapt out to join Stuart's support fire team. Tiger and I lifted off to orbit over the villa and suppress any guards we saw on the other, or 'black' side of the villa. For ease of description, I had designated each side of Escobar's villa with colours. We had dropped the teams on the front, which I called the 'white' side, 'black' was the back, 'red' was the right and 'blue' was the left, the end nearest the guards' quarters. The three main parts of the villa were numbered: 1 (the blue or left wing), 2 (the centre) and 3 (the red or right wing).

"Moving to red side, to start point for white 1," said Billy Potts, the assault team leader, his voice crackling in my earphone. I looked down and saw the dark figures with the distinctive yellow crosses on their heads running forward through the trees to begin the house clearance from the right-hand end, opposite the guards' quarters at the red end.

"Entering white 1!" panted the assault group commander beneath me as his team lobbed grenades inside the building. They waited for the explosion and then burst through the door. This was no hostage-saving situation, when assault teams use stun grenades and aim to save lives. We were attacking an enemy stronghold where we expected every man we met to be armed and we were using high explosive military grenades.

As the assault team worked its way through the rooms of the villa beneath us, Tiger kept the Hughes swinging round over the target area, on the black and blue corner of the villa, furthest from the teams, where I could fire the GPMG at any guards trying to escape or reinforce those in the villa.

Our backers had given us good intelligence about Escobar's guards. There were seventy of them and they were well armed, with a variety of useful weapons such as the GPMG and M60 machine guns, AR-15s, Israeli Uzis and American Ingrams. However, they did not expect to be attacked, so no one carried more than three magazines. Like all bullies, they were used to throwing their weight about, their morale was high and they drank heavily all day. They wore civilian clothes. This is why the yellow crosses on the guys' black balaclavas were vital, so I could see in the smoke where they had got to and fire at everyone else with my machine gun in the Hughes.

With the support team on the white side of the villa, and the assault groups working their way through inside from red to blue, my job in the Hughes was to use my GPMG to eliminate anyone

I saw outside the villa on the black and blue sides. Between us, we had all sides of the villa covered and no one could escape.

While Tiger kept the Hughes moving back and forth above so we couldn't be hit by ground fire below, the explosions of grenades and firing were almost continuous. The support group by the LZ were firing 66mm anti-tank rockets, the G3s on automatic fire and rifle grenades fired from G3s.

The assault team's breathless voices marked their relentless, explosive progress through the villa. "White 1 cleared, moving to white 2!"

They advanced in two smoothly practised four-man groups, under Billy Potts and Don Milton, taking each room in turns, throwing grenades, waiting for the explosions, running inside firing their M16s at anyone left alive, reforming while the second group followed through to attack the next room without pause.

"Guards trying to enter at black 3," I called on the radio, to warn the men inside the villa and at the same time opening up with my GPMG at the new targets. The assault teams knew the black side was a free fire zone for me in the Hughes flying above them and let me get on with it. "Guards suppressed at black 3," I radioed after several minutes firing.

"White 3 cleared," crackled Billy Potts' voice. "Jose killed and mission complete."

"Phase Three," I ordered.

The assault team ran back through the shattered house and regrouped with the support group while I called in Pablo with the Bell. He had been to check an airstrip about 4 kilometres away which we were going to use on the withdrawal, and I wanted him to make sure there were no barbed-wire obstacles across it.

When he arrived, Tiger dropped down to hover-land and we picked up Dave, Almaro and Ramon. Then Tiger pulled the Hughes back into orbit so we could maintain suppressive fire at opportune targets while Pablo swooped in to hover-land and pick up the ground teams. I fired a last long burst and pulled away after them.

We landed and I debriefed the men. I was delighted with the dress rehearsal. The weeks of making them practise over and over and over again had produced a viciously efficient assault. The men really worked hard, they fought through with the confidence of knowing they had everything tuned to a degree they had never experienced before in their army lives and now it was all coming together.

Every morning when we started work, I went through the whole plan. I tested them over and again, firing questions at one after the other as we practised, "What callsign is the support group?"

"What callsign is assault group 2?"

"What side of the door are you?"

"Who's the entry man?"

"Who's the bomber?"

"Who's the lookout?"

"Stop!" I would say, and switch my questions to another man. "You take over. What happens next?"

We did that live rehearsal five times and by the end I can tell you that after eleven weeks of training, these men knew every answer to every possibility we might meet on the attack, every single drill, action and emergency. We had the Principles of War tuned to a fine pitch. This was no bunch of mercenary idiots. Whatever stupidities these men may have committed elsewhere in their lives, whatever foolishness they may be accused of in the future, they had become excellent, highly motivated fighting men on that task in Colombia. They were all raring to go, and I was proud of them.

Then Ricardo, who as our intelligence officer had been keeping in touch with our backers in Bogotá, told me that Escobar was at the villa. The attack was on!

I started the attack sequence, warning the men, our support and the pilots that we were ready to go the following morning. Our Colombian helicopter mechanic worried me by saying there was something wrong with the Bell 204 engine, but he grafted hard for several hours up to his elbows in oil and cured the problem.

That last evening in the cabin at La Gagua, the fury of our recent live rehearsals gave way to calm, deliberate preparations for the fight. Outside in the darkness of the jungle, we could hear the soft rustle of the river beyond the sandbank, the endless call of night insects and see the glowing trails of fireflies. Inside the cabin, the men moved about singly in their rooms or on the verandah making their final adjustments to their equipment. They were quiet because they knew the next morning, they were going to risk their lives. They were thinking of the seventy guards, not in panic, but calmly going through in their minds the way our attack plan would eliminate them from the contest, in the towers, in the guard house, the other side of the fences, in the guards' quarters. They looked at Escobar's photographs. They cleaned their weapons for the last time, minutely checking them so there would be no malfunctions. They examined the grenades and the 66mm rockets, and Dave Tomkins checked over his explosives. There was the occasional joke, which showed bravado and the fire of confidence, and then we went to bed early. I slept well.

We were up at five o'clock. I stepped off the verandah out onto the sandbank to look at the weather. The sky above was brilliant blue and promised a perfect day. Immensely cheered, I went back into the cabin for a cup of tea. After breakfast, we dressed and loaded the helicopters, making a few small adjustments to equipment. Ricardo called on our radio net from Cali and, using coded language, told me to hold on until he had received definite confirmation that Escobar was at home. We waited tensely for several hours, and then the radio crackled again with Ricardo's voice. "He's there! Go!"

I checked my watch. Eleven o'clock. We all paraded by the helicopters, weapons to hand and ready to board. After so much time together training, there was an unspoken bond between us which prompted me to say something to them before we set out. "Good luck" seemed insufficient. The words sounded too much like a weak English understatement which did nothing to match the excitement boiling inside me. I have always been fascinated by military history, famous generals, their battles and leadership, and I tried to recall the lines from *Julius Caesar*, when Marcus Brutus leaves Cassius before the Battle of Philippi, knowing he might never survive. In tones which rang with my own feelings of total commitment, I told them, "Whether we shall meet again I know not. Therefore, our everlasting farewell take. For ever and ever farewell. If we do meet again, why, we shall smile! If not, why then, this parting was well made!"

Some of you may think this was pretentious, but we were a close band of men at that moment, with fire in our bellies and gladness in our hearts for the fight, and that sensation is the finest drug in the world.

The adrenalin rushed again as the rotors began thumping round before take-off, and then we were airborne, swinging over the green jungle towards our target. I was convinced we had done everything possible to make this operation work; I had been given everything I asked for and we had trained like never before. There was nothing to stop us.

Tiger flew our Hughes low over the trees, swinging back and forth up the curving valleys, north-eastwards into the foothills, while Pablo kept the Bell slightly higher on our left, so we could both see each other in case anything went wrong. We had two high mountain ranges to cross, between peaks rising to 15,000 feet, the first to cross back into the Río Cauca valley and then the second beyond that into the Río Magdalena valley to our final refuelling point at Kiko just short of the target.

I looked round at the guys squashed in the back and I could see morale was high. Tiger was piloting us with elan, sitting cheerfully on a tiger skin I had given him for good luck, and Almaro passed sweets around from the back next to Dave Tomkins and Ramon. Once again, I felt that great sense of closeness which is only experienced among men about to risk their lives. I really felt a surge of delight, like the last of the great adventurers.

We crossed the first mountain range without incident, though I was concerned to notice wisps of cloud thickening over the peaks.

We left Manizales in the Cauca Valley on our right and began to gain altitude to cross the last range. I looked ahead through the Hughes' glass bubble and could see cloud hanging round the peaks on all sides. It was not encouraging.

I asked Tiger if he had contact with the other air support. Three other aircraft were vital to my plan. They were flying to join us from Cali and a secret airstrip near Bogotá. Soon, precisely on cue, we heard the pilots answering calls crackling on our earphones. Everyone was on station. One of the newcomers was a Bell JetRanger helicopter ambulance flying low, which would hold off the target ready to fly in if there were wounded or dead, or to ferry us out if the Hughes or the Bell 204 were shot down. The second plane was a small Cessna flying high above us with Ricardo aboard, which was loaded with Telstar radio rebroadcast equipment. All the men on the ground had a radio, earpiece and mike. They were on Channel 1 to me in my Hughes. All the aircraft were on Channel 2 also linked to me, through the other side of my headphones. The Cessna Telstar had rebroadcast equipment to boost the signals to make sure we would all get through to each other. On tests, it had punched through loud and clear from as far away as 40 miles. I could listen and control both channels through my earpiece and mike in the Hughes, switching from one to the other to speak, though most conversation was going to be with the men on the ground. Ricardo was on hand in the Cessna to translate my commands for the Colombian pilots if required. Lastly, an empty fixed-wing Caravan was circling further away ready to pick us all up from an airstrip about 4 kilometres away from the target after the attack.

Everything was in place. The operation was set to go. We had one range to cross, then a refuelling at Kiko, and then the attack!

We flew up towards Cuchilla del Tigre, the last obstacle before the target. The clouds swirled round the peaks above us and hung thickly on the dense trees covering the steep mountain slopes below. The sun occasionally burst through with a flash of light on the white clouds. Tiger followed the valley up to the top, hugging the contours. Our Hughes was sandwiched between the white clouds and green slopes as he searched for a way through, while Pablo circled higher in the Bell on our left.

Tiger pointed at his fuel gauge. We were running low, with less than 100 pounds of fuel in the tanks.

I glanced at the altimeter: 8,900 feet.

Suddenly, Tiger pointed to a bright shining gap in the clouds ahead, where the sun lit the pass and the way through. I was confident we could make it. Kiko was only a couple of minutes' flying time away. Tiger tilted the Hughes forward and headed fast for the bright hole in the clouds.

However, just as quickly, the sunlight then vanished, the gap disappeared like the reflection it was, and we flew into the clouds and straight at the trees. Tiger shouted and pulled at the controls but it was too late. The Hughes smashed into the green canopy, the Perspex bubble shattered and the whole aircraft twisted upside down with the power of the rotor blades caught in the branches. I felt a terrible wrenching pain in my back. The noise of the screaming engine, breaking metal and splintering branches filled my ears. Disorientated, I remember falling, the momentum of the helicopter plunging it through the trees towards the steep slope beneath.

Eventually, we stopped moving. I was hanging head down, held in my seat by my safety belt. I looked around. The Hughes was wedged near the ground between jungle trees, broken palms and thick undergrowth, creaking as it settled. The engine stopped. Almaro and Ramon clambered out past me and dropped to the ground beneath. Tiger was hanging beside me in his seat, covered in blood, mortally wounded by one rotor blade, which had half-severed his left arm and cut off his leg at the hip. He was alive, but grey-faced with shock and hardly conscious.

I unclipped myself and fell painfully to the ground, which

sloped away steeply. I stood with difficulty and searched in the upside-down Hughes for the emergency shock packs we had prepared. Dave and Almaro lifted Tiger from his seat and laid him on the ground, but when I tried to give him a drip, I could not find a vein strong enough to take the needle of the giving set. Instead, I jabbed him with morphine. Dave also tried to get a drip in Tiger's remaining arm but he was too far gone.

My own pain washed over me and I began to suffer shock myself, losing balance. Dave, Almaro and Ramon had survived the crash better in the back and they sat me down among the leaves and undergrowth on a small ledge on the steep slope. Dave made me drink from another drip. Pain seemed to well up in every part of my body. It was worst in my back and chest. I had severely wrenched my spine and broken five ribs, and we had all sustained superficial cuts and bruises from the chunks of metal and Perspex flying round when the Hughes broke up on impact.

I knew Tiger was dying and I tried to persuade the others to give him the last rites, but they did not know how. Maybe they felt it was pointless, but I felt bad about leaving him alone as he died. Perhaps in retrospect, the extreme disappointment at being stopped dead in our tracks was beginning to affect me. We had been only eight minutes' flying time from the target. However, regret was a luxury for later. For the moment, our problems of survival were paramount. We could hear the other helicopter still above us, out of sight, circling over the canopy and, when Dave and Almaro found two personal radios from the wrecked Hughes which still worked, we called up Pablo. Our flares had been smashed in the crash but they found a red panel which they hung out on the trees and Pablo announced he could see it.

He radioed that there was a landing site about 700 metres away to the east and that we should make for that, but I was in agony. I had stiffened up fast and found I simply could not move at all.

I told Dave to set out on the bearing Pablo had given and try to find the landing site, where he could be picked up by the

Bell. He left with his weapon and one radio. Meanwhile, Almaro found another jacket to cover me and keep me warm against the dangers of shock.

"D'you want an Uzi?" he asked. "There's one in the chopper." I shook my head, confused.

Then he set out too, with Ramon and the other radio.

As I heard him sliding down the steep slope below me after Dave, I realised that in my confusion I had broken the rules. I had let both radios go and had no weapon handy either. I felt no anger about it but, as the day passed and I lay cocooned in pain, unable to move, I knew I was in for a hard time.

We had crashed at 8,900 feet and the mountain was cold. The wind got up at dusk, buffeting the trees around me and making the wreck of the Hughes grind on the tree trunks. Painfully, I wrapped myself in my jackets, but I could not help shivering. The grey light faded quickly under the dripping leaves of the canopy and the temperature dropped. I lay in total darkness, cold, filthy, covered with Tiger's blood and exhausted, while images drifted through my consciousness, of Jane looking clean and beautiful at home, of the men during our training and the tough figure of my grandfather, Old Miles. It was a long, miserable and sleepless night.

Eventually, thankful, I glimpsed the first translucent grey light of dawn through the trees above me and I decided it was no good just lying there and freezing to death. I was struck by the idea that I might be on my own for days. We had been flying over a very remote part of the mountains and I knew from my map study that it might take a long time for a search party to find me.

In the meantime, I had to survive. I struggled to my hands and knees, which I think took more than half an hour, and crawled my way by inches over to the wreck, holding on to roots and trees, desperate not to slip on the steep slope and trying to keep my breathing shallow against the pain of my broken ribs and back.

Like a modern-day Robinson Crusoe, I wanted to find what might be useful in the shattered ship. I tried to ignore Tiger's stiffened corpse under the broken Hughes and the terrible grey pallor of his blood-spattered face. Maybe my own problems made me oversensitive, but he had been a Catholic and I still felt badly about his lonely death without the comfort of the last rites. For the living, debris was scattered all around. I picked up the Uzi Almaro had offered me, with three full magazines, Almaro's sweets, several drip bags to drink and plastic explosives which I burned to warm my hands. Most irritating, my watch had come off in the crash and I could not find Tiger's. It had probably been flung off his wrist into the jungle somewhere when the rotor blade ripped into his arm.

Later in the morning, I heard a fixed-wing plane, probably the Cessna, circling above and it was intensely frustrating not being able to speak to them on a radio. However, at least they were there to encourage me. After the Cessna left, some time passed till I heard a helicopter which I thought landed not far away. Soon afterwards, I was sure I heard two shots. I fired several times in reply but the sounds of the low-velocity rounds seemed to be absorbed in the cloud and the wet foliage of the trees. No human noise disturbed the damp silent jungle around me. I guessed I had imagined it.

The clouds filled the trees with mist and it began to drizzle. I realised the planes would be unable to fly any more that day.

"The game is on!" I said to myself fiercely. I was on my own. The fight was on for my life.

I made another slow, painful trip to the helicopter, concentrating hard not to slip on the steep wet earth. This time I found a packet of biscuits, two tins of tuna, part of the escape rations we had prepared, some shell dressings, plastic bags and a torn flak jacket.

Back on my shelf, I tried to warm myself up as the shivering wrenched my broken ribs. I wrapped the torn flak jacket and the shell dressings round me as best I could, and stuck my legs in a plastic bag. My feet had gone numb and felt as though I had broken bones there too, but I dared not take off the boots to see.

They had swollen badly and I was sure I would never be able to pull them on again. This was Escobar's territory and I could not shake off the fear that his men might find me before my own. I had to keep on my boots in case I was forced to fight.

Darkness fell again and I began to dream, visualising extraordinarily real scenes against the black jungle night. I found myself on a brightly lit stage and dreamed all my pain and injuries were drug-induced. The drugs had somehow caused me to make up the crash in my mind. I began to float deliciously free of my painful body and relegated the whole plan of attack on Escobar's house to a dream. I sat comfortably on the roof of a railway building and observed that the jungle had shrunk to a number of small theatrical bushes dotted around on the stage.

"I'm just going to change the videos," announced Dean Shelley helpfully, appearing with extraordinary realism right in front of me. At this, I noticed I was being filmed and Shelley explained that I was not in the jungle at all, but in a railway station and the bushes were props. "All your actions and words are being taped," he added, and vanished into the shadows behind the stage.

The dark night and shivering took over for a moment but my mind was soon away again and this time I was standing in a very luxurious apartment surrounded by bemedalled and moustachioed Colombian generals in splendid uniforms, who were saying, "Bueno! Bueno! The experiment was excellent. Mucho gusto. Bueno!"

I was puzzling over what the experiment might be, when another man, wearing a suit, butted in to say that no one was really concerned about the two hostages still being held under a bridge in Cali. The generals took no notice of this at all.

The inconsistencies seemed quite normal. Nor was I in the slightest surprised to see Frederick Forsyth walk over in a bright, shiny, white mackintosh. Rather casually, he said the generals had taken my watch and added that he intended to use the material for his new book.

"But what about the hostages?" I demanded furiously.

He turned up his mackintosh collar, stuck his hands in his pockets so that he looked like something out of *The Third Man*, and declared throatily, "Don't worry! There'll be no problems anymore!"

They say dreams are a release, and I suppose for a short while mine gave me respite from the cold and the painful restlessness. I shifted position in the dark, unsuccessfully trying to find a comfortable position on the wet leaves. I was so tired I drifted off again.

I heard Dave Tomkins speaking in a seductive voice, saying, "Walk to the railway lines, Peter, and when you've crossed them, the drug will wear off. One of the other guys who is injured will come and pick you up in a truck and it'll all be OK."

I walked to the railway in broad daylight and waited, looking up and down the empty track which disappeared down straight lines into the distance in both directions, but no one came for me. The daylight of my dream faded.

I opened my eyes and was enveloped in the black jungle night. Water dripped on me from the unseen leaves above, cold seeped round my aching back and shivering tremors racked my ribs. Halfconscious, I swore out loud, "Dave, if this is your idea of a joke, it's gone a bit too fucking far!"

Dawn gradually lightened the tall trees over my hide on the steep slope and I set my mind to another day. I crawled again to the helicopter and found some more plastic explosives from Dave's satchel bombs which I burned for a little warmth. I also hoped the smoke would filter up through the trees and show the others where I was.

I heard a plane above me, but the grey mist was still thick in the trees and I doubted they could see anything through the clouds over the mountain.

I crawled back to my ledge, made a fire with plastic explosives and heated a tin of sausages I had found. I had little appetite but forced myself to eat. I began to wonder if I could get out myself. Maybe I could drag myself to Kiko, our refuelling point. We had been so nearly there. I had only three drips left, of 1½ litres each, but I reckoned I could catch water dripping from the trees using the plastic bag. The trouble was that I had not found my compass and had no idea which direction to take. I told myself that at least if I stayed where I was, the others knew where to look. I spent the rest of the day shivering painfully, and drifted exhausted in and out of fitful sleep during my third, long, cold night on the mountain.

I dreamed again that night, that I lived in a graveyard and I had a friend. His name was Death.

In the morning, I was terribly stiff and cold, but I crawled again to the helicopter, more for something to do than in the hope of finding anything. I saw Tiger's dead body again, lying under the wreck. He was a ghastly grey-blue colour and his awful wounds looked glazed and unreal. I guessed the altitude and cold were delaying his decomposition as I saw no flies on him. I felt no anger that he had crashed the chopper and stopped us reaching our target. He had been a good pilot and he had paid the final price for his mistake. It was small satisfaction that we had at least softened the pain of his death with the morphine, though there was nothing we could have done to save him.

Scrabbling about on my hands and knees in the undergrowth beneath the chopper, I found a drip, which I drank at once, and some more of Almaro's sweets among the dead leaves. I suppose all that took up most of the morning, as I only moved for short moments before having to rest and stop the surges of pain round my chest. Later, I dragged myself back to my ledge and prepared to wait out the rest of another day.

Suddenly, I heard the sharp hard sounds of a machete chopping wood below me. Voices speaking Spanish carried clearly up the slope. I came fully alert with a jolt of terror, like Robinson Crusoe finding the footprint in the sand. I pulled myself round to the edge of my hide, rolled on my stomach and held my Interdynamic ready. We had crashed only eight minutes' flying time from Hacienda Nápoles, in the centre of Escobar's territory, and for all

I knew these were his men. I gripped myself and prepared to fight.

Like most people in a tight corner, I prayed. Fervently and urgently, I asked God to guide me and to let me give a good final account of myself before departing this world.

I tried to ignore the pains in my chest and back, and concentrate on the men advancing steadily up the slope towards me. Determined to 'die with honour', I waited, gripping my Uzi tensely, as the chatter of Spanish voices and the steady clipping of machetes drew closer and closer.

I peered carefully over my ledge and could just see the black hair of the nearest man. As he topped the rise, I shouted hoarsely to release my own fear, and thrust my Uzi forward, ready to fire.

"Amigo! Amigo!" screamed the man, staring terrified into the muzzle of the Uzi. "Colonelo Ricardo, Ricardo!"

Even in the state I was in, I recognised Colonel Ricardo's name, which the man was using as a codeword.

The rescue party had arrived.

There were two men and three campesinos whom they had employed to cut a path through the thick jungle. None of them spoke English, but one gave me a can of Pepsi which I poured straight down my throat. Another cut off my boots and my trousers, to see the extent of my injuries. I lay back for a moment with the sheer relief of being found and then painfully pulled on some civilian trousers and running shoes which one of them brought from the helicopter. Then I asked, "How far have we got to go?"

One of the Colombians shrugged and replied, "Five hours' walk."

I stared at him. It took all my energy and resolve just to crawl to the helicopter, let alone to then 'walk' for five hours.

I had no option. We were nearly 9,000 feet up at the top of a mountain and thick clouds meant they could not winch me out by chopper. I had to walk down. Leaning on a succession of men in the rescue party for support, I set out down the slope, tottering and sliding downhill terribly slowly, like an old, old man. Each step was agony. I thought that once I got moving, the pain would ease off or soften to a dull ache. It did not. I could hear my ribs grind and stab my chest every step of the way down the endless slope.

The campesinos did their best to cut away the undergrowth, leaves, hanging vines, lianas and clinging, spiny fronds, but I have never seen such thick, overgrown and dirty jungle. We followed the same stream bed they had walked up to find me and I lost count of the waterfalls we had to descend. The ground was so steep the only way down was to climb the rocks beside the tumbling water. When they realised I couldn't climb, they cut down a tree, skinned off the leaves and lowered me down by a rope tied round my chest as I hung onto the pole, gasping and speechless, gritting my teeth.

When I had recovered, the men below me stepped confidently down the slope again, while I followed, hanging on to one of the campesinos, slithering uncertainly on wet leaves and stones, afraid to hurt myself by slipping, and hurting anyway. Every time I heard another waterfall below, my morale sank.

Hours later, it seemed to me, I asked in desperation, "How much longer?"

"About one hour," called back the man in front of me, turning round and looking up the slope.

Looking at the watch of the man helping me, I asked an hour later, "How far?"

"Maybe two hours," the man called up casually, ignoring the shock on my face, and carried on slashing right and left at the lianas hanging in our way.

The light faded, dusk turned into night and the campesinos never stopped. In utter darkness, lit only by the swinging, flashing beams of their torches turned back to light my path, I slithered on downhill. I have never felt so sorry for myself, or so desperate and helpless.

I made a pact with my God not to complain of the pain. I kept my side of this unequal bargain and I guess he kept his because eventually they stopped walking on a sandy island in the middle of the stream.

They had sheets of polythene which they wrapped round me and we settled down to sleep, one on each side of me to keep me warm. At least, I think that's what they were doing, because in the darkness I heard them rootling through my small zip bag, and whispering to each other as they divided up my goods. I felt disappointed, as they had been so helpful. Next morning, I found they had stolen all my 30,000 pesos escape money.

I slept fitfully, unable to turn properly against the hurt in my chest, and was awake as the grey light of dawn filtered through the trees hanging over us. Gesticulating at their watches and pointing downhill, I asked them, "How far to go?"

"One hour," they said. "One hour."

Cheered, I gritted my teeth and we set off down the slope again. Fighting every step of the way, I became obsessed with time, looking at their watches as we walked. Five hours later, we were still going.

The jungle canopy was thinning in places overhead and when I saw a hosepipe running out of the stream, I asked again, "How far?"

"One kilometre."

This took another hour. I was totally exhausted. The last several hundred yards towards a small wood and atap leaf hut seemed to last forever. They took me inside, laid me down to rest on the floor and covered me with hessian sacking. I passed out at once.

Only an hour later, they shook me awake from deep sleep to say the helicopter was coming. Needless to say, the landing site was not right outside the hut. They patiently dragged me to the top of a hill further away where I sat in long coarse grass, drugged by exhaustion and the warmth of the late afternoon sun.

I heard the steady beat of a helicopter. A Bell 204 swung round above the hilltop and came in to land. I stood up, swaying on my feet. As its skids touched the grass, the down-draft from the rotors washed over me and I collapsed flat on my back, like a puppet without strings. Several of the campesinos helped me to my feet and into the helicopter.

After a short flight to an airstrip, they transferred me to a light Cessna and we took off at last light. An hour later, we circled over the lights of Cali spread out in darkness below, and landed. A Land Cruiser came onto the airfield to take me straight from the Cessna into Cali, to the apartment block we had used weeks before when we had first arrived.

Four days had passed since I had seen the guys. We had all been fully committed to the attack, we had trained like brothers, as closeknit a team as any I have prepared in all my experience, but when I walked in, filthy, bruised and exhausted, I found them all there, showered and changed, arguing with each other about money.

We tried to go again. We assessed that security was not so badly breached by our crash that Escobar knew, so I moved the men out of the way to Panama while our backers found another helicopter and a new pilot. I stayed in Cali, recovering from my injuries, planning a new attack and liaising with Almaro and Ricardo. However, the momentum of the operation had gone. I had wound up the men over weeks of detailed, intensive training and they had willingly followed me over the top. Only crashing into a mountain had stopped us killing Escobar. The galling anti-climax was too much for them. The fire in their bellies had died.

Ned Owen skipped off of his own accord. He flew back to England from Panama and promptly sold his story to television.

On August 13, James Adams published his story, telling us on the phone that he had to do it because our mission was virtually public knowledge. He said Washington knew all about us, which was true enough because we had told the CIA ourselves, but our assessment in Colombia, with Ricardo, was that Escobar had no inkling. Adams may have been sensitive about this, as his article was headlined that our plans were "known – even to their enemy". He rather condescendingly described me as a "simple soldier" and then proceeded to blow the whole plan with an artist's impression of our attack and plenty of speculation to fill in the gaps.

Terry Tagney added his two pennyworth by going on television from South Africa.

Security was in shreds, our backers agreed it was pointless to continue and I disbanded the team after a farewell dinner in Panama.

CHAPTER 14
ESCOBAR
SURRENDERS

Bombings and high-profile kidnappings pressured the government into accepting the terms set by Escobar for his surrender. Behind the scenes, lawyers for Escobar and the government distilled the terms down to three issues: the location of the prison, the guards and the involvement of the police and army. As the prison holding his crime partners, the Ochoas, could have been car bombed, Pablo had refused that suggestion. Out of forty staff, Pablo would get to select half of the guards, all of whom would come from his region, Antioquia. He would decide who would attend the surrendering, and whether there would be any journalists. He would surrender by helicopter, accompanied by the two men who had brokered the peace: Father Garcia and the politician, Alberto Villamizar.

"I do not go alone without the priest in that helicopter," Pablo told Alberto. "They knock him down and I'll go hand in hand with him to the heavens. No, you give me cover too."

"Carlos Gustavo Arrieta and the Director of Criminal Instruction, Carlos Eduardo Mejía, also wanted to attend."

"Let them come, but we'll put them in another helicopter." He told his men that he would use the officials as a decoy: "If there is an attack, we will shoot them first."

He had wanted to convert a convent in El Poblado into a prison, but the nuns had refused to sell it. A proposal to reinforce a Medellín prison was also rejected. The remaining option was the Municipal Rehabilitation Centre for Drug Addicts on property called La Catedral del Valle, stationed on a mountainous slope over the Honey Valley, 7,000 feet above sea level, which would give the guards and the occupants a bird's-eye view of any threats. The area was foggy in the evening and at dawn, which made a surprise raid from the air more difficult and provided a means for the occupants to slip away unnoticed. They could easily lose their pursuers in the surrounding forest, which was teeming with wildlife, such as armadillos, sloths and huge, iridescent butterflies.

The building and 30,000 square metres of land had been registered in the name of one of Pablo's friends, a trusted old ironmonger. Pablo wanted only local guards and for the police and army to have nothing to do with it. The Mayor of Envigado approved the transfer of the building into a prison called the 'Cathedral'.

It had cement floors, tile roofs and green metal doors. Formerly a farmhouse, the administration section included three little rooms: a kitchen, a courtyard and a punishment cell. It had a big dormitory, library, study and six cells with their own bathrooms. The large dayroom included four showers, a dressing room and six toilets. Motivated by Father Garcia's blessing of the project, seventy men had been working around the clock, remodelling it. Due to its inaccessibility, mules had brought furnishings: water heaters, military cots, tubular yellow armchairs, potted plants ...

Despite its secure location, Pablo wanted bodyguards inside the prison, just in case anything happened. "I won't surrender alone." He wouldn't abandon his associates to be slaughtered by the Elite Corps, while omitting to say that by keeping his network close, he could continue to run his operation. As added insurance, he and Roberto buried weapons near their designated cells. "One day we'll need them," he told his brother.

The night of his kidnapped wife's release, Alberto stayed up until dawn chatting with Maruja. After an hour's sleep, he set off for Medellín. At La Loma, he met Monkey, one of two men, including Jorge Ochoa, whom Pablo had authorised to finalise the negotiations. Monkey was tall, blond and had a golden moustache.

The phone rang. "Dr Villa, are you happy?" Pablo asked Alberto. "I thank you for coming. You're a man of your word and I knew you wouldn't fail me. Let's start to arrange how I'll turn myself in."

Monkey and Alberto visited the Cathedral. They found a wide comfortable construction with a view that Pablo didn't want obstructed. From outside, it appeared austere, with only concrete visible. They discussed security concerns as they examined a double fence over 9 feet high, with fifteen rows of electrified barbed wire. Out of the nine watchtowers, the two at the entrance were being reinforced. The cells had grills, tables, tubular beds and a cement area with an electric grill and a two-burner gas stove. Alberto frowned upon the Italian tiles in Pablo's bathroom, so he issued an order: "Remove that shit! This is a prison!" After the inspection, he said, "It seemed to me a very prison-like prison." He hadn't noticed that the switch for the 10,000-watt fence was in Pablo's room.

An arrangement with Pablo had been made whereby Alberto would receive an anonymous call: "In fifteen minutes, Doctor." Then he would go to his upstairs neighbour, Aseneth, and take a call from Pablo. As her house was a jumble of writers and artists, who came and went throughout the day and night, it was considered a safe place to call.

One evening, Alberto didn't get to the phone on time. Aseneth answered, "He doesn't live here."

"Don't worry about that," Pablo said. "He's on his way up."

Alberto tried to tell Aseneth what was going on. She covered her ears. "I don't want to know anything about anything. Do whatever you want in my house, but don't tell me."

At La Loma, Maruja thanked the Ochoas for facilitating her release. Alberto mentioned that her emerald and diamond ring, taken by the kidnappers, had not been returned as promised. Monkey's offer to buy a new one was declined because Maruja wanted the original due to its sentimental value. Monkey promised to refer the matter to Pablo, who successfully ordered it to be returned.

The president's fear of the priest saying a word that might threaten the negotiations at the last minute was realised during a broadcast of *God's Minute*. The 83-year-old called Pablo an unrepentant pornographer and an abuser of minors, and demanded that he remove himself from the hands of the devil and return to God's path. The about-face astounded the viewers. Enraged, Pablo believed that something seismic had occurred. As the priest's blessing had cajoled his men into surrendering, he was now faced with a rebellion. Demanding an immediate public explanation, he refused to surrender.

Perplexed, Alberto wondered whether the priest's mind had started to crack, as it wasn't the first time that he had rambled out of reality. Or maybe Pablo's enemies had encouraged the outburst to torpedo the surrender. Alberto hustled the priest over to the Ochoa family at La Loma to speak to Pablo on the phone. Out of the various explanations he offered, the acceptable one was that an editing error had made him appear to say pornographer. Having recorded the conversation with the priest, Pablo played it to his men, which satisfied them.

Government demands presented the next challenge. The politicians wanted more say in the selection of the guards. They wanted army and National Guard troops to be on patrol outside the Cathedral. They wanted to cut down trees to make a firing range adjacent to the Cathedral. Citing the Law on Prisons, which prohibited military forces from going inside a jail, Pablo rejected the idea of combined patrols. Cutting down trees would permit helicopter landings and a possible assault on the prison, which was unacceptable. He changed his mind after it was explained that the removal of the trees would provide greater visibility, which would give him more time to respond to an attack. The National Director of Criminal Investigation was insisting on building a fortified wall around the prison in addition to the barbed wire, the prospect of which infuriated Pablo.

On May 30, 1991, newspapers began reporting the terms of surrender. What caught the public's attention the most was the removal of General Maza and two prominent police leaders. After meeting the president, Pablo's nemesis, Maza, sent him a six-page letter, saying he was in favour of Pablo's surrender: "For reasons known to you, Mr President, many persons and entities are intent upon destabilising my career, perhaps with the aim of placing me in a situation of risk that will allow them to carry out their plans against me." He suspected that the government had negotiated his position away, even though there was no official evidence of it.

Pablo informed Maza that their war was over. There would be no more attacks. His men were surrendering. He was turning in his dynamite. He listed the hiding places for 700 kilos of explosives. Maza was sceptical. He asked the president, "How come you're going to take him to his natural refuge, where he hid for at least five months while we chased him?"

Losing patience with Pablo, the government appointed an outsider as the director of the prison, not a local person as had been requested. They assigned twenty National Guards to the prison, who were also outsiders.

"In any event," Alberto said, "if they want to bribe someone, it makes no difference if he's from Antioquia or somewhere else."

Not wanting to make a fuss, Pablo agreed that the army could guard the entrance. The government offered assurances that precautions would be taken to ensure that his food wasn't poisoned. Policies and procedures for the prison were determined by the National Board of Prisons. Prisoners had to wake up at 7 AM. At 8 PM, they had to be locked in their cells. Females could visit on Sundays from 8 AM until 2 PM. Men could visit on Saturdays. Children could visit the first and third Sunday of each month.

On June 9, 1991, Medellín police started to implement security measures, including removing people from the area who didn't live there. Two days later, Pablo asked for a final condition: he wanted the prosecutor general to be present at his surrender.

Pablo lacked the official ID necessary for a surrendering person. To get citizenship papers, he was supposed to go to an office at the Civil Registry, which was impossible for a man with so many enemies. His lawyers asked the government to issue citizenship papers without him having to make an appearance. The solution proposed was for him to identify himself with his fingerprints and to bring an old notarised ID, while declaring that his new ID had been lost.

On June 18, Escobar's kids, Juan Pablo and Manuela, called their father from Miami and excitedly described the recent days they had spent in Los Angeles, San Francisco and Las Vegas.

"Tomorrow I'm going to turn myself in," Pablo said to his son, "because I already know that the new constitution will not include extradition." Juan Pablo expressed concerns that his father might be falling into a trap. "Don't worry, son. Everything has gone to plan. I can no longer be extradited." After Manuela got on the phone, he said, "Soon my problems will all be in the past. Before long, we'll all be living together again. On the news, you will see that I'm in a prison, but don't worry, because it is my decision to go there."

On June 18, Monkey called Alberto at midnight, waking him up. Alberto took the elevator to Aseneth's apartment, where a party with accordion music was in full swing. Wrestling his way through the revellers, he was stopped by Aseneth. "I know who's calling you. Be careful, because one false step and they'll have your balls." She escorted him to her bedroom, where the phone was ringing.

Above the ruckus, Alberto heard, "Ready. Come to Medellín first thing tomorrow."

On June 19, Pablo asked his wife to go home and get her things in order, so that she could meet him at the Cathedral.

At 5 AM, Alberto appeared at the dwellings of Father Garcia, who was in the oratory finishing mass. "Well, Father, let's go. We're flying to Medellín because Escobar is ready to surrender."

Aboard a Civil Aeronautics plane were representatives of

the government. Travelling with the priest was his nephew, who assisted him. They were met at the Medellín airport by Marta Ochoa and Jorge Ochoa's wife, Maria Lia. The officials went to the capitol building. Alberto and Father Garcia headed for Maria Lia's apartment.

Over breakfast, the surrender arrangements were finalised. The priest was told that Pablo was on his way, employing his usual evasion techniques, travelling sometimes by car and at other times walking around checkpoints. His imminent surrender unnerved the priest so much that one of his contact lenses fell out and he stood on it. To remedy his despair, Marta took him to an optician to get a pair of glasses. On the way there and back, they were stopped at numerous checkpoints, where the guards saluted Father Garcia for bringing peace.

At 2:30 PM, Monkey showed up and said to Alberto, "Ready. Let's go to the capitol building. You take your car and I'll take mine."

At the building, the women waited outside. Putting on dark glasses and a golfer's hat, Monkey disguised himself. Misidentifying Monkey, a bystander called the government to report that Pablo had just surrendered at the capitol building. About to leave the building, Monkey received a call on a two-way radio notifying him that a military plane was heading for the city, carrying injured soldiers. To keep the airspace open for Pablo, Alberto had the military ambulance rerouted and repeated his order to keep the sky clear.

"Not even birds will fly over Medellín today," the defence minister wrote in his diary.

After 3 PM, a helicopter lifted from the capitol building's roof with the National Director of Criminal Instruction, the prosecutor and a cameraman. Ten minutes later, an order was despatched to Monkey's radio.

Boarding a second helicopter with two of his closest men, Otto and Mugre, Pablo appraised the passengers: Monkey, Alberto, Father Garcia and Luis Alirio Calle, the only reporter invited because Pablo admired his honest daily TV broadcast about solidarity, peace, hope and religion. He offered his hand to Alberto. "How are you, Alberto?"

"How's it going, Pablo?"

Smiling, he thanked Father Garcia. "What are you doing here?" he said to Monkey. "Do you want to get yourself killed?" He sat next to his bodyguards. His friendly tone left the passengers wondering whether he had praised or chastised Monkey. Smiling, Monkey shook his head. "Ah, Chief." Turning to the reporter, he said, "At last I meet you, man, Luis Alirio."

Based on Pablo's tranquillity and self-control, Alberto's first impression was that Pablo possessed a dangerous level of confidence bordering on the supernatural. Monkey was unable to close the helicopter door, so the co-pilot did it.

"Do we take off now?" the pilot asked.

"What do you think? Move it!" Pablo said, briefly dropping his polite mask. As the helicopter ascended, Pablo said to Alberto, "Everything is fine, isn't it?"

"Everything's perfect." For five minutes, Alberto braced for the helicopter to be blown out of the sky as it flew towards Treasure Hill, between El Poblado and urban Envigado. Gazing at the view, Pablo saw the La Paz neighbourhood and the mountains where he had played as a child, and from where years later he had commanded a war against those officials whom he felt had unjustly persecuted him.

"I'm going to turn on my tape recorder," the reporter said.

"Tell the president that I'm not going to disappoint him in anything," Pablo said for the tape. "He knows that people are going to start slandering about me committing crimes from here and all those things …"

As they flew, a radio broadcast announced that the government's position on extradition had been defeated in the Constituent Assembly by a vote of fifty-one to thirteen, with five abstentions. It was official confirmation of Pablo's demand for non-extradition. The reversal on extradition had come about at a time when President George HW Bush was mustering support for the invasion of Iraq. Colombia had used its seat on the United Nations Security Council to vote against the attack. Bush had wanted the Colombian government to reverse its position. Dozens of traffickers had been extradited to America, which was still providing arms and soldiers to Colombia. In a quid pro quo fashion, the Colombian president had voted to attack Iraq, while reversing its policy on extradition.

After 3 PM, Pablo's wife Victoria, his mother Hermilda and Aunt Ines set off for the Cathedral, while listening eagerly to radio stations for any news. Approaching the prison, they prayed that everything would go smoothly.

Monkey directed the helicopter pilot to a soccer pitch by tropical flower gardens. "Put it down over there. Don't turn off the engine."

As it descended, 100 armed bodyguards arranged by his brother Roberto formed a circle to protect Pablo, who helped the priest get off the helicopter. Wearing tennis shoes, faded blue jeans, a blue checked shirt and a light-blue jacket, the bearded boss with long hair walked with a carefree stride. Overweight and tanned, he said goodbye to and hugged the nearest bodyguards, some of whom were crying. He told Otto and Mugre to come with him.

Pablo spotted a government cameraman recording him. "Turn off your equipment!" Fifty nervous guards in blue uniforms were brandishing guns. He responded like thunder: "Lower your weapons, damn it!" They were lowered before their commander issued the same order. They walked to a house containing the official delegation, more of Pablo's men who had surrendered and his wife and mother, who was sobbing. "Take it easy, Ma," he said, patting Hermilda on the shoulder. He guaranteed Victoria that her suffering was over. He said that he lived for her and the kids, all of whom deserved to live peacefully. She gazed at him with hope that he had left his past behind.

"Nothing wrong is going to happen to Pablo," Alberto said. "Don't worry. I give you my word."

"Thank you," Pablo said.

The prison director shook Pablo's hand. "Señor Escobar. I'm Lewis Jorge Pataquiva."

Pablo pulled up a trouser leg, revealing a SIG Sauer 9mm pistol with a gold monogram inlaid on a mother of pearl handle. The spellbound crowd watched him remove each bullet and throw the gun to the ground. The gesture was designed to show confidence in the warden, whose appointment had worried him. "Will you sign the delivery certificate as a witness?" he asked Alberto. On a portable phone, he told his brother that he had surrendered, and then he acknowledged everyone. Addressing the journalists present, he said his surrender was an act of peace. "I decided to give myself up the moment I saw the National Constitutional Assembly working for the strengthening of human rights and Colombian democracy."

The journalists wrote about Pablo:

"I had thought that he was a petulant, proud, disciplined man, one of those who is always looking over his shoulder. But I was wrong. On the contrary, he is educated. He asks permission if he walks in front of a person and is agreeable when he greets someone."

"You can see that he is someone who worries about his appearance. Especially his shoes. They were impeccably clean."

"He walks as if he had no worry in the world. He is very jovial and he laughs a lot."

"He had a bit of a belly, which makes him look like a calm man."

"I'm here," said Carlos Arrieta, the attorney general, taking Pablo's hand, "Señor Escobar, to make certain your rights are respected."

Pablo thanked him. Having seen a TV show about Arrieta, Pablo had enjoyed the part which showed the attorney general playing with his daughter. Having been at war with the government since his daughter's birth, he had been unable to spend any proper time with her. "Do you know why I did this?"

"No. Why?" Arrieta asked.

"One day I want to play with Manuela just like you did with your Camila." The attorney general was impressed by his desire to be a good father.

Pablo took Alberto's arm. "Let's go. You and I have a lot to talk about." In an outside gallery, they both leaned against a railing. "I apologise for what I have done to you. Know that neither I nor any of my men will ever touch you or your family again. I apologise for the pain I've caused your family. Both sides have suffered much in this war."

The words surprised his bodyguards. "You are the only person to whom the boss has ever apologised," one said.

"Leave us alone to talk," Pablo told them.

"Why was Luis Carlos Galán killed?" Alberto asked.

"The fact is that everybody wanted to kill Dr Galán. I was present at the discussions when the attack was decided, but I had nothing to do with what happened. A lot of people were involved in that. I didn't even like the idea, because I knew what would happen if they killed him, but once the decision was made, I couldn't oppose it. Please tell Doña Gloria that for me," he said, referring to Galán's widow. Alberto later stated that Pablo had told him that a unanimous decision had been made to kill Galán by politicians, members of the Colombian Congress, paramilitary groups and the Cali Cartel.

"Why was an attempt made on my life?"

"A group of your associates in Congress had convinced me that you were uncontrollable and stubborn, and had to be stopped somehow before you succeeded in having extradition approved. Besides, in that war we were fighting, just a rumour could get you killed. But now that I know you, thank God nothing happened to you."

"Who in Congress said those things about me?"

"No. I'm not going to give you names, but you know who they were."

"Why did you kidnap my wife and sister?"

"I was kidnapping people to get something and I didn't get it. Nobody was talking to me. Nobody was paying attention, so I went after Doña Maruja to see if that would work." He said that the negotiations had convinced him that Alberto was a brave man of his word, and he was eternally grateful for that. Even though he was not expecting them to ever be friends, he assured Alberto that nothing bad would ever happen to his family. "Who knows how long I'll be here, but I still have a lot of friends, so if any of you feels unsafe, if anybody tries to give you a hard time, you let me know and that'll be the end of it. You met your obligations to me, and I thank you and will do the same for you. You have my word of honour." He asked Alberto if he would give further reassurances to his mother and wife, who were having sleepless nights as they suspected the government had arranged for him to be murdered in prison.

After leaving the Cathedral, Alberto went to Itagüí prison to thank the Ochoas. Some of Pablo's men took him on a tour of the city. Fascinated by how they lived and their belief systems, he stayed out drinking with them until 7 AM. For an entire day, he slept at the Ochoas' La Loma property, before returning to Bogotá.

Forty-one-year-old Pablo underwent the medical examination required for new prisoners. Courteously, he answered questions asked by Marta Luz Hurtado, the Director of Criminal Instruction of Medellín. After his fingerprints had been taken, she asked, "Did you have surgery on your fingers?" Displaying his hands, he laughed. The actor from *The Godfather*, Gianni Russo, has described in his book and interviews witnessing Pablo burning his fingertips. His health was documented as that of "a young man in normal physical and mental condition". He said that the scar on his nose was due to an injury from playing football as a child. The only abnormality found was congestion in the nasal mucous membranes.

To obtain imprisonment, he cited a crime that he had been found guilty of by the French authorities: acting as a middleman in a drug transaction arranged by his cousin, Gustavo. He issued a statement: "That country's penal code ... gives one the right to apply for a revision of their case, when they appear before their national judge, in this case a Colombian judge. This is precisely the objective of my voluntary presentation to this office, in other words, to have a Colombian judge examine my case." Rather than plead guilty to a crime, he had surrendered to appeal the French conviction.

In court in Bogotá, he declared his job title as "livestock farmer," and added, "I have no addictions, don't smoke, don't drink." He said he had done an accounting course and, while incarcerated, he was going to obtain a college degree. "I wish to clarify that there may be people who might try to send anonymous letters, make phone calls or commit actions in bad faith under my name in order to harm me. There have been many accusations, but I've never been convicted of a crime in Colombia."

"Do you know where they got the 400 kilos of cocaine?" the judge asked, referring to his conviction in France.

"I think Mr Gustavo Gaviria was in charge of that."

"Who is Mr Gustavo Gaviria?"

"Mr Gustavo Gaviria was a cousin of mine."

"Do you know how Mr Gaviria died?"

"Mr Gaviria was murdered by members of the National Police during one of the raid-executions, which have been publicly denounced on many occasions."

"Let's talk about your personal and family's modus vivendi, and the economic conditions you've had throughout your life."

"Well, my family is from the north-central part of Colombia, my mother is a teacher at a rural school and my father is a farmer. They made a great effort to give me the education I received, and my current situation is perfectly defined and clear before the national tax office ... I have always liked to work independently and, since my adolescence, I have worked to help sustain my family. Even when I was studying, I worked at a bicycle rental shop and other less important jobs to support my studies ... Later,

I got into the business of buying and selling cars, livestock and land investments. I want to cite Hacienda Nápoles as an example of this – that it was bought in conjunction with another partner at a time when these lands were in the middle of the jungle. Now they are practically ready to be colonised. When I bought land in that region, there were no means of communication or transport, and we had to endure a twenty-three-hour journey. I say this in order to clarify the image that people have that it's all been easy …" Asked if he had originally started in business with other people, he said, "No. It all began from scratch, as many fortunes have started in Colombia and in the world."

"Tell the court what disciplinary or penal precedents appear on your record."

"Yes, there have been many accusations, but I've never been convicted of a crime in Colombia. The accusations of theft, homicide, drug-trafficking and many others were made by General Miguel Maza, according to whom every crime that is committed in this country is my fault."

After he denied any involvement in the cocaine business, the judge insisted that he must know something about it. "Only what I see or read in the media. What I've seen and heard in the media is that cocaine costs a lot of money and is consumed by the high social classes in the United States and other countries of the world. I have seen that many political leaders and governments around the world have been accused of narco-trafficking, like the current Vice President of the United States [Dan Quayle], who has been accused of buying and selling cocaine and cannabis. I have also seen the declarations of one of Mr Reagan's daughters, in which she admits to taking cannabis, and I've heard the accusations against the Kennedy family, and also accusations of heroin dealing against the Shah of Iran, as well as the Spanish president. Felipe González publicly admitted that he took cannabis. My conclusion is that there is universal hypocrisy towards drug-trafficking and narcotics, and what worries me is that from what I see in the media, all the evil involved in drug addiction is blamed on cocaine and Colombians, when the truth is that the most dangerous drugs are produced in labs in the United States, like crack. I've never heard of a Colombian being detained for possession of crack because it's produced in North America." He had a point: the journalist Gary Webb discovered that the CIA had facilitated the importation of tons of cocaine into America, some of which had contributed to the crack epidemic. George HW Bush and other senior politicians had been deeply involved in covert drug activity while using Pablo's operation as a smokescreen.

"What is your opinion, bearing in mind your last few answers, on narco-trafficking?"

"My opinion, based on what I've read, I would say that cocaine [will continue] invading the world ... so long as the high classes continue to consume the drug. I would also like to say that the coca leaf has existed in our country for centuries and it's part of our aboriginal cultures ..."

"How do you explain that you, Pablo Escobar, are pointed out as the boss of the Medellín Cartel?"

Avoiding the question, he referred the judge to a statement he had submitted on videotape. "Another explanation I can give is this: General Maza is my personal enemy ... [He] proclaimed himself my personal enemy in an interview given to *The Time* on the eighth of September, 1991. It is clear then that he suffers a military frustration for not capturing me. The fact that he carried out many operations in order to capture me, and they all failed, making him look bad, has made him say he hates me and I am his personal enemy ..."

The court heard a list of traffickers who had claimed that Pablo was their boss. "I don't know any of these people," he said. "But through the press, I know about Mr Max Mermelstein. I deduce that he is a lying witness, which the US government has against me. Everyone in Colombia knows that North American criminals negotiate their sentences in exchange for testifying against Colombians ... I would like to add to the file a copy of *Semana* magazine, which has an article about Max Mermelstein, to demonstrate what a liar this man is: 'Escobar was the chief of chiefs. The boss of cocaine trafficking wore blue jeans and a soccer shirt, was tall and thin.'" Pablo stood to display his short, stout body. "I ask you to tell me, am I a tall and thin person? For a gringo to say that one is tall, you would suppose that man to be very tall."

From all over the world, hundreds of journalists requested direct interviews, which Pablo refused, while allowing a few to post him questions. *The Colombian* published his first prison photo, which showed him with a white poncho and a long beard, but hid the almost 50 pounds weight gain from over two years. He wrote a public statement for the media, which he recorded on tape: "After seven years of persecution, abuses and struggles, I wish to serve all the years of jail necessary to contribute to the peace of my family, to the peace of Colombia, to the strengthening of respect for human rights, to the strengthening of civil power and the strengthening of democracy in my beloved Colombian homeland."

The attitude of the police was expressed by Colonel Naranjo: "When Escobar imposed conditions and surrendered, the police did not feel totally mocked. Some people believe that the police did not receive that surrendering well. But we understood that if he turned himself in, it was the same as having captured him and, in any case, the institution was not able to contain terrorism anymore. The police rested when he surrendered. In general, that first phase was irrational. People believed that only the police – and not the whole of society and institutions – was responsible for combating Escobar."

CHAPTER 15
ESCOBAR'S
PRISON LIFE

With some of the millions he had smuggled into prison, it wasn't long before Escobar started to modify his surroundings. Cash was stored in milk cans inside containers of salt, sugar, rice, beans and fresh fish, which were classified as food rations and permitted. Extra money was buried near the soccer field and in underground tunnels accessible by trapdoors in the cells. When his men needed paying, helicopters transported cash out.

With officials from the Ministry of Justice and the National Bureau of Prisons frequently visiting, Pablo lobbied them for a doctor, a social worker and areas to work. He wanted to study law or journalism, but universities kept rejecting his applications.

With his own surveillance system, he monitored the four sides of the prison and installed checkpoints on the access road. With the approval of the prison director, he converted the prison into a club.

The centre of the prison became a room with two pool tables, fitness equipment, a roulette table, games such as chess and parquet, and several motorbikes. The cabins had double beds, bookshelves, lavish bathrooms, stereos, refrigerators, TVs and VCRs. He added a bar, lounge and disco, where he hosted parties and weddings. Famous people, models, politicians and soccer players danced and cavorted. He installed a sauna in the gym, and jacuzzis and hot tubs in the bathrooms.

A sliding door replaced his cell's bars. Two adjacent cells became an office. The only people allowed into his quarters without pre-authorisation were his family and his delivery driver Limón, who had worked for his brother, Roberto, for twenty years, and was an official employee of the municipality of Envigado. Those waiting to see him remained outside of his office. When not receiving visitors, he spent hours hand-writing responses to letters, using a cheap pen and with a dictionary on hand. A journalist in Medellín received a letter of safe passage to be presented to anyone threatening him. A communications student with an illustrious surname who wanted to write a biography about Pablo was instructed to postpone the book idea and establish a friendship. To a TV presenter who detailed her emotions, he described how excited he became upon receiving drawings from his daughter, about the long trips that Manuela had taken to visit him. He admitted writing poetry and requested some books by Tolstoy. The hundreds of letters he wrote had his signature, fingerprint and a request for forgiveness for any spelling mistakes. They were posted in traditional envelopes with red and blue stripes.

Years later, one of the freed hostages, Pacho Santos, commented on Pablo's writing ability: "As an editor, I would not have changed a comma in his letters and communications. He wrote in a simple, direct, perfectly coherent language and did not say another word than what he had to say."

An area in his living room included a bar, fridge, small table, kitchenette and valuable paintings. His paperweights were little soldiers, a gift from a guard's wife, and a rumour circulated that the soldiers represented people he wanted to kill. His bedroom included a double bed below a gold-framed portrait of the Virgin Mary, a library, a fireplace, a closet that looked out onto a cove (which stored weapons) and a bathroom with a tub. In a corridor, on one side of the cell, was the control panel for the lights, the foghorn and the electricity of the outer fence. At the front was a terrace with a view of Medellín – and a few metres away was Manuela's dollhouse. One of the biggest benefits was the time he could spend with his family.

Large items such as computers and big-screen TVs were

smuggled in by Limón's truck, with crates of soda disguising the contraband. Limón brought women in too. Despite rules restricting visits to official days, people were always sneaking in, as the checkpoint guards had been bribed. Vans with fake walls held up to twenty people. It was ideal for those who wanted to keep their visits a secret, such as criminals and politicians. Allocated a $500,000 a month budget, one of his sicarios, Popeye, paid the guards at the six checkpoints. Coloured pieces of paper convertible into cash at a bank in Envigado were used for bribery at the rate of approximately $100,000 monthly. As well as cash, the guards received gifts, including refrigerators and TVs, which Limón shopped for.

Pablo's extensive record collection was there, including albums signed by Frank Sinatra from when he had visited him in Las Vegas, and Elvis records purchased during a Graceland trip. His books ranged from bibles to Nobel Prize winners. He had novels by Gabriel García Márquez and Stefan Zweig, a prominent Austrian writer from the 1920s. His movies on videotape included *The Godfather* trilogy and films starring Chuck Norris. He also loved the series *The Untouchables*, which he watched three times.

Most of the prisoners had posters on their walls, whereas Pablo hung valuable paintings. His closet was full of neatly pressed jeans, shirts and Nike sneakers, some with spikes on in case he had to flee. He never tied the laces of his sneakers – it was said that if he did, then an emergency was imminent. In case of danger from above the prison, a remote control allowed him to turn off all the internal lights. Communications were a priority. He had cellphones, radio transmitters, a fax machine and beepers. Roberto has denied claims by authors that Pablo used carrier pigeons. Further up the slope, cabins were built for privacy with females and as hideouts in case the prison was attacked. They were painted brightly and had sound systems and fancy lamps. Paths were made into the forest to allow a quick getaway and to enable the prisoners to enjoy fresh air.

Women with high heels, silicone enhancements and their hair

dyed blonde were smuggled into the Cathedral for sex, with Pablo preferring teenagers. He would order his men to find the most beautiful girls from the colleges of Medellín. Working for Pablo, a madam who helped to coordinate beauty pageants provided him with catalogues of girls, naked and clothed, and if he saw somebody attractive on TV, she would send the person an invitation to the Cathedral with the enticement that they would earn a car or something valuable.

She would tell Pablo, "I have this girl who will have sex for this amount." The girls earned up to 3 million pesos for each visit, and those who performed well were invited back. For each girl provided, the madam earned 1 million pesos. TV models and divas received up to 12 million pesos per visit. Money made inaccessible women available to men from poor backgrounds, who delighted in soiling the privileged and beautiful. During the police raid of the Cathedral, sex toys and blow-up dolls were photographed. Popeye claimed that the sex toys had been used by girls performing lesbian shows and the orgies were compensation for a workforce who needed to destress.

As the location included a direct sightline to his family's home, he mounted a telescope so he could see his wife and children while talking to them on the phone. His daughter's playhouse was filled with toys. He said that his main motivation to come to an agreement was to restore his family life.

His mother was the first to attend the regular Sunday family visits. She arrived with the family's favourite food, including tamales, and religious images for the chapel. At noon, Father Garcia arrived. Delighted with the chapel, he heard confessions from Pablo, Roberto and some of Pablo's men. All the prisoners and their families attended the mass hosted by the priest. Pablo made a large donation for his social projects.

The visits exposed Pablo's drug habit to his sister. While tidying up the library in his room, Luz Maria discovered cannabis. After she asked what it was, Popeye constantly joked that the boss's sister didn't know what weed was. For Christmas, Pablo's wife brought lobster and caviar, and his mother contributed fritters and custard sweets. To satisfy them both, he put the caviar inside of the fritter and ate it. Popeye pointed out that Pablo's solution would be good material for a book he hoped to write. The boss responded that dead people don't write books.

The soccer field was renovated, night lights installed and wires positioned above it to sabotage helicopter landings. Despite having a bad knee, Pablo played centre forward. His men made tactful allowances such as passing him the ball to score winning goals. The professional teams who came to play were careful never to win. He had a replacement on standby in case he grew tired. When he regained his energy after resting, he would resume. The guards served the players refreshments. Sometimes his lawyers had to wait hours to see him if he was on the field.

The introduction of two chefs known as the Stomach Brothers addressed his concerns about getting poisoned. They prepared his favourites, including beans, pork, eggs and rice. He had installed exercise equipment such as weights and bikes for the prisoners to get in shape, but as they were no longer on the run and had access to endless food and alcohol, they started to gain weight. The Cathedral became known as "Club Medellín" or "Hotel Escobar." *Hustler* magazine published an illustration of Pablo and his associates partying in prison, throwing darts at a picture of President George HW Bush. Pablo obtained the illustration and hung it on his wall.

As if it were a religious kingdom, the Cathedral was full of Catholic images and symbols in which Pablo acted as a feudal overlord. His drive for family, hierarchy and lavish rituals contrasted with the debauchery of his underlings. Surrounded by yes men, he began to believe that he was the future of Colombia.

With the government protecting him instead of hunting him down, his cocaine business thrived. Father Garcia tried to broker peace with the Cali Cartel. He arranged for Pablo to speak to its leaders, who were offering a peace agreement plus $3 million to end the war. Some of his men urged him to accept the deal, but others opposed it. "No, boss, you are already safe, and you can bend them from here," Big Gun said. Those in support of total war knew that Pablo's money would flow faster to them because they would perform more hits. He was the chief of a clan of warriors who died with detachment, a course that would eventually swallow him. As if he were defending his pride, he demanded $5 million from Cali. No progress was made because he found them too stubborn.

"I don't believe a word of those two," Pablo told Roberto, referring to the brothers Gilberto and Miguel.

News of Pablo's resumption of criminal activity and affairs with women devastated Victoria, who withdrew from him and decided that her role as a mother was only to maintain the relationship between Pablo and their children. To her, it was incomprehensible that he would violate his promise for those he had claimed were the most precious to him. The increasingly chaotic atmosphere at the prison made his family uneasy. Hermilda warned that if he kept having so many visitors, he would not last there another year. After he insisted that nothing bad would happen, she called him hard-headed.

To address legal problems, he had thirty lawyers working almost full-time. He was facing an indictment for being the intellectual author of the murder of the presidential candidate, Galán. During a raid of one of his properties, paperwork was found linking him to the assassination of the journalist, Guillermo Cano. On September 25, 1991, one of his men, La Quica, was using a payphone in New York when he was arrested for travelling with a fake passport. La Quica and two others offered no resistance to the police. Although he gave the fake name Esteban Restrepo-Echavarria, his fingerprints matched records provided by Colombia. Believed to be in the country as part of a hit on Max Mermelstein, a witness in the Barry Seal murder case, La Quica was held without bail and accused of being a player in the bombing of Avianca Flight 203. Eventually, charged with "conspiracy to import and distribute cocaine, substantive importation of cocaine, participating and conspiring to participate in a racketeering enterprise, engaging in a continuing criminal enterprise, various offenses relating to the bombing of a civilian airliner and the extraterritorial murder of two citizens of the United States," he would receive ten life sentences plus forty-five years, all to be served consecutively.

When he wasn't meeting his lawyers, Pablo was usually on the telephone or reading. He tried to learn Mandarin. He received endless letters from people asking for help, business advice and money. If their stories checked out, he often sent cash to them. People gathered at the prison gate with notes, seeking his assistance. At nights, he sat in a rocking chair and watched the lights come on in Envigado, while thinking about his family.

Near the end of September, President Gaviria summoned one of the men who had brokered Pablo's surrender, Alberto Villamizar, to the Palace of Nariño. He arrived with his wife Maruja and his sister Gloria Pachón. "Escobar confessed his participation in the export of cocaine to France," the president said. "It's an offense that has already been condemned by a Paris court, and we're going to need to release him within a few months because there is no way to condemn him for other crimes. There is no evidence. The only person who can convince him to confess to a crime of such importance is you."

Realising that the president needed to give Pablo a bigger sentence to improve his standing in the international community, Alberto went to the Cathedral. "If you do not stay in jail for at least ten years, this cannot stand. If you get out, everything will ignite because neither the United States nor Colombia can stand for you to be freed quickly."

"For how many presidential terms do you think I'll have to stay?"

"At least three. Gaviria, the next and another, at least ten years."

Silently, Pablo contemplated. "With you leaving as an ambassador for the Netherlands, who will liaise with Gaviria for me?"

"I'll find someone in the government to fill that role."

Days later, Pablo's legal team arrived at the Cathedral. At noon, he rose and dressed in jeans, a short-sleeved shirt and a sapphire-coloured watch. In the early afternoon over breakfast, he considered what to say about his sentence. During the meeting, a lawyer offered him an ostentatious gold bracelet. Studying it, Pablo calmly said, "First of all, I have nothing to buy it with, but also, if I buy it from you, they will all say that I am a Mafioso." Everyone laughed.

After complaining about his unfair treatment, the lawyers measured their words because they were aware of Pablo's knowledge of the laws, the codes of procedure and the jurisprudence of the Supreme Court of Justice. The meeting ended with him accepting that he would serve a minimum of ten years, after which he would live a normal life with his family and enjoy his wealth.

After dismissing the lawyers, he received the ELN guerrilla Lucho, who had been smuggled into the Cathedral in the double bottom of a small truck. Lucho found Pablo with cardboard boxes full of cash. "You see, the commander of this military base changes every month, and when the new one arrives, I'll buy him for 30 million pesos." Holding up a photo album with a tree diagram of his hit men, Pablo said, "This was a gift from an army officer. It's the album that the military was using to decide who to allow in and out of the Cathedral."

A feast prepared by the Stomach Brothers arrived: beans, rice, chorizo, egg, ground beef, salad, ripe plantains and black pudding. In-between devouring food, Pablo talked: "Look, Lucho, man, you are well placed. The most loyal people are the people of the communes. On the other hand, the most corrupt class in Colombia are the politicians, and there they have the power to protect themselves and their interests." The two agreed that it was necessary that their men in the neighbourhoods kept the peace. "I have twenty Uzi sub-machine guns for you."

After Lucho left, Pablo was informed that some men acting suspiciously outside of the Cathedral had been detained. "Bring them in for questioning." During a speedy kangaroo court, Pablo determined that they were spying for Cali and sentenced them to death. After they had been executed, he sent one of his US-trained pilots to bomb Cali Godfather Miguel at his house in the Ciudad Jardín neighbourhood, but the helicopter crashed outside of Cali.

He sent Tyson to kill Cali Godfather Pacho: "I hope he dies with his soccer shoes on." Fifteen men stormed the Villa Legua estate near Cali and killed twenty-two people, but Pacho was absent. Pacho would eventually die with his soccer shoes on, but five years after Pablo's death.

On the day of the Feast of the Virgin of Mercy, the patron saint of the traffickers, Pablo rose at 1 PM and read newspapers. He was told that a woman called Claudia was waiting, but he couldn't see her yet, because he had scheduled a soccer game: the traffickers versus members of Atletico Nacional and Deportivo Independiente Medellín. Despite the rain, he dressed in the German national team's colours, with a number nine on his back, and tied the laces of his black and white Nikes. On the pitch, he greeted some World Cup players, including the goalkeeper René Higuita, a.k.a. El Loco (The Madman), whom he admired for his flamboyant style and behaviour. While goalkeeping, he had developed a unique scorpion kick, which wowed the world at Wembley Stadium, England, two years after Pablo's death.

Despite his weight and a bad knee, Pablo regularly played soccer for hours. The visiting team scored three goals in fifty minutes, but after an hour and a half it was drawn. The visitors scored two more goals, but the traffickers pulled back by the end. From outside of the penalty area, Pablo scored the equalising goal, so that the game finished 5-5. Years later, when the goalkeeper was asked whether he had dived the wrong way on purpose to allow the ball into the net, he responded with a smile and said that Pablo had scored the goal and what more could he say.

After the game and spending an hour in the bathroom, Pablo emerged wearing a wool poncho given to him by Father Garcia. He stopped in a corridor to play with a bird that had landed on his shoulder. At 4 PM, Claudia spotted him with the indigo bunting. She had come to ask Pablo – the godfather of her marriage – to intervene on her behalf. She wanted to leave her trafficker husband without getting killed for doing so. So that he could hear her story, they sat.

"You know I married at 15 years old. I lived a horrible life. I never spent eight days in my house. It was from here to there, fearing we would get detained, accompanied day and night by an entourage of bodyguards. It was a year and a half that felt like fifteen years. Every day, I lived so many things that it seemed forever. Every day, I saw a gang of men coming over, one day with their wives and the next day with their lovers. They had weapons coming and going. The phone rang constantly. The TV could not be turned on because the ads about rewards appeared. No, no, I was like, uh, impregnated with it all. That hectic life, that bustle. How tired I became! Like a skeleton.

"I became pregnant. I had an abortion because I felt it was the right step after an argument, but for him there were no reasons. The idea of divorce made him angry, and he went from threats to actual attacks. One night, they broke into my mother's house, then I decided to go to Cali and there I suffered another attack."

Although he had listened attentively, Pablo was reluctant to go against his men in family matters. "And so? What am I supposed to do? Do you want to stay here and live with me?"

"It would be foolish for me to come here to act as a maid."

"So, what can I do? Why are you so upset?"

"If I leave this place, I will die like a chicken in a corner."

"What chicken? Stop talking like that! Don't you think you should go back to him?"

"No. I'm not crazy."

As one of Pablo's men had vouched for Claudia's plight, he took pity on her. "You have faced things very well. You have the ability to get out, and there are very few people who have that ability. I'm going to help you." Pablo sent her husband his reasons for helping her, and her life was spared.

Years later, after Pablo's death, Claudia said, "Pablo was only bad because he thought he was God. He once said, 'I dispense justice myself, period.' But there was a balance in how he was not entirely good or bad. The bad guys liked him because he was bad and the good ones because he was good. Everyone liked him. I liked him because of the sense of justice he had."

On December 1, 1991, he celebrated his 42nd birthday with a party in the Cathedral, where his guests ate stuffed turkey, caviar, pink salmon, smoked trout and Russian salad while live music played. His gifts included a red jacket bought by his wife in Spain and a Russian fur hat from his mother. Photographed wearing it, he declared it would be a symbol of his identity "like Che Guevara's beret". At night, the kids watched the inmates release coloured balloons.

In early 1992, the attorney general's office published photos taken at Hotel Escobar, including of waterbeds, jacuzzis, big-screen TVs ... The embarrassed president commissioned an investigation, but the justice minister found that the furnishings were legal because each prisoner was allowed a bed and a bathtub, and TVs were permitted for good behaviour.

"I want all of these things taken out immediately!" the president said. "Tell the army to go in there and take everything. Escobar has to know we're not kidding."

No government department wanted the job. "No way," the minister of defence said. "I cannot do it because I don't have the people." When it was pointed out that he had 120,000 troops, he still refused. Due to the deal struck with Pablo, the police couldn't do it. The DAS said that they couldn't act because they were only allowed inside the prison in the event of a riot.

In the end, a lawyer was told to take a truck and some workers, and to go to the prison and get the goods. "What have I ever done to you?" the lawyer responded. "Why'd you give me this assignment?" Banking on the truck not being allowed to enter the prison, so he could turn around and go home, the lawyer set off. When he arrived, the prison gate opened and Pablo waved them in.

Upon being told why they were there, Pablo said, "Certainly. I didn't know these things bothered you. Please, take everything out." He and his men helped them carry the goods until everything was gone. The lawyer rushed to his boss with photos of the bare prison. While the president was examining the photos, the goods were heading back to the Cathedral.

When the government tried to build a maximum-security prison based on the American model to transfer Pablo to, no construction company would accept the job. One said, "We're not going to build a cage with the lion already inside." Finally, a company owned by an Israeli security expert attempted to do it, with supposedly incorruptible workers from afar. Watching the work crew, Pablo's men wrote down their licence-plate numbers, and eventually attacked them, causing many to quit. The project was abandoned.

While in the Cathedral, Pablo wanted to preserve his legend in literature and the media. An illustrator from Medellín, Guezú, had a cartoon published in *The Time* called "The Epistle of Pablo", showing him heaven-bound with wings and a saintly aura, with his characteristic lock of hair, and with Father Garcia blessing him. Pablo commissioned Guezú to make a book containing the cartoons and caricatures of him published since 1982, the year he made a splash in politics. While talking, Guezú noticed that on the walls of Pablo's room hung his wanted posters.

Travelling to Colombia, Guezú gathered pictures, which were brought to the Cathedral. Lying on the bed, barefoot, Pablo examined the cartoons one by one and even accepted some lampooning him. In *Hustler* magazine, he is in a prison called the Medellín Club, holding a joint with three half-naked women and firing a gun at President George HW Bush on TV. Its caption says, "Pablo, leave something for us." He admired a cartoon by Velezefe published in *The Colombian* in 1984, which shows a raid on his estate, and one of his giraffes with its neck in a cloud, and on its head above the cloud is Pablo. In July 1991, *Semana* published a drawing of him on its cover, which he asked to be replicated.

Obsessed with archiving everything about him, he had his private secretary maintain a room full of articles and books. In a video, he narrated the history of his social projects, which she claimed had been obstructed by people he had assassinated: Lara Bonilla, Guillermo Cano and Galán. Besotted by the idea of his own greatness, he studied the Republican history of Colombia in the nineteenth century. He saw himself as a chieftain and believed that he was similar to the early guerrillas.

The publication of the book was taking too long, so he asked Guezú to accelerate things. When it was done, one of his assistants went for the book, only to have it confiscated by DAS agents, whom Pablo had to bribe to retrieve it. Finally, he ordered the publication of 500 copies of *Pablo Escobar Gaviria en Caricaturas 1983–1991*. He chose a leather binding, added his signature and thumbprint to the cover, and included some text criticising the War on Drugs. His family and friends received them as gifts with personal messages. He intended to write an autobiography that would clarify his role in the biggest crimes he was blamed for, including the assassination of Galán.

Some of what he wrote ended up in the hands of the authorities:

"The sale of cigarettes, aguardiente, weapons, votes, public sector jobs, patronage, import licences, grants, joints of cannabis or a dose of cocaine, is something totally and completely obscene."

"Neither divorce nor prostitution are bad in themselves. They are simply rather undesirable solutions, but necessary to avoid greater problems. Personally, I could not care less, because I do not have a problem with them."

"All those stimulants and antidepressants, all those synthetic drugs from the big legal laboratories abroad – legal they say, because they believe that they solve everything – they are more dangerous and more deadly and more addictive than cocaine. But they don't say that, because they are produced by them. The problem is that, speaking in monetary terms, cocaine left them behind and it is the first time in history that they haven't had the guts or the imagination or the power that we have."

He told *Semana*, "Legalisation is the solution to finish with the drug traffickers. Education is [necessary] to finish with drugs."

The elite police sent the government reports warning about his activity: "In the Cathedral, people enter without authorisation, and from there, Escobar has reorganised the drug trafficking network, kidnapping and attacks are ordered, and it is even insistently rumoured that Pablo leaves his cell to go to the city." The government paid no attention. With the nation fixated on economic turmoil and power outages, Pablo in jail was the president's only trophy.

Years later, the former hostage, Francisco Santos, said, "The great mistake of the Gaviria government was forgetting who that gentleman was, not appreciating who it had in jail, not realising that Escobar was the most villainous of the bandits, and that he was always going to be like that. Pablo was a criminal by heart. He could not do anything different from what he did, and it was inevitable that the situation would end up like it did. Crime was his reason for being, it was a natural thing, his genetic essence, or I do not know … but that was what took him to the grave. It is just that not only was he a great bandit – which is a very nice term – but also a murderer and a shameless person who could steal candy from a child. That is what it means to be a bandit: being able to do the smallest thing and jump to the biggest thing without any problem."

CHAPTER 16
BOMBING ESCOBAR'S PRISON

The Cathedral provided the Cali Cartel with a direct target to strike, so the godfathers decided to bring us back. Dave Tomkins and I were in England. After the Galán assassination on August 18, 1989, the authorities went from house to house hunting down anyone associated with the traffickers. Tomkins and I had been in an apartment outside Cali, which had housed Gilberto and Miguel's mother. With a raid imminent, we moved by jeep around security checkpoints to another apartment.

Worldwide media had published the story of the failed raid on Hacienda Nápoles, putting our lives in danger, so we had fled to Panama. Disturbed by the developments, our entire team had wanted to complete the mission but couldn't return to Colombia. Two team members who had dropped out of the mission made the mistake of doing media interviews. One appeared on CNN. In a UK interview, a team member's face and voice were disguised, but when the interview was shown overseas, his disguises were lifted. He ended up shot in both knees and in a wheelchair. Due to the government's crackdown and the media leaks, the godfathers postponed the operation. In September 1989, we had eaten a farewell meal with Cali's head of security, Jorge Salcedo, at a restaurant in Panama, and then flew home.

On September 13, 1989, the US government went public about a witness who had defected from the Medellín Cartel. It was a doctor whom Tomkins had trained during the previous mission financed by Gacha of the Medellín Cartel. On a screen,

Tomkins' picture was identified by the doctor, who said that Tomkins was an explosives instructor and one of the leaders of the foreign mercenaries. He blamed Tomkins, the mercenaries and our training programmes for the surge in violence in Colombia.

Two US Senate investigators showed up at Tomkins' house in England, insisting that both of us testify in Washington. After refusing, Tomkins notified Jorge. The next year, the US Senate investigators asked Tomkins to testify again at a hearing about arms trafficking, mercenaries and drug cartels. The Cali Cartel didn't object, so on February 27, 1991, Tomkins testified about his missions in Colombia in 1988 and 1989, including the failed attempt to kill Escobar. At the hearing, Tomkins refused to name anyone involved.

In 1991, at the US Embassy in London, the DEA told Tomkins that the US government was offering millions for Escobar's assassination. They wanted him to get hired by Pablo, so that Tomkins could kill him. With the failed attempt to kill Escobar in the news, Tomkins replied that Pablo would certainly hire him only to torture and murder him. The DEA said they hadn't thought of that.

In July 1991, Tomkins met Jorge in Panama City. Tomkins learned that the prison consisted of concrete blocks, a roof of asbestos sheeting, dorm-style cells and Pablo's wing. The surrounding area was booby-trapped with anti-helicopter wires, which were only removed to allow permitted landings. Cyclone-wire netting under the roof prevented anyone getting in from above. The biggest threat, though, Jorge told Tomkins, was the machine-gun posts.

As Pablo had built defences against the standard methods of assault, Tomkins was going to decline the mission. Then he considered bombing the Cathedral, using helicopters dropping drums of C-4. By remote-control detonating 200 kilos of C-4 at both sides of the building, the men inside would be cooked alive. After listening to the plan, Jorge offered his own. He knew a pilot keen to sell some 500-pound bombs.

Jorge flew to Guatemala to negotiate a price for four Mk 82 bombs from an El Salvadoran colonel who knew Cali's intentions. He wanted $600,000. Jorge said it sounded expensive, but he would relay it to the godfathers. Before leaving, he asked whether $500,000 was acceptable.

In need of a small ground-attack bomber, Tomkins asked a former CIA contractor whom he had met before. He was put in touch with a seller and all three met in Miami. A deal was agreed, and Jorge arrived. The seller and his technician showed Jorge and Tomkins the plane. The lack of ID markings and the presence of a six-barrel machine gun in its nose made Tomkins and Jorge suspicious. Intending to do background checks, they filmed the seller and the technician.

Jorge left the country. Tomkins stalled the seller by paying a $25,000 deposit and arranging a meeting the next week to conclude the transaction. The seller offered Tomkins a range of military equipment, which he had not requested. The cartel established that the technician was a former DEA agent and that the seller had been compromised.

On Sunday, Tomkins fled to England. He contacted the seller and apologised for his sudden departure. Working with US Customs as part of Operation Dragonfly, the seller tried to entice Tomkins back to inspect another plane. The former CIA contractor who had brokered the deal was a confidential informant who received a percentage of the money or assets confiscated in sting operations.

While the bombing mission stalled for lack of a plane, Escobar was planning further attacks against Cali. A cousin of godfather Miguel's fourth wife revealed that a family member visiting from Medellín had asked too many questions about the godfathers, including the names of their children's schools, how they were transported to school and details of other properties that the fourth wife frequented. The godfathers sent a Cali police sergeant to arrest the inquisitive family member. The police brought the man to one of Pacho's ranches. He was escorted into a living room full of bodyguards. Sitting around with glasses of orange juice, the godfathers ceased their small talk to examine the new arrival, whose face glossed over with sweat.

In a hostile tone, Gilberto asked why the man had so many questions about their children. The captive acted clueless. Gilberto cited the questions about their children's schools. The captive claimed that he was just curious and apologised for being rude. Cutting him off, Gilberto accused him of working for Pablo. With a pained expression, the captive protested that he would never betray them. Softening his voice into a coaxing tone, Gilberto said he understood that the captive had no choice because he lived in Medellín, and Pablo had threatened him and his family. The captive fell back on his curiosity excuse.

Gilberto sprang up, demanded a fork and yelled that he was going to remove the captive's eyeballs as punishment for lying. The bodyguards grabbed the captive and dragged him to a dining room, where there was a large polished table. Everybody moved into there. Standing at the end of the table, Gilberto demanded that the captive be brought to him. The bodyguards shoved the captive forward. One handed the godfather a fork.

Sobbing, the captive broke down. He had done it because Pablo had threatened to kill his entire family, including his baby. He would rather die than hurt the godfathers and their families. After resting the fork on the table, Gilberto ushered everybody back into the living room. In the dining room, the captive revealed everything to the godfathers and Jorge. He had been to the Cathedral a few times, which is where Pablo had sanctioned the fact-finding mission. Not wanting the captive to die, Jorge asked Miguel to intervene, because the man had details about the interior of the Cathedral, which could help the bombing mission. Miguel instructed the captive to tell Jorge everything.

Although the information provided was helpful, Jorge wanted to get inside the Cathedral to get a feel for the place. The cartel bribed the authorities to allow Jorge onto a police helicopter that had been scheduled by Pablo to transport a judge and a court reporter to the prison.

On the morning of the flight, Jorge put on a disguise: a green jumpsuit with three stripes, indicating the rank of police captain; a big helmet with a radio microphone, sunglasses and well-polished boots. He asked the pilot to approach the Cathedral in a way that would provide the best view of the grounds. He noticed the absence of a section of perimeter fence behind the building, which the prisoners could easily escape through if attacked. Seeing the playhouse for Pablo's daughter and imagining the underground bunker, Jorge sensed that Pablo had prepared countermeasures for a raid.

Jorge believed that the bombing mission would end in disaster, but the godfathers were so gung-ho that he had to tread carefully. He alerted Miguel to the risks: dropping bombs on a mountainside target was far more imprecise than on flat land. Bombing what was technically classified as government property would be classified as terrorism, which could result in a government crackdown on the cartel. Miguel told Jorge to get on with the damn job.

With the civil war in El Salvador ending, the bombs had to be purchased quickly, otherwise the transaction would become impossible. Jorge visited Miguel to get the cash. He found the godfather meticulously wrapping $500,000. Taking possession of a red and gold box, Jorge said that it was heavy. The godfather's frown silenced him.

Jorge put the box in a shopping bag and boarded a plane. In El Salvador, a sergeant who was an aide to the colonel selling the bombs got on board. After the passengers had left, Jorge gave the sergeant $500,000. The sergeant left with the money, took $20,000 and gave the rest to the colonel.

Jorge told the airport officials that he was in El Salvador on business-seeking opportunities in orange-juice exportation. He checked into a hotel, where his passport was photocopied. The following morning, he met Nelson, the local leader of Cali's interests in El Salvador. The massive man had arranged a crew to help the mission.

At an airbase, the sergeant who had received the money got in a forklift and moved four green bombs onto a small red truck, with straw disguising the bombs. Accessories such as detonators were moved separately. The sergeant drove out of the compound with the bombs.

The truck travelled 2 miles and stopped at a busy restaurant, where Jorge was waiting with Nelson shadowing him at another table. The sergeant entered, and while walking past Jorge, he placed a key on his table. Jorge moved the key to the side of the table and left the restaurant. Nelson stood. While walking past Jorge's table, he grabbed the key. Outside, he got in the truck and left.

On the way to his hotel, Jorge stopped at a payphone. He told Miguel that they had possession of the articles. Miguel sent a plane to a landing strip on the border of El Salvador and Guatemala. Jorge told the pilot assigned to pick up the bombs that it had to be done in less than ten minutes to give the El Salvadoran authorities no time to respond to the breach of their airspace. They took note of the radio frequencies that they would use to guide the landing.

Anticipating military roadblocks, Nelson and his crew painted the bombs yellow to match the colour of the trucks and the Caterpillar equipment often seen in the area. For extra camouflage, straw and four pigs were added. The mess made by the pigs would hopefully deter any thorough searches.

It was night when Jorge boarded a canoe on the Zapote River, which took him to Nelson's house. Although pleased to learn that the bombs had been transported without any setbacks, he was disheartened to discover that Nelson had broken a leg, making him useless for bomb loading. The pilot radioed Jorge that he was thirty minutes away. On a motorbike, he headed for the landing strip, followed by Nelson's men in the truck with the bombs. Jorge arrived at a dirt runway barely visible in the darkness.

The men hushed to listen for the plane. A call came on the radio that the plane was minutes away. A light from the plane on the runway exposed a variety of farm animals. On a motorbike, Jorge sped to the runway to scare them away to prevent them from damaging the plane and destroying the mission. As he revved the engine, pigs and chickens scattered.

The plane was much smaller than he had expected. Was it big enough to transport the bombs? It landed and the passenger door opened. To make space, the pilot threw out empty fuel cans. The plane was so small that the crates holding the bombs wouldn't go inside. The men extracted the bombs from the crates. It took twenty minutes to pack three bombs. Attracted by the commotion, local people started gathering.

The third bomb had been placed so precariously that it slipped. Not wanting to risk taking any more bombs, the co-pilot insisted that they take off. Extracting a fourth bomb from a crate, Jorge told them to wait, but the door was slammed shut. The pilot pledged to return the next day. Jorge handed the bomb accessories to the pilot, while warning him to be careful with them.

Jorge was left with a bomb, farm animals and local witnesses gazing at him, while empty fuel cans littered the runway with MADE IN COLOMBIA written on them. Fearful of getting caught, he told Nelson to sink the bomb in the river.

Detecting a small plane invading their airspace, Colombian air traffic control alerted the military. Two fighter planes were despatched to intercept the plane with the three bombs. When the cartel pilot noticed the fighter planes homing in, he hid in the clouds over mountain peaks and radioed for assistance. The cartel despatched a second small plane. While the fighter pilots forced the decoy to land, the plane with the bombs escaped.

Jorge called Miguel from a payphone. To ensure that the landing strip had not been compromised, they decided to wait a few days before sending a plane for the fourth bomb. Feeling unsafe, Jorge flew to Panama and called El Salvador. Distressed, Nelson's sister said that everybody had been arrested because the authorities had discovered the fourth bomb. She asked him where he was. Assuming that her phone had been tapped, he said that he was in Costa Rica.

While he sought sanctuary in a friend's house in Panama, the international news reported the discovery of a 500-pound bomb intended to kill Pablo Escobar. The news in Panama and Colombia described Jorge as a Colombian Army reserve captain involved in the plot and revealed his real name. Instantly, he went from operating invisibly to landing on Escobar's most wanted list. Not only Jorge, but also his family were now targets for Pablo and the authorities he controlled.

The bomb plot ended the party atmosphere at the Cathedral. The prisoners evacuated the main building and moved to wooden cabins hidden in the forest beyond the soccer field at the limits of the perimeter fence. Pablo hid in a cabin in a mountain cleft, where his location was disguised from above and from the forest. It was so cold due to a spring below that he moved to another well-concealed cabin. He instructed his men to watch the sky and to shoot down anything that violated their airspace. He ordered anti-aircraft artillery and he commissioned an architect to sketch anti-bombing designs. The architect drew a building with individual pods, with bombproof insulation consisting of concrete and steel. Camouflaged by earth, the building would be undetectable by spy planes or satellites. Preferring to be disguised by nature, Pablo rejected the plan.

Media attention had ruined the plot, so the godfathers hid the bombs in a storage facility and tried to keep a low profile.

CHAPTER 17
ESCOBAR'S DEATH

At the Cathedral prison, Escobar made the mistake of executing the heads of the Moncada and Galeano factions of the Medellín Cartel, which turned other Medellín Cartel factions against him, including the paramilitary Castaño brothers. Don Berna, formerly the head of security for the Galeano faction, brokered a deal between the Castaños and the Cali Cartel, whereby Cali financed the death squad Los Pepes, which targeted everyone who had anything to do with Pablo and his property.

After escaping from prison, Pablo relaunched his war against the government, but quickly burnt through his cashflow. At war with the Cali Cartel and with Los Pepes closing in by increasing the threat to his immediate family, Pablo found himself stripped of his organisation and weaponry. Most of his workers and bodyguards were dead, in prison or had defected to his enemies, and his health was failing.

In late November 1993, he moved to building number 45D-94 on Street 79A, a two-storey home. He lived with Limón and a cook, his mother's cousin, Luzmila. When he wanted to make calls, Limón drove him around in a yellow taxi, which had given him a false sense of security. He spent his time reading newspapers and watching TV, mostly remaining quiet except for the occasional outburst about him getting blamed for every crime in Colombia. Clinging to the hope of finding refuge in the jungle with the guerrillas, he didn't want to leave until he had guaranteed his family's safety.

The family considered fleeing to Germany, because his brother Roberto's eldest son had stayed there for three years without any problems, and Pablo's sister had been there for three months. From a travel agent a neighbour had recommended, Pablo's wife Victoria sent an anonymous person to buy four business-class tickets to fly to Frankfurt on November 27, 1993, which was less than a week away, so rapid preparations were required. She informed attorney general de Greiff that they were leaving and requested protection for the trip to the airport and at the connecting airport in Bogotá, before the transatlantic flight.

When they were all packed, they learned that the attorney general wanted Pablo to surrender before they would be allowed to fly. A female official arrived and announced that criminal charges had been filed against Pablo's son, which would prevent him from leaving, including transporting illegal drugs and sexually assaulting young women. Summoned to the room by Victoria, Juan Pablo angrily protested his innocence. He pointed out that he had a girlfriend and as Pablo's son he had never lacked female interest.

The official responded that Juan Pablo matched the description of a rapist, who had claimed to be Pablo Escobar's son. Even though she had no evidence for the sexual-assault charge, a witness had reported seeing him bring a box of guns into the building, which backed up the weapons offence. Juan Pablo granted her permission for the building to be searched. He said that the only thing they would find was a shotgun left by his bodyguard. The conviction in his voice convinced the official that he had told the truth, so she departed.

The attorney general's office dropped the charges but issued a death sentence by stating that the CTI cops were going to be removed from their building in the next few days. Pablo's wife screamed at the officials for threatening to leave her children unprotected and at the mercy of Los Pepes. On November 27, the family were informed that ten SUVs with CTI bodyguards had arrived to escort them to the airport. In the living room, they hugged each other and prayed not to be intercepted on the way to the airport.

A decoy SUV led the convoy, with Victoria and Pablo's daughter Manuela in one vehicle, and Juan Pablo and his girlfriend Andrea in another. In helicopters, cops with machine guns escorted the speeding convoy as it roared along the road, with other cars screeching out of its way and keeping their distance. A few hours later, they arrived at Rionegro, where dozens of police assigned to protect them swarmed around.

Boarding the plane first, Victoria wondered about Pablo, whom she had not heard from in days. At least the absence of any new stories meant that he was still alive. Waiting for take-off, she braced for agents to storm inside and remove the family. Unable to take her eyes off the door, she was momentarily distracted by Juan Pablo and his girlfriend playing a game of spot the undercover cop posing as a passenger. Two men particularly stood out.

During the flight, her eyelids slid, but she forced them open, terrified that if she dozed, she would lose her children. Ten years of dodging raids, assassination plots and adrenalin spikes had taken their toll. After a few hours in the air, a reporter approached her, revealed that someone high up in Bogotá had tipped him off and requested an interview. Figuring that his presence would reduce the risk of anything bad happening to them, she agreed to speak to him in Frankfurt.

With his family airborne for Germany, Pablo learned that they were going to be denied entry. The attorney general had tricked him. Plans had been made to return his family, because their safety had been guaranteed by the Colombian government. Infuriated, he made a call: "This is Pablo Escobar. I need to talk to the president."

"OK. Hold on. Let me locate him," an operator said, and contacted the National Police.

A policeman got on the phone. "We can't get in touch with the president right now. Please call back at another time." He hung up.

Pablo called again. "This is Pablo Escobar. It is necessary that I talk to the president. My family is flying to Germany at this time. I need to talk to him right now."

"We get a lot of crank calls here. We need to somehow verify that it is really you. It's going to take me a few minutes to track down the president, so please wait a few more minutes and then call back."

The president refused to speak to Pablo. Gaviria wanted Pablo's family members back in Colombia because he believed that if they were safe overseas, El Patrón would be free to regroup and unleash unlimited terrorism. With them in the country, they would all be at risk, which meant that Pablo would have to be more careful.

After the phone rang, a policeman told Pablo, "I'm sorry, Mr Escobar, we have been unable to locate the president." Enraged, he threatened to bomb the presidential palace and the German Embassy if his family couldn't stay in Germany. Colonel Martínez and his team were listening to the calls. The longer the incident could be extended, the better chance they had of tracing him.

At 6 AM on November 28, the plane landed. The pilot announced that some people needed to be removed before it could dock at the terminal. The two obvious undercover cops sprung up and approached Victoria. After stating that they were from Interpol, they claimed that they wanted to protect the family. Outside of the window, police cars surrounded the plane.

When a cop grabbed Manuela's arm, Victoria rushed over. "Please don't take her. She's only 8 and still drinking from a bottle." While Juan Pablo and Andrea were escorted to police cars, Victoria started screaming. Speaking in Spanish, a cop explained that Pablo had threatened to bomb every German airport. Not knowing whether that was true, Victoria insisted on accompanying Manuela in the same car.

With the family classified as undesirables and detained in Germany, the attorney general announced, "If Escobar gives up, we will try to mediate to get protection for his family in Germany."

Pablo's desperate calls enabled the Search Bloc to narrow

down an area with a radius of 1,500 metres. Out of five properties identified in the zone, two were believed to be owned by Pablo, one by Roberto, and two others were classified as hideouts. Martínez ordered the perimeter of the area to be secured. He didn't want to repeat previous mistakes such as sending in helicopters and hundreds of troops, which Pablo had always been able to dodge. This time, Martínez wanted to be certain of trapping his nemesis.

With everyone expecting a massive raid, only twenty-two men had been selected to launch a targeted attack. Stationed at Pueblito Paisa, the men had an excellent view of Medellín, including the Los Olivos district they had narrowed down as Pablo's area, a neighbourhood in Medellín near the football stadium, consisting mostly of two-storey homes. The men waited for information from surveillance units discreetly posted, including one led by Frequency – the son of Colonel Martínez – only ten blocks away from Pablo's hideout.

At Frankfurt airport, the family were searched, and interviews started in separate rooms. Pablo's emergency phone number remained concealed in one of Juan Pablo's shoes. For over thirty hours, with Manuela next to her on a sofa, Victoria was questioned about Pablo's whereabouts, finances and crime partners. Why had they come to Germany? Who were their local contacts? How much money had they brought? Holding her bottle, Manuela fell asleep, so Victoria extracted a blanket from her handbag and covered her daughter.

A sympathetic Spanish-speaking lawyer listened attentively to Victoria's pleas to stay in Germany, rather than risk death in Colombia. After Victoria broke down crying, the German lady confessed that she had been instructed not to help them. They had to go back. Before a harsh German voice ordered her to leave, she wished them luck. The typist departed. The lead interrogator said that Victoria needed to get her family ready for departure. They were leaving immediately. With the family in police cars heading for the plane, she protested that they were being condemned to die.

As soon as they were aboard the delayed flight, the door was closed. The passengers scowled at the family for stranding them on the runway for hours. After an hour, the pilot announced a further delay because France had denied access to its airspace because Pablo Escobar's family was on board. The tension rose. An attractive dark-haired woman approached Victoria, gave her a bible and said that Psalm 23 would help her situation. In case Victoria ever needed anything, she wrote down her contact details. Relieved by the act of kindness, Victoria thanked her and read Psalm 23:

"The Lord is my shepherd; I shall not want.
He maketh me to lie down in green pastures: he leadeth me beside the still waters.
He restoreth my soul: he leadeth me in the paths of righteousness for his name's sake.
Yea, though I walk through the valley of the shadow of death,
I will fear no evil: for thou art with me; thy rod and thy staff they comfort me ..."

As Victoria was a practicing Catholic, the words resonated. With so many forces operating against her, God was her only hope. (In the coming years, she would treasure that bible.) Approaching Colombia, she grew terrified about what might happen when the plane landed.

In Bogotá, Major Gonzalez lobbied the government to relocate Pablo's family to the Residencias Tequendama Hotel, where the Search Bloc had installed hidden microphones and had tapped the phone lines, which were monitored from an electronic-surveillance office.

At 8 PM, the flight landed in Bogotá. As if refusing to risk confronting danger, Victoria's legs froze and she remained seated with breathing difficulties. Hugging her kids, she said that their fate was in God's hands.

Three government agents boarded. "Everyone needs to remain

seated while we remove some passengers." Promising to get them stamped, they took the family's passports. They emerged to a cold night and numerous cops with rifles. An agent stated that the government could only guarantee their protection if they stayed at a hotel owned by the Colombian Armed Forces Retirement Fund. Feeling uneasy, Victoria requested transport to a regular hotel with a high level of security, only to be told that she had no choice.

Travelling in an armoured SUV, they were escorted by over 100 cops in a procession of vehicles to the Residencias Tequendama Hotel, which included an apartment building. In an elevator with armed bodyguards, they got out on the unoccupied twenty-ninth floor, and were escorted to two mundane apartments at the end of the hall. Exhausted, they tried to sleep, but kept waking up terrified. Outside, over 100 cops patrolled with bomb-sniffing dogs. Using mirrors, they inspected the undercarriages of cars.

The following day, Victoria heard nothing from Pablo, but received a call from her sister. Assuming that the phones had been tapped, she was unable to ask about her husband or even mention his name. With the building surrounded by guards, she fretted over the impossibility of him getting a message in to them. To let him know that they were OK, she asked Juan Pablo to give a radio interview, describing what had happened in Germany.

After hearing the radio interview, Pablo started to call his family using fake names. On the phone, he told them to stay at the hotel, to lobby the government to allow them to go to another country and to contact the United Nations. Pablo contacting his family raised the hopes of the authorities, but as the calls were short and sporadic, pinpointing their origin was difficult. For fifteen seconds, he stayed on the phone to his brother-in-law Alonso, but after that, the calls stopped for thirty-five hours, which tested the patience of the surveillance team, Colonel Martínez and Los Pepes. While waiting for them to resume, the men in the surveillance units, including Frequency, ate and slept in their vehicles.

To test the tracking equipment, Frequency sent a man to Medellín with instructions to call Don Berna. By sheer chance, and completely unknown to everybody involved, the man ended up near Pablo's hideout. The tracking system worked, and Pablo failed to notice the unusual activity in his area.

When the receptionist at the Residencias Tequendama Hotel received a call from a man identifying himself as Pablo, she was sceptical. The electronic-intelligence agents instructed her to delay him on the phone. The team in Medellín determined that the caller was in a moving vehicle.

Lonely and surrounded, Pablo's health deteriorated further. Weight gain reduced his mobility. Due to gastritis, he kept jars of Mylanta in every room of the safe house. Staying longer on the phone out of concern for his family, Pablo was dropping his guard. Slowly closing in after years of hunting El Patrón, the team in Medellín set up headquarters in a car park at the Atanasio Girardot Sports Complex. But a few setbacks occurred: several key members of the police had to leave for promotional courses, a new Search Bloc senior member arrived who was a desk worker, and pressure from the government to produce a result exacerbated the fatigue shared by the pursuers. In newspapers, the attorney general stated that the failure to capture Pablo was not just due to his talent at remaining invisible, but also because of the corruption of the Search Bloc. The government announced that Colonel Martínez would be replaced.

"An unusual noise is obstructing the clarity of the signals received," Frequency said.

"It sounds like water," Don Berna said, after listening to the call.

"Maybe a stream." The signal indicated that Pablo was moving between America Street and a sports complex. The only ravine in the area was the Bone.

Los Pepes announced that their cessation of operations – which had been initiated at the government's request – was over and they were resuming hostilities with Pablo. Fearing a bombing, the other guests checked out of the hotel housing his family. Walking around the hotel, Pablo's daughter, Manuela, sang about Los Pepes coming to kill her and her family.

On November 30, 1993, Pablo sent a letter to the suspected leaders of Los Pepes, including Colonel Martínez, the leaders of the Cali Cartel, the Castaño brothers and members of the Search Bloc. "I have been raided 10,000 times. You haven't been at all. Everything is confiscated from me. Nothing is taken away from you. The government will never offer a warrant for you. The government will never apply faceless justice to criminal and terrorist policemen."

On the same day, he met Milton Hernandez, one of the leaders of the National Liberation Army, and the second most powerful guerrilla in Colombia. Over the years, Pablo had remained close to the organisation, even recruiting its members to help fight Cali. Milton offered to protect him, provided that authorisation was granted by the guerrilla group. Pablo planned to leave on December 3 to eastern Antioquia, where he would be protected by the Carlos Alirio Buitrago Front. He thanked Milton and gave him a gift: a SIG Sauer pistol.

The tracking signal indicated that Pablo's vehicle had just driven by the car park the team was using as headquarters. Immediately, Don Berna and his men took to the roads, but due to congestion, he escaped yet again. They returned confounded that the most wanted man in the world was driving around like an ordinary person.

Now they had him worried about his family's safety, the authorities relied on him calling the hotel. Returning early from a rest period, Frequency resumed the hunt. In a dangerous neighbourhood, Frequency's van was spotted. A child on roller-skates approached the van and gave Frequency a note: "We know what you're doing. We know you are looking for Pablo. Either you leave or we're going to kill you."

Roberto Escobar received a note from someone on his payroll to warn Pablo to stop talking on the phone or else he would be caught. Immediately, he sent a note to his brother urging him to stop using his phone, because it had been compromised. Other sources told Pablo that if he surrendered, he would be killed.

Aware that the end was near, Pablo left a recording for his daughter telling her to be a good girl and that he would protect her from heaven. He bought his brother a copy of *The Guinness Book of Sports Records*, wrote a personal note to Roberto – who he described as his soul brother – and put it in the book. In one of his final letters, he wrote: "Mother, don't believe everything you read in the newspapers. I was good. They turned me bad."

With Los Pepes closing in, he spent his 44th birthday, December 1, 1993, at his hideout. Pouring champagne, Luzmila knocked a glass over which didn't break. "What good luck," she said.

"No," Limón said. "It means something bad is going to happen."

Breaking his usual abstinence, Pablo drank some and lit a cannabis joint. With Limón and Luzmila, he ate one of his favourite Antioquian meals, chicken sancocho, a hearty soup similar to a stew, including potatoes, yucca, corn, plantains and meat. Due to gastritis, he hardly ate any of his dessert: a chocolate cake sent by Victoria. With his mood low, he took Mylanta medicine.

With tunnel vision, he read the mail sent by his family, including drawings and a card from Manuela: "Daddy, I love you very much. I want to give you a big kiss on your birthday. You are my heaven. You are my dove. You are my heart. I wish you lots of luck on your birthday. I adore you. Your little girl." A card from a sibling read: "Even though you try to hide ... You cannot escape another birthday."

A fly persistently attempting to land on Pablo upset Limón, who viewed it as an omen, because flies are attracted to corpses.

Birthday congratulations kept Pablo on the phone longer than usual with his family. Frequency picked up a signal and sped to the location, which brought him to a roundabout with nobody there. Convinced that he had just missed Pablo, he was disappointed. The next day, he returned to his apartment to rest.

Getting so close to El Patrón spread apprehension among the troops, some of whom were instructed to write letters to their families because of the likelihood of death. With Pablo determined not to be captured alive, the troops expected a heavy battle. They were awarded extra money to buy clothes and send presents to their families in case they never saw them again.

Pablo wanted to say goodbye to his mother first, so he risked going to her apartment in the early morning. He told her that it was the last time he would see her in Medellín. His plan now was to form a new group, establish an independent country and be its president. Without crying, his mother said goodbye.

On December 2, 1993, the cloudless sky and peaceful weather held a promise of good things to Don Berna, as his driver took him and his bodyguards to a wine cellar that had been converted into a Los Pepes headquarters. Afterwards, he moved to a car park by the Atanasio Girardot Sports Complex, where Major Hugo Aguilar was waiting with tracking equipment.

"I'm very worried," Aguilar said. "The government wants results, so they have increased pressure on the police. The president is close to finishing his term, and he doesn't want that to happen with Escobar still on the loose. He has given us one week to find him. If we don't achieve that objective, we're all going to be dismissed."

"You need to remain calm," Don Berna said. "His days are numbered, so don't be getting anxious. He is unable to detonate bombs. He is unable to set up police to be murdered. He is unable to kidnap people from the government. His power is totally reduced. As his brother and Popeye have surrendered, he can receive no support from them. On top of that, we've intercepted a letter from one of his men, Marlboro, in which he told Pablo that he had to sell his motorbike and gun to buy food. His men have run out of money and out of work, or else they are in jail. The message issued by Los Pepes – whoever helps El Patrón must die – has been extremely effective."

"I hope things get sorted out," Aguilar said, "because even

though I'm happy to be assigned to the promotional course to acquire the degree of lieutenant colonel, I find it disappointing that after so much effort and time, we haven't been able to finish Escobar. In the entire history of Colombia, nobody has endured such a strong and relentless enemy, where forces so dissimilar have had to combine to be able to eliminate him." At 12:30 PM, Aguilar said, "I'm tired of eating chicken and potatoes, so for lunch, I'm heading to the Carlos Holguín School." He departed. In the car park, Don Berna ordered restaurant food and waited with twenty of his men, his brother Seed, Frequency and a sergeant called Toño.

"The teams are ready," Frequency said on a call to his father, "but Pablo hasn't used his phone."

Pablo woke up around noon. He dressed in jeans, a polo shirt and flip-flops instead of sneakers. He learned that the police had raided and killed Gustavo's son at 11 AM. After eating Italian food, he sent his cousin to buy the supplies he would need in the jungle: stationery and toiletries.

News broadcasts of his family's ongoing suffering at the hotel disturbed him. In a taxi, he called them while they were in a meeting with some generals, but upon recognising his voice, Juan Pablo kept referring to him as 'Grandma,' stating that the family was OK, and hanging up out of concern that the calls would be traced. Noticing the strain in her son's voice, Victoria knew that 'Grandma' was her husband. The generals said that the hotel had authorised them to bring an additional 100 soldiers to protect the building and that they were going to completely lock down the twenty-ninth floor. After Pablo called a third time, Juan Pablo told 'Grandma' that they were doing fine, but instead of hanging up, he handed the phone to his mother. While he said goodbye to the generals, she rushed into another room.

Excitedly, Victoria spoke, but was quickly interrupted by Juan Pablo, who insisted that she needed to hang up, because the call was being traced. She told Pablo to take care of himself because they all needed him. He said his only motivation in life was to fight for his family, and that he was safely hiding out in a cave and that the hardest part of their struggle was over. Despite the warnings, he called two more times, and Juan Pablo hung up on him. A call to Manuela and Victoria was interrupted by Juan Pablo yelling that Pablo would be found and killed if they didn't hang up.

While putting on a brave face to his family, he was adjusting mentally to the reality of death. He wrote a prayer, which he put in his wallet:

"Pray for us
O Lady Saint Marta
To the mountain work you entered
With the serpent that you found
With your great symbol
You tamed it, defeated it and dominated it
I ask you Queen and Lady
That you defeat, calm and dominate all the enemies that challenge me
Be they men or women, tigers or lions
I have to defeat all of them and make them tremble with fear in front of me
The Lord played the seven horos
Being horo and game
They are the seven arts that killed him with poison
And now they are my companions
There are seven of them, eight including me
I call on all of them to defend me from fist, bullets, force and every type of firearm that could come against me
Christ gives me courage and gives me faith
The Holy cross travels with me
In front and behind"

By adding himself to the seven horos or evils, he appeared to be taking responsibility for his crimes in the face of his mortality.

Martínez notified his son that Pablo was talking. Frequency rushed back to his team. After 1 PM, pretending to be a radio journalist, Pablo called his family. The house receptionist unsuccessfully attempted to delay the call. The noise of the creek convinced Don Berna that Pablo was nearby. His wife, Victoria, was crying. Numerous of their family members and associates had been killed by Los Pepes. The family was distraught.

"So, what are you going to do?" Pablo asked.

"I don't know," she said, still crying.

"What does your mother say?"

"It was as if my mother fainted," she said, referring to a few days earlier at the airport, when the family had unsuccessfully tried to flee to Germany. "I did not call her. She told me bye, and then—"

"And you haven't spoken to her?"

"No. My mother is so nervous …" Victoria said the murders committed by Los Pepes had traumatised her mother.

"What are you going to do?" Pablo asked softly.

"I don't know. I mean, wait and see where we are going to go, and I believe that will be the end of us."

"No!"

"So?"

"Don't you give me this coldness! Holy Mary!" Pablo said.

"And you?"

"Ahhh."

"And you?"

"What about me?" Pablo asked.

"What are you going to do?"

"Nothing … What do you need?"

"Nothing," Victoria said.

"What do you want?"

"What would I want?"

"If you need something, call me, OK?"

"OK."

"You call me now, quickly," Pablo said. "There is nothing more

I can tell you. What else can I say? I have remained right on track, right?"

"But, how are you? Oh my God, I don't know!"

"We must go on. Think about it. Now that I am so close, right?" Pablo said, referring to his proposal to surrender.

"Yes," Victoria said. "Think about your boy too, and everything else, and don't make any decisions too quickly. OK?"

"Yes."

"Call your mother again and ask her if she wants you to go there or what …" she said. "Ciao."

"So long."

Concerned about his family due to the death of Gustavo's son, Limón called his wife, who had been registering their children for school. "Take care, because you know things are going to get very hot." Upon learning that one of his daughters had slept at his sister-in-law's house, he said, "Get her back to the house. Things are going to get ugly."

"Are you OK?" his wife asked.

"I'm fine. Give my love to the kids."

At 2:52 PM, Pablo called his son.

"We have narrowed down the location," Frequency told Don Berna. "Let's go with all of your men." With the Search Bloc attempting to find the precise location, Juan Pablo asked his dad to help him formulate answers to questions from a journalist. Pablo got out of the taxi and returned to the apartment, making the mistake of speaking for longer than five minutes.

"Look, this is very important in Bogotá," Pablo said, hoping to present his case favourably through the media. He wanted to hear the questions first. "This is also publicity. Explaining the reasons and other matters to them. Do you understand? Well done and well organised."

"Yes, yes." Juan Pablo began with the first question from the magazine *Semana*: "'Whatever the country, refuge is conditioned on the immediate surrender of your father. Would your father be willing to turn himself in if you are settled somewhere?'"

"Go on," Pablo said.

"The next one is, 'Would he be willing to turn himself in before you take refuge abroad?'"

"Go on."

"I spoke with the man and he told me that if there were some questions I did not want to answer, there was no problem, and if I wanted to add some questions, he would include them."

"OK. The next one?"

"'Why do you think that several countries have refused to receive your family?' OK?"

"Yes."

"'From which embassies have you requested help for them to take you in?'"

"OK."

"'Don't you think your father's situation, accused of X number of crimes, assassination of public figures, considered one of the most powerful drug traffickers in the world …?'" Juan Pablo stopped reading.

"Go on."

"But there are many. Around forty questions."

Pablo said he would call back later in the day. "I may find a way to communicate by fax."

"No," Juan Pablo said, concerned about a fax being traced.

"No, huh? OK. OK. So, good luck."

Frequency traced the call to the Los Olivos neighbourhood. They waited for him to make another call.

At 2:57 PM, Pablo called his son for the last time. Juan Pablo said that the journalist wanted to know what conditions Pablo would be satisfied with in order to surrender. Members of the Search Bloc started to go street to street, hoping to detect his location. Frequency's scanner led him to an office building. Convinced Pablo was inside, the troops stormed in, but Pablo was still conversing as if nothing had happened.

"Tell him, 'My father cannot turn himself in unless he has guarantees for his security.'"

"OK."

"'And we totally support him in that,'" Pablo said.

"OK."

"'Above any considerations.'"

"Yep."

"'My father is not going to turn himself in before we are placed in a foreign country, and while the police in Antioquia—'"

"The police and DAS is better," Juan Pablo said. "Because the DAS are also searching."

"It's only the police," Pablo said.

"Oh, OK."

"'While the police—'"

"Yeah."

"OK," Pablo said. "Let's change it to, 'While the security organisations in Antioquia …'"

"Yeah."

"'—continue to kidnap—'"

"Yeah."

"'—torture—'"

"Yeah."

"'—and commit massacres in Medellín.'"

"Yes, all right."

"OK," Pablo said. "The next one."

Due to the amount of time Pablo had spent on the phone, Frequency's scanning equipment had narrowed down the location to a street in the América sector, on the right beyond La Hueso Creek. He led Don Berna's men to a stream by Pablo's house.

Juan Pablo asked why so many countries had refused to allow their family in.

"'The countries have denied entry because they don't know the real truth,'" Pablo said.

"Yes."

"'We're going to knock on the doors of every embassy from all around the world, because we're willing to fight incessantly. Because we want to live and study in another country without bodyguards and hopefully with a new name.'"

"Just so you know, I got a phone call from a reporter who told me that President Alfredo Cristiani from Ecuador, no, I think it is El Salvador—"

"Yes?" Pablo went to a second-floor window and scanned the street, checking for cars.

"Well, he has offered to receive us. I heard the statement. Well, he gave it to me by phone," Juan Pablo said.

"Yes?"

"And he said if this contributed in some way to the peace of the country, he would be willing to receive us."

"Well," Pablo said, "let's wait and see, because that country is a bit hidden away."

"Well, but at least there's a possibility, and it has come from a president."

"Look, with respect to El Salvador …"

"Yeah?"

"In case they ask anything, tell them, 'The family is very grateful and obliged to the words of the president, that it is known he is the president of peace in El Salvador.'"

"Yeah."

The length of the call had exceeded his safety limits. When asked about how the family had felt about living with government protection, he said, "You respond to that one."

"'Who paid for maintenance and accommodation? You or the attorney general?'" Juan Pablo asked.

"Who did pay this?" Pablo asked.

"Us. Well, there were some people from Bogotá who got their expenses paid … but they never spent all of it, because we supplied the groceries, mattresses, deodorants, toothbrushes and pretty much everything."

After two more questions, Pablo said, "OK, let's leave it at that."

"Yeah, OK," Juan Pablo said. "Good luck."

"Good luck."

The call had lasted for so long that Frequency and Los Pepes were on Pablo's street, driving up and down. Frequency stopped studying his equipment and started observing the houses. He noticed a bearded man behind a second-storey window, phone in hand, watching the traffic. After a few seconds, the man disappeared into the house.

Frequency leaned out of the window. "This is the house!" he yelled at the vehicle behind him. Suspecting that Pablo had noticed his white van, Frequency told the driver to keep going. He radioed his father: "I've got him located. He's in this house." Assuming that Pablo's hit men were on their way, Frequency wanted to leave.

"Stay exactly where you are!" Colonel Martínez yelled. "Station yourself in front of and at the back of the house. Don't let him come out!"

"Are you sure Pablo is there?" Aguilar asked.

"I'm absolutely sure of it."

"Do not make any movement. We're on our way with backup."

With the roads congested with holiday traffic, the journey would take approximately forty-five minutes. To Don Berna, waiting was eternal and too agonising.

With all units of the Search Bloc on their way, Frequency parked in a back alley and got his gun ready.

At 3:15 PM, Frequency asked Don Berna, "Do your men have the area completely secured?"

"Yes. It's impossible for El Patrón to escape."

"Let's go in. I assume responsibility."

Seed put on a bulletproof vest, grabbed an M16 rifle loaded with ammunition provided by the Americans and approached the residence accompanied by two of Don Berna's men and the policeman, Toño. A sledgehammer demolished the door.

"Boss, they found us!" Limón yelled.

"Limón, run, because you still have time," Pablo said. "It's easier for you to run. I know what I have to do."

Limón dashed out of the back door. Charging up the stairs, one man fell as if shot, startling the rest, but he had only slipped.

Limón went through a window onto an orange-tile roof. As he fled, Los Pepes members behind the house sprayed gunfire. Shot multiple times, he careened off the roof onto the grass. Limón's wife believes that they shot him in the forehead to ensure his death.

After tossing his sandals, Pablo ran upstairs to a small window that faced a neighbour's roof. Not wanting to end up like Limón, he stayed against a wall, which blocked clear shots at him even though gunmen were everywhere. Aiming to escape down a back street, he hastened along the wall.

Shots erupted.

The gunfire was so intense from all sides of the house that it tore up the bricks and the roof, and some members of Los Pepes thought they were under attack by Pablo's bodyguards and radioed for help. As he ran across the roof, Seed fired at him. A bullet entered Pablo's thigh above the knee and emerged below his kneecap. Another went in below his right shoulder and got stuck between two teeth in his lower jaw.

Pablo fell.

Having pledged to never be captured or killed, he most likely shot himself in the head to deprive the government of being able to claim that they had killed him. The bullet went into the right ear at the centre and came out in front of his left ear, slicing through his brain, killing him instantly. To this day, Roberto and Juan Pablo believe the wound above the ear was a suicide shot.

The shooting stopped.

"It's Pablo! It's Pablo!"

Los Pepes members approached the blood-soaked corpse and flipped it over. One used his foot to remove Pablo's gun and lifted his hair. Stood at a window, Don Berna felt no emotion gazing at the obese, shoeless corpse leaking blood.

"Viva Colombia! We've just killed Pablo Escobar!"

"We won! We won!"

Amid men shooting in the air and yelling "Viva Colombia!" Aguilar arrived and hugged Don Berna, Seed and Frequency.

"You and your men need to leave immediately," he told Don Berna, "because the media are coming."

The police shaved a Hitler moustache onto Pablo's face and posed for pictures with him.

After leaving the death scene, Don Berna called Fidel Castaño: "Commander, Pablo is dead."

"Are you sure?"

"Yes, Commander, I saw it myself."

"Meet us at Montecasino as soon as possible."

With Los Pepes gone, the police told the media, "We advanced rapidly up the stairs and we saw two subjects throw themselves from the window at the back of the house, falling onto the roof of the house next door, where they were neutralised by the personnel located in this part of the house, who fired back after coming under attack."

It was the birth of the official version of Pablo's death: "The two fugitives attempted to escape by running across the roofs of adjoining houses to reach a back street, but both were shot and killed by Colombian National Police. Escobar suffered gunshots to the leg and torso, and a fatal gunshot through the ear."

Years later, the chief of intelligence for the Colombian National Police claimed that Pablo had been killed at close range because nobody had wanted to risk the disaster that would have arisen if he had been captured alive. He compared Pablo to a trophy at the end of a long hunt. According to the official reports, a sergeant who died two years after Pablo's death, along with a police chief and another officer, were the only three who shot Pablo.

After Don Berna wrote his book, *Killing the Boss*, describing his brother, Seed, as one of the main shooters, Colonel Martínez denied the role of Los Pepes: "It's not true what he says. I was constantly communicating by radio with Colonel Aguilar and Lieutenant Hugo Martínez Bolivar, my son, and the operation was carried out entirely by policemen."

In 2014, Juan Pablo wrote in *Pablo Escobar: My Father* that he had overheard Pablo on a radiophone to a bodyguard stating that he would never allow the authorities to take him alive. He had told his son that when the inevitable showdown came, he would fire fourteen rounds from his Sig Sauer pistol, leaving one to shoot himself in the right ear. The photo of Pablo on the roof shows him next to his Sig Sauer pistol, which had been fired. His Glock was still holstered. Even *Narcos* showed Pablo's blood-soaked right hand, which would have arisen from the spray if he had shot himself in the ear. Juan Pablo is convinced that his father honoured the suicide pledge.

Shortly after Pablo's death, his mother and two sisters arrived. At first, they thought that only Limón was dead. "It's Pablo," one of his sisters said, identifying the second corpse.

Hermilda later described what happened when she found his corpse: "I felt something I have never felt in my life. It was terrible. Since then, my soul has been destroyed because there will never be anyone like Pablo again."

Manuela was singing in the shower and Victoria was on the phone with her sister when Juan Pablo yelled that his father had been killed. Pablo's wife told her sister to find out what had happened. The return call confirmed his death and added that helicopters were circling his hideout.

At Montecasino, Don Berna described what had happened to Carlos and Fidel Castaño. "Now that Pablo is dead," Carlo said, "we must start a new fight against the guerrillas. With all the weapons, contacts and resources we've obtained from the war against Pablo, we'll build a political military organisation. From this moment on, Don Berna and Seed, you belong to this new anti-subversive organisation as members of the general staff."

From his personal wine cellar, Fidel brought a 1948 French vintage. After the celebration, Don Berna asked for permission to rest because he was exhausted. Back in his apartment, he reminisced about the friends he had lost. Feeling alone, he wondered if it had all been worth it and finally fell into a deep sleep.

The American president wrote a letter:

"December 2, 1993
Dear Mr President:
I just learned of the success of your long struggle to bring Pablo Escobar to justice. I want to offer my congratulations to you and the Colombian security forces for your courageous and effective work in this case. Hundreds of Colombians, brave police officers and innocent people lost their lives as a result of Escobar's terrorism. Your work honors the memory of all of these victims. We are proud of the firm stand you have taken, and I pledge to you our continued co-operation in our joint efforts to combat drug trafficking.

Sincerely,
BILL CLINTON"

At 6 AM on December 4, 1993, in Bogotá, at a military airport, a plane scheduled to collect the president stopped by the VIP lounge. Inside, Gaviria was on a natural high, celebrating and toasting glasses of orange juice with Rafael Pardo, his advisors and the police and military commanders. A dozen journalists accompanied them on the flight to Medellín. After Gaviria removed his coat, grey suit and shirt to put on a bulletproof vest, a cameraman started recording, but was quickly instructed to stop. The festivities continued for the duration of the short flight.

In Rionegro, officials at the airport expanded Gaviria's entourage, some of whom yelled "Hurray!" and congratulated the Search Bloc. At an army facility, the policemen whom the government had credited for eliminating Pablo were paraded to the media. The journalists were instructed to turn the men into heroes. A presidential advisor coached the journalists on what to report: the bravery of the Search Bloc and the security forces, the historical importance of Pablo's demise ... "This is the beginning of the end of narco-trafficking in Colombia," the advisor boasted.

EPILOGUE

In England, I went back to Birmingham and took over the Gunmaker's Arms. I enjoyed running this red brick Victorian pub in Small Heath and produced an impressive turnover, so much so that over the next two years the brewery persuaded me to take on another thirteen other pubs in the area. I know they thought I was ideal, as poacher turned gamekeeper, but I have never come across so much unprincipled lying, begging and stealing in my life. Once, I walked into one of my bars to find that every man there owed me money, and they were all drunk.

However, I set up all the memorabilia of my military life in the Gunmaker's, in the snug bar on the first floor, and numerous friends visited.

One cold spring day in April 1992, I was sitting in the snug when Liz in the bar phoned up to say there were two men to see me.

"Who are they?" I asked.

"They look like toffs, Peter," she said in her strong Brum accent.

"Send them up," I told her, puzzled. I was not expecting any trade visitors.

I met them at the top of the stairs and saw at once Liz was right.

These two were definitely out of place in Small Heath.

One was tanned, in his mid-40s, wearing a grey suit and highly polished brown brogues. He was of medium build and fit-looking, his jaw was strong – with a small dimple in the chin – and he had the air of one who spends a lot of time abroad.

The older man was in his 50s, of medium height with grey hair brushed straight back. He wore glasses, a white shirt with a silver club tie, and a long black Melton overcoat over a dark pin-striped suit and expensive black brogue shoes. I rather thought his outfit lacked a red carnation in the lapel of his coat. He was the very image of a man who works in the City of London, but he was certainly no stockbroker.

There was something about the manner of both these men, concerned, benign, but with eyes that were implacably hard, and I was suspicious at once.

"Are you Mr McAleese?" the older man asked politely.

"I am," I nodded.

"May we talk with you in private, please?" It was more a command than a request.

I invited them into my snug bar and noticed them both looking round my photographs and badges which I had mounted on the walls.

"What can I do for you?" I asked. Plainly, they were not paying a social call, but I offered them drinks. The older man chose a half-pint of Black Label lager, while the younger man in the grey suit opted for a safe glass of Perrier water. Oh dear, is there no style anymore? James Bond, where are you?

We sat at one of the round tables and the older man began with, "You don't know who we are, but we know who you are."

His voice rang with all the righteous conviction I recognised from the past, when my actions have been so often prejudged by men who have never met me or even bothered to ask for my side of events. I watched them, I listened carefully, but I said nothing.

He went on, "We know all about your recent trips to South America, and we feel we should advise you of the following." At this, a tiny smirk appeared on his face.

I looked up at the wall beside our table. We were sitting between a framed copy of the McAleese family motto written in calligraphic style, "Touch not the cat but a glove," and a photograph of fearless Old Miles in his Argyll kilt. I looked back at my two suave visitors and said nothing.

The knowing smirk again. He said, "You've been out there again, to set up another operation to kill Escobar, haven't you?"

I said, "Yes."

"You were told to recruit your own pilots, weren't you?"

"Yes."

"Do you realise you were going to be shot down, after the job?"

I said nothing. Did he take me for a fool? Did he not wonder how I had survived so far? Plainly, these men had read a file on me in their office in London, an incomplete file in all likelihood, as the security services never trouble to ask us for an interview when they make up a file on us, and there was nothing I could say. Plainly, their view of me stemmed more from my time as a wild young trooper in the British SAS and from lurid nonsense about me in the tabloid press, always thirsty for talk of 'mercenaries', rather than any knowledge of my time in Africa. I was certain they had not read my confidential reports as a sergeant major in the South African Army. They had come with their minds made up. There was nothing I could say which would make any difference and I did not want to upset them. They were powerful, because they represented the establishment, and I did not want to buck the system.

I have had plenty of fights in my time, I've been in jail, but I have never fought the system, because I believe in it. Through years of service and operations for three governments and in three regular armies, I rose to the rank of sergeant major, and I did my best to uphold in that rank all the traditions of the finest airborne regiments anywhere in the world: the British Army, the Rhodesian Army and the South African Army. If these men saw nothing of that in me, then there was no point talking.

They spoke to me for a few more minutes while I politely answered "Yes" or "No", and then they stood up to go. The older man paused a little theatrically at the door and said, "My final piece of advice to you, Mr McAleese, is don't do it!"

The younger man gave me a hard look and then they stamped downstairs and out into the street.

I expect they thought they had done rather well, and it was nice of them to take the trouble.

But who knows what I might do next?

AFTERWORD: MEETING PABLO ESCOBAR

From a conversation with Shaun Attwood (YouTuber and author of five books on Escobar):

Peter: You mentioned the assassination of Galán, the Colombian presidential candidate. I was accused of planning that assassination and training the guys for it. There was a newspaper man called Andrew Neil who I think was working at the *Mirror* at the time and he came to see me and he asked me different questions. "Did you train anybody in mini-Uzis?" I said I never saw a mini-Uzi out there. I saw some Uzis but there weren't many Uzis. I said I saw an Interdynamic MP9, which is a small weapon, but not many Uzis. "Oh, come off it, Peter," he said, "come off it!" This guy had it in his head, and nothing I could say could change it, that I was involved in that. So, he went ahead and printed it all.

I went to a book launch after and he come up to me and said hello. One other thing he did: he had a company van there and he said, "Let's get away from this van and go for a walk." I lived in Telford at the time and we walked along the top of a motorway bridge and I still had a microphone on and he said, "Come on, Peter, there's nobody here now, tell me, were you involved?" I couldn't convince the guy. I didn't even know who Galán was until he was assassinated. But he got it and he printed it and then he tried to do the same to Dave Tomkins and Dave was very quick tempered. He said, "Right, you want to print stuff, go on and print it and give me the money and fuck off."

There was nothing I could do to convince this man that I didn't do that assassination. Nowadays, the way things are going, even with your background and mine, it is beneficial to have a little bit of a villain in your background. Even now I can turn around and say yes I was involved, but I wasn't. I lost any respect I had for the man. I think he'd done a little bit of a search, picked a few things up and got on with it.

Shaun: So, you saw Pablo's prison, the Cathedral?

Peter: Now, I was the one who was gonna attack the place along with Dave. Dave came up with the idea of bombers and helicopters, and at one stage we considered actually going into the place and assassinating him, Pablo, and that's why I know so much about the fence. Pablo ran that place totally. The army was there. There was a token gesture by the police because even then people spoke about, you know, he's running the place. Also, I met Pablo Escobar.

Shaun: What was that like? Was that when you were working for Gacha?

Peter: Yeah, Pablo was funny and he had a presence. I didn't know who he was at the time and I found out afterwards, but he seemed to command the whole thing. He laughed and joked, and within the area of Medellín, nobody had a great deal to say about him. OK, if you crossed him, you were getting the message, but a lot of people didn't. They had the whole area sewn up. Every corner had a little shop and, you've got these dirt roads. You come … there's a dirt road there and there's a little crossroads and there's a little shop there selling sweets and pop and that, and of course you notice the guy has got a radio. So, he had the whole place sealed off pretty well. But he knew everything that was going on in the area. If there was a strange car, a strange person, it went straight back to him.

Basically, what happened, I was with Jorge Salcedo, you know, Jorge, but they called him George. He took me there and I was training the guys and the whole thing was based on doing an attack on a villa at Casa Verde which was a mountain, and they wanted me to take out some communists up there. It was nothing to do with the West against communism. The communists were taxing people and the drug lords wanted to get their market share.

The communists were taxing the emerald people and the … although there was cocaine involved somewhere along the line, it would be wrong for me to say so because I have no proof of it. It was mainly … what was spoke about was the emeralds, controlling the emerald people who were probably swapping emeralds for cocaine. I don't know that part. I just know that I had to attack this hill and there was a place on top of it called Casa Verde.

Shaun: Did you attack it?

Peter: No, we planned it. That's why we trained Gacha's guys. Now I was training them and they had a little sort of patrol and the guys were going around shooting at things, little targets popping up here and there. And up comes – I didn't know who it was at the time – a couple of heavy-duty vehicles tooled up to the eyeballs, and they sort of looked at what was going on. And then I went away … when we finished the day's training, I went to this little hacienda somewhere with Jorge and that's when I met Pablo.

Shaun: Did Pablo speak to you in English?

Peter: No, he didn't, no, Jorge was translating all the time. He was actually a funny guy. Entertaining, you know?

Shaun: Yeah. I mean, from writing about him he had a really funny sense of humour.

Peter: Yeah, and I never found out who he was till we left. Jorge said, "That was Pablo." And there was quite a few of them fairly heavily tooled up.

Shaun: How long did you speak to him for?

Peter: I think probably about an hour and a half.

Shaun: What did he say?

Peter: He just asked me how it was going, how I liked Colombia, how the training was going. Bearing in mind, I didn't know who he was. It was all business and, you know, how would we handle it based on the fact that Jorge had done a heavy sell, you know – these people are Falklands guys and used to fighting and hitting heights and what not. And, of course, Jorge was in it for the money himself.

Shaun: Did Pablo come across as someone who had good military knowledge?

Peter: No, I wouldn't say that. He had gangster knowledge. He was in total control of the group. Don't get me wrong, there was leadership there. Whether the leadership was based on natural leadership or the amount of money he had, I don't know, but when he said to them jump, they jumped.

Shaun: There's not many people out there who can say they had a conversation with Escobar, wow.

Peter: To be honest with you, I took a liking to the guy. Then I got to know that he had done more for the people of Medellín than the actual government. Well, that's what I heard, but I couldn't prove it. I've never been in Medellín. I've been in the outskirts of it to buy a car, which was roughly in the area of Pablo's hacienda.

Shaun: He did a lot: he built houses, soccer fields, put lights in soccer pitches, built hospitals, schools, roads, everything …

Peter: The thing was, he had the place sewn up the whole time we were there. You can't get this through to people. Army guys with us, military people, they were with us the whole time. They were on the payroll with Pablo. No aircraft could get near his place without being reported by the people on the radar. He had them on the payroll, the air force guys. And that was partially the reason I crashed because we were hedgehopping to try and get a good run at the target.

Shaun: Yes, the local air force base alerted him to anyone coming.

Peter: A lot of people, whatever they said about him, Pablo responded, "You're holding me responsible for what an American wants to throw up his nose."

Shaun: So true, and it's still going till this day.

Peter: Money didn't seem to be an object with him. At one stage, I saw a roomful of money, not a safe, a roomful.

Shaun: It's great that through your first-hand experience, you've got all of these accurate details.

Peter: Yeah, if you talk to a DEA guy, all you would get is, we had them lined up, we had this and we had that. Then why the hell didn't they do something about it?

Shaun: I know. Some American authors tried to claim their forces killed Pablo and everything, their special forces killed him. It's all bullshit.

Peter: One of the Americans was speaking to me. I spent a lot of time doing special forces work, and this guy was talking to me like I was someone just recruited ... just left the military depot. And I never said a word to him. Behind him was a photograph of my grandfather, and I just stared at the photograph and I looked at a man who had been through it all – First World War – and he got himself wounded on three different occasions. And I looked at these guys there with the Milton overcoats on and the only thing missing was the carnation in the lapel, you know?

Shaun: I've been researching Escobar now for about five or six years, so it's a real honour to finally chat with you about your story because I've been so fascinated by it.

Peter: Yeah, I mean, can I say something here? I am not a special guy. I'm an ordinary guy who got himself into a lot of extraordinary situations. People say to me, "You're some man." No. It's like, you write books, and when a book gets finished, you start thinking about another one, and you carry on from there. It's the same with soldiering, it's just a job. I came out of the British Army, I then went to Rhodesia and in the process I got myself in an awful lot of shit.

Shaun: It's a bloody miracle you're alive, Peter.

Peter's two-hour interview with Shaun Attwood is available on YouTube: Killing Escobar For Cali Cartel: SAS Soldier Peter McAleese | True Crime Podcast 218

OTHER ESCOBAR BOOKS BY GADFLY PRESS

Pablo Escobar's Story (4-book series)

By Shaun Attwood

"Finally, the definitive book about Escobar, original and up-to-date" – UNILAD

"The most comprehensive account ever written" – True Geordie

Pablo Escobar was a mama's boy who cherished his family and sang in the shower, yet he bombed a passenger plane and formed a death squad that used genital electrocution.

Most Escobar biographies only provide a few pieces of the puzzle, but this action-packed 1000-page book reveals everything about the king of cocaine.

Mostly translated from Spanish, Part 1 contains stories untold in the English-speaking world, including:

The tragic death of his youngest brother Fernando.

The fate of his pregnant mistress.

The shocking details of his affair with a TV celebrity.

The presidential candidate who encouraged him to eliminate their rivals.

Pablo Escobar: Beyond Narcos

By Shaun Attwood

The mind-blowing true story of Pablo Escobar and the Medellín Cartel beyond their portrayal on Netflix.

Colombian drug lord Pablo Escobar was a devoted family man and a psychopathic killer; a terrible enemy, yet a wonderful friend. While donating millions to the poor, he bombed and tortured his enemies – some had their eyeballs removed with hot spoons. Through ruthless cunning and America's insatiable appetite for cocaine, he became a multi-billionaire, who lived in a $100-million house with its own zoo.

Pablo Escobar: Beyond Narcos demolishes the standard good versus evil telling of his story. The authorities were not hunting Pablo down to stop his cocaine business. They were taking over it.

American Made: Who Killed Barry Seal? Pablo Escobar or George HW Bush

By Shaun Attwood

Set in a world where crime and government coexist, *American Made* is the jaw-dropping true story of CIA pilot Barry Seal that the Hollywood movie starring Tom Cruise is afraid to tell.

Barry Seal flew cocaine and weapons worth billions of dollars into and out of America in the 1980s. After he became a government informant, Pablo Escobar's Medellin Cartel offered a million for him alive and half a million dead. But his real trouble began after he threatened to expose the dirty dealings of George HW Bush.

American Made rips the roof off Bush and Clinton's complicity in cocaine trafficking in Mena, Arkansas.

"A conspiracy of the grandest magnitude." Congressman Bill Alexander on the Mena affair.

The Cali Cartel: Beyond Narcos

By Shaun Attwood

An electrifying account of the Cali Cartel beyond its portrayal on Netflix.

From the ashes of Pablo Escobar's empire rose an even bigger and more malevolent cartel. A new breed of sophisticated mobsters became the kings of cocaine. Their leader was Gilberto Rodríguez Orejuela – known as the Chess Player due to his foresight and calculated cunning.

Gilberto and his terrifying brother, Miguel, ran a multi-billion-dollar drug empire like a corporation. They employed a politically astute brand of thuggery and spent $10 million to put a president in power. Although the godfathers from Cali preferred bribery over violence, their many loyal torturers and hit men were never idle.

Printed in Great Britain
by Amazon